Exploring Agrodiversity

Issues, Cases, and Methods in Biodiversity Conservation

ISSUES, CASES, AND METHODS IN BIODIVERSITY CONSERVATION

Series Editor: Mary C. Pearl
Series Advisers: Christine Padoch and Douglas Daly

This series combines two earlier Columbia University Press series: *Methods and Cases in Conservation Science* and *Perspectives in Biological Diversity.*

BOOKS IN THE SERIES

Thomas K. Rudel and Bruce Horowitz, *Tropical Deforestation: Small Farmers and Land Clearing in the Ecuadorian Amazon,* 1993

Holmes Rolston III, *Conserving Natural Value,* 1994

Joel Berger and Carol Cunningham, *Bison: Mating and Conservation in Small Populations,* 1994

Jonathan D. Ballou, Michael Gilpin, and Thomas J. Foose, eds., *Population Management for Survival and Recovery: Analytical Methods and Strategies in Small Population Conservation,* 1995

Susan K. Jacobson, ed., *Conserving Wildlife: International Education and Communication Approaches,* 1995

Gordon MacMillan, *At the End of the Rainbow? Gold, Land, and People in the Brazilian Amazon,* 1995

David S. Wilkie and John T. Finn, *Remote Sensing Imagery for Natural Resources Management: A First Time User's Guide,* 1996

Luigi Boitani and Todd K. Fuller, eds., *Research Techniques in Animal Ecology: Controversies and Consequences,* 2000

Exploring **Agrodiversity**

Harold Brookfield

COLUMBIA UNIVERSITY PRESS   NEW YORK

Columbia University Press
Publishers Since 1893
New York   Chichester, West Sussex

Library of Congress Cataloging-in-Publication Data

Brookfield, H. C.
     Exploring agrodiversity / Harold Brookfield.
       p.   cm. — (Issues, cases, and methods in
       biodiversity conservation series)
     Includes bibliographical references (p.   ).
     ISBN 978-0-231-10232-2 (cloth : alk. paper)—
     ISBN 978-0-231-10233-9 (pbk. : alk. paper)
   1. Agrobiodiversity.   I. Title. II. Series.
S494.5.A43 B76 2000
306.3′49—dc21   00-045175

Casebound editions of Columbia University Press books
are printed on permanent and durable acid-free paper.
Printed in the United States of America

Contents

Introduction

Introducing an Exploration   *xi*
The Plan of the Book   *xvi*
Acknowledgments   *xviii*

Part I. Presenting Agrodiversity

Chapter 1. Presenting Diversity by Example: Mintima and Bayninan   3

Mintima, Chimbu, Papua New Guinea   *3*
Bayninan, Ifugao, Philippines   *16*
Comment: Dimensions of Diversity   *20*

Chapter 2. Diversity, Stress, and Opportunity   23

Three Contrasted Examples   *23*
Threats to Crop Biodiversity: Paucartambo, Peru   *23*
A People Resettled Again and Again: The Zande of the Southern Sudan   *27*
The City in the Village: Four Villages Around Kuala Lumpur, Malaysia   *32*
Comment Arising from the First Two Chapters   *38*

Chapter 3. Defining, Describing, and Writing About Agrodiversity   40

Summarizing the Elements   *40*
Defining Agrodiversity   *42*

Describing and Classifying Agrodiversity   *46*
Following What Farmers Do   *50*
Analyzing and Writing About Agrodiversity   *52*
Themes for a Structured Argument   *54*
Two Cautions   *56*
The Way Forward   *57*

Chapter 4. Learning About the History of Agrodiversity   59

Two Very Relevant Questions   *59*
Selection of Favored Sites   *68*
Diversity in Early Management: Evidence from the Ground Surface   *69*
Evidence from Within the Soil   *73*
Toward Answers to the Questions   *75*

Chapter 5. Understanding Soils and Soil–Plant Dynamics   80

Introducing Soils   *80*
Soil Taxonomy and Its Problems   *82*
Soil-Forming Processes   *88*
Introducing Nutrients and Soil–Plant Relationships   *90*
The Human Factor   *96*

Part II. Diversity Within Land Rotational Systems

Chapter 6. Analyzing Shifting Cultivation   103

Introducing Part II   *103*
Farming in the Forests of Borneo   *105*
Borneo in Perspective   *116*
The Forces of Change   *119*

Chapter 7. Alternative Ways to Farm Parsimonious Soils   123

Citemene and *Fundikila:* Northeastern Zambia   *123*
Farming Systems Across Space and Through Time   *132*
Some Concluding Remarks About Work on Shifting Cultivation   *138*

Chapter 8. Managing Plants in the Fallow and the Forest   140

Introducing the Management of Plants   *140*
Managing the Successional Forest in Latin America   *141*
Managed Successional Fallows in Amazonia and Southeast Asia   *144*
Complex Multistory Agroforests in Southeast Asia   *149*
What Is Natural and What Is Human-Made?   *151*

Using Plants and Soil in Conjunction   *154*
Conclusion   *156*

Chapter 9. Coping with Problems: Degraded Land, Slope Dynamics, and Flood   157

Degraded Land   *157*
Coping with Degradation in Southeastern Ghana   *161*
Managing the Dynamics of Steep Slopes   *165*
Managing Water   *170*
Discussion   *173*

Part III. Paths of Transformation

Chapter 10. Who Has Driven Agricultural Change?   179

Introducing Part III   *179*
Bursts of Innovation and Incremental Change   *181*
Two Completed Experiments   *183*
Agricultural and Social Change in Japan, 1700–1950   *188*
Japan and Java   *194*
Conclusions   *196*

Chapter 11. Farmer-Driven Transformation in Modern Times   198

A Focus on Spontaneous Change   *198*
Management and Investment in a Sahel Village   *201*
Management and Migration Among the Kofyar of Northern Nigeria   *206*
Interference and Invention in Machakos, Kenya   *211*
Intensification, Revolution, and Agrarian Transformation: A Review   *214*

Chapter 12. The Green Revolution   218

Science and Public Policy as the Drivers of Change   *218*
North and South India   *222*
Farmers and the State in Java   *230*
Back to Diversity   *233*
Conclusions   *235*

Part IV. The Future of Agrodiversity

Chapter 13. Recent Trends in Agriculture   241

Economy and Ecology Hand in Hand   *241*
The Background: Genetic Erosion and Conservation   *244*

Innovations in Plant Breeding   *249*
Alternative Agriculture in the North   *253*
The Special Case of Cuba   *257*

Chapter 14. Science, Farmers, and Politics   262

A Check to the Seed–Chemical Juggernaut   *262*
Progress in Wider Biotechnology Fields   *265*
Biosafety and Ethical Issues   *268*
Understanding the Scientific Basis of Agrodiversity   *270*
Biophysical Diversity and Its Management: Alternatives to Herbicides   *275*
Diversity in Farm Management   *276*
The Organizational Domain: An Area of Weakness   *278*
The Conditions for Success of Diversity   *279*

Epilogue: Looking at the Future   281

Allies of Agrodiversity   *281*
In Conclusion   *284*

References   287

Index   325

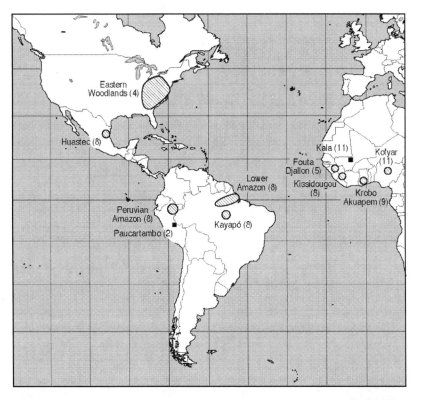

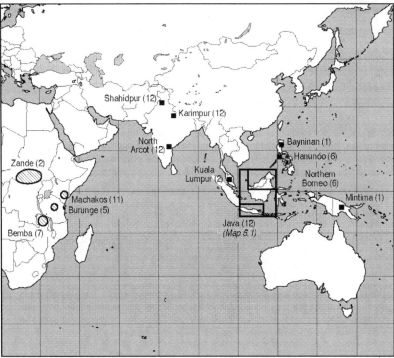

Figure 0.1 & 0.2 Location of places, areas, and peoples discussed in the text, with chapter number of the principal text reference.

■ Introducing an Exploration

HOW THIS BOOK HAPPENED

This book was started in 1993, but in its present form most of it has been written only since mid-1997. During the same period of time, I have been involved as principal scientific coordinator in a United Nations University (UNU) international project on People, Land Management and Environmental Change (PLEC). This began in a small way as a cooperative research project on small farmers' practices. PLEC has grown since 1993. After a long struggle, it gained approval as part of the Global Environmental Facility (GEF) portfolio in 1997, beginning its 4-year period of more ample funding in February 1998. It is a project within the Biodiversity Programme of the GEF, under the aegis of the United Nations Environment Programme (UNEP). It is now a demonstration project, focusing at chosen sites on ways in which farmers manage their land and its biodiversity. The project brings together about 200 scientists, together with their students and, most importantly, cooperating farmers, in Africa (Ghana, Guinea, Uganda, Kenya, and Tanzania), the Asia–Pacific region (China, Thailand, and Papua New Guinea), and the Americas (Brazil, Peru, Mexico, and Jamaica). This book does not arise out of that project, which will produce its own book later, but the PLEC job and this book have evolved hand in hand. Each has influenced the other, and to some extent each has held up the progress of the other.

The real origin of the book was at a meeting in Washington, D.C., in August 1992, which brought together several people who were interested in setting up PLEC. At one point, the discussion turned to the question of managing biodiversity. I responded that PLEC would be concerned primarily with diversity in the manner in which farmers used all their resources, and I described this as agrodiversity. Having said this, I went on to remark that "I have just coined a term, and I intend to use it." Within a short time, *agrodiversity* became a central concept for PLEC. Christine Padoch, who was present at the Washington meeting and later became a scientific coordinator in PLEC, suggested toward the end of 1992 that she and I write a book on agrodiversity, and we began to do this in 1993 and 1994. Christine's view of the book was that it should be based primarily on our field experience, of which I had less that was of direct relevance than she had, so my view involved a heavier use of the literature. Working in a little-charted field, I found it hard going. During 1995 it became obvious that what I was writing was not what she wanted to see, and it was different from the book that I was coming to want to see. The year 1996 was taken up much more with fundraising for PLEC than with writing, delaying inevitable abandonment of the original design, which we did amicably in early 1997. Beginning again in the middle of that year, I rewrote all that I had written before, which amounted to a disconnected set of draft chapters, and have completed this book on my own.

The enterprise survived for three reasons. First was PLEC. The PLEC project not only kept me in touch with what my collaborators were doing with the idea of agrodiversity but also gave me opportunities to widen my field experience. Between 1960 and 1990 this experience had been confined largely to Papua New Guinea, small islands in the Pacific and elsewhere, and, mainly after 1980, island and peninsular southeast Asia. Although only for short visits, PLEC enabled me to renew an acquaintance with Africa, where my working career began, and to spend some time in the field with colleagues elsewhere. It introduced me to China and Amazonia. The second reason is that from 1994 until the end of 1998 I was supported in research for this book by the Australian Research Council (ARC), making possible the appointment of Helen Parsons, an agricultural scientist with valuable developing country experience, as part-time research assistant. She greatly enhanced my possibilities for finding scarce gems among a large and very scattered literature. I am grateful to UNU–PLEC for continuing to support Helen's appointment beyond the end of the ARC grant period. Most work on the large list of bibliographic references has been hers. Search possibilities were further enhanced by the insightful efforts of Muriel Brookfield, working in such time as was not occupied by her work as joint editor of the twice-yearly PLEC periodical *PLEC News and Views*. Both have gone through the whole manuscript, making many helpful comments and corrections. Their support has been basic. It has encouraged me in the third reason why the book did not perish, perhaps the best: I found writing it so interesting.

WHAT THE BOOK IS ABOUT

This book is an exploration into agrodiversity, not an exposition. I have been the explorer and the book has developed and changed as it has gone along, with successive plans and many drafts discarded along the way. My concluding argument, when I came to it, surprised even me. The varied and changing ways in which small farmers in the developing world use their resources are presented by example in two introductory chapters before any attempt is made to define the content of agrodiversity or to discuss how it should be examined. The dynamic diversity of small farmers' practices has only the most limited formal literature, and principal titles draw heavily on material from west Africa (Richards 1985; Netting 1993; Mortimore 1998). This book explores the topic of agrodiversity among small farmers over a large part of the world and through as long a span of time as possible. It draws on a selection from a large literature based on field research in small areas and on a purposive sampling of the historical literature covering specific areas and periods of time. Here and there, it is illuminated by my own fieldwork. Except for the case presented in depth in chapter 1 and to a lesser extent one case presented in chapter 2, most of this has been short-period fieldwork of a type derided by some as "academic tourism."[1] I use these latter experiences to help me understand the writings of others rather than for their own sake.

It is much easier when this is possible because the agriculturally managed landscapes of the small-farming regions of developing countries differ enormously from one another. At one extreme are the huge extents of carefully tended rice fields in the lowlands of southern China or the northern plain of Java. At another is the temporary occupation of a part of a forest or tall grassland by a field that is quickly overwhelmed by secondary growth once the main crops are out of it. In some areas, a short arable farming period is merely the most total transformation of land cover in a sequence that goes from one state of managed forest to another. The only generalization that is possible is that all their landscapes are unlike the wide arable fields, managed with machinery, chemicals, and few workers, that have arisen in modern times across wide areas of North America and northern Eurasia. Yet until 1950 there was greater diversity in some of those regions, too. My first paid employment was as a temporary laborer on an English upland farm of a type that has now vanished.

I confine attention almost entirely to arable and tree crop farmers. Although many also fish, hunt, and collect wild produce, sometimes on a substantial scale, these are not their main activities. Many also have ruminant livestock, but none of the people discussed in this book are primarily pastoralists, using rangeland. Farmers making extensive use of irrigation enter the story in some places, but the main concern is with dry land farmers. Collection and sometimes cultivation of medicinal plants are important activities in many systems, and in several instances many hundreds of such plants can be identified even on the land of a single community and are well known to specialists in their use.[2] Medicinal plants are a special field of study in their own right, and the topic is not discussed in this book.

All peoples described are small farmers, although there may be big differences in the scale of operation, even within a single community. I do not confine attention to the people whom Netting (1993:2) described as smallholders, "rural cultivators practicing intensive, permanent, diversified agriculture on relatively small farms in areas of dense population." By no means do all farmers discussed in this book practice what could be described as intensive, permanent agriculture, although all practice diversified agriculture. Furthermore, several of these latter peoples hold land individually even under low population densities. The gradient Netting (1993) and others have proposed, in which land rights become more strict and explicit in higher-density than in lower-density areas, cannot be supported as a universal rule. Other important considerations intrude. To avert any possibility of confusion, I avoid the term *smallholders* and write only of small farmers under whatever conditions of population density and land tenure they may live.

This minor difference with Netting notwithstanding, all farmers described in this book are, in his sense, householders, and few make extensive use of hired labor. Almost all participate in production for the market, whether as major or minor activity, but most also produce a large part of their own subsistence. Few have significant farm machinery, and most do the greater part of their work by hand or with use of farm animals. Many have adopted crop varieties bred on research stations, but it is characteristic of small farmers that they also cultivate varieties that are locally bred or have been acquired through informal channels, on journeys, or in the marketplace. Most save their seed or other germplasm material from the previous harvest. Above all, they experiment with almost any potentially useful new planting material that comes their way, and sometimes they do this in a manner not greatly different from formal experimental method. Paul Richards (1985) has demonstrated this very clearly.

Contrary to an old but still common belief about small farmers' conservatism, there is a very widespread willingness to try out new ideas but also to reject ideas and methods that seem to be unsuitable for their needs or expose them to risk of crop failure. Confronted with an innovation or new advice from the authorities they carry out an informal cost–benefit analysis to determine its value before deciding to adopt or reject. Also contrary to a still-common belief, they plan their activities over a long time horizon, investing in improvements while continuing production. This incremental pattern of agricultural change can easily be lost from sight in a quick and casual view of what they do, but it is widespread. It seems to have been the manner in which farmer-driven agricultural revolutions have taken place, both in sixteenth- to nineteenth-century Europe and in modern Africa. The evidence for these statements emerges throughout the book but especially in the later chapters.

PLACING THE BOOK IN CONTEXT

The source material used in this book is drawn from a wide range of disciplines, especially from anthropology, agricultural science, and geography. Both the anthropology and geography are of a type that in the Americas would be called cultural ecology or sometimes human ecology. This interdisciplinary field evolved in the 1950s and 1960s, and I was a part of its evolution (Butzer 1989). It suffered competition for students' attention from successive waves of logical positivism, neo-Marxism, and postmodernism and from the inroads of a politicized environmentalist ethic that emerged strongly in the 1970s, offering a different meaning for the term *ecology*. Yet it has survived. Cultural ecology has continued to appeal to a minority of students and to command a wider multidisciplinary attention than some of the mainstream work in geography and anthropology. As a field, it has changed with the incorporation of a developmental perspective and adaptation to modern concern with biodiversity conservation. Its essential characteristics are a strong base in empirical research and in exploration, in depth, of the long-lived people–environment paradigm in the social and natural sciences. Necessarily, it remains eclectic, but it has a clear focus in the management by people of their biophysical resources. This book is written as a contribution to cultural or human ecology per se, not to the mainstream areas of any of the several mother disciplines of modern cultural ecologists.

SOME CONVENTIONS, ONE OF MAJOR IMPORTANCE

Before I move on to present what agrodiversity is about, this is the place in which to touch on some conventions I have decided to adopt in writing. I write of agriculture, not horticulture, *pace* even some of my own former students who have written of horticulture because the fields of small farmers often are described as gardens. Horticulture is practiced by many agricultural farmers discussed in this book, as the care of specific plants in true gardens, but it is best to avoid the term for general use. The second convention is perhaps somewhat idiosyncratic in that it concerns rejection of a term very common in the cultural ecology literature: that of *swidden* for a temporary field. For describing burned temporary fields, this old north English dialect word was introduced into the literature by Izikowitz (1951), and its use was popularized principally by Conklin (1954, 1957).[3] Not all temporary fields are cleared by use of fire, and fire may not be the major tool used. Because it seems specific but in fact is applied very loosely, I avoid the term *swidden* except in quotations and write only of fields, cleared fields, temporary fields, or, where appropriate, gardens.

The third convention is of greater importance. It addresses a more serious issue than just terminology. The problems involved in using the writings of others are discussed more fully in chapter 3, but I must address a general one before I begin. The literature contains many excellent accounts, but they have been written over a 50-year period during which a great deal has changed. Only a few studies have tracked systematically this pattern of change in any detail for particular areas, peoples, and their agricultural systems. In some cases follow-up studies provide valuable information, but in other areas even the latest writing, from as recently as the early 1990s, describes a political, social, and economic context that is no longer valid. In any case, even 1999 titles generally are based on work done a few years before.

I therefore decided to put everything that does not clearly relate to ongoing, mainly biophysical conditions into the past tense. This is especially important because I put such emphasis on change and adaptation in this book and because so much modern change has been very rapid and often stressful. If this decision seems pedantic, it also avoids a good deal of inaccuracy that could arise by failing to recognize the rate and nature of change. Though less exclusively, I continue it even in the final chapters where, depending on which present-day cliché we might prefer, I am writing of "near real-time" or the "virtual present." Both the science and the politics of present-day developments are changing so rapidly that even what I have written as 1999 gave way to 2000 will be history before the book is published.

The problem of tense is compounded by a former practice among anthropologists, whose work provides some of the best material on agrodiversity. This is the use of the ethnographic present. Using this device, everything is written in the present tense, and the effect is that change through time is suppressed, and observations over the period of fieldwork are treated as observations of a continuing state. The practice was developed by anthropologists between the two world wars. It facilitated writing based on a single extended period of fieldwork in which so many data were assembled that they could be written up only over several years. It also rested on a general belief that traditional societies were essentially static and unchanging, a belief dead in anthropology but not in some other social science disciplines or in agricultural science. The ethnographic present continued to be used by some anthropologists into the 1970s, but it is little used now in view of an emphasis on the forces of change that came into all the social sciences, initially in the form of Marxist thought. A lot of the writing on which this book has to rely has made use of the ethnographic present, which creates a problem in writing an interpretation that stresses dynamism.

Other conventions of minor importance include use of the term *manioc* rather than *cassava,* which, properly, was originally the flour derived from manioc. Although the term *cassava* is much more commonly used than *manioc* in Africa and Asia, I follow J. D. Sauer (1993) in using the American-region term for a plant of American-region origin. Because its main genus is Asian–Pacific, taro is always taro, never the coco-yam, even though the latter is an ancient term in Africa. Maize is always maize, never corn. *Wet-rice field* and sometimes *pond field* are always used in preference to *paddy.* The literature is full of the use of regional terms, and it is simpler to standardize. For botanical terms, I have tried to standardize use of the most up-to-date sources wherever possible, but there remain a few variations. Finally, all measurements given in this book are in metric units, and all temperatures are in degrees Celsius.

■ The Plan of the Book

FOUR UNEQUAL PARTS

Broadly, I divide the book into four unequal parts. Part I, the longest, is introductory, and consists of five chapters. The first two present agrodiversity by example, then the third discusses it as a concept and as a field of study. There follow two chapters that are more specific in content, one concerning the place of diversity in agricultural history and the second concerning diversity in the ground, principally in the soil. Part II contains four chapters, all concerned in one way or another with the large group of agricultural systems that are often generalized under the term *shifting cultivation* but in fact exhibit great variation. This stress on shifting cultivators has the effect that an important part of the book is concerned with people who farm their land at only low to medium population density, thus distinguishing this book more clearly from Netting's important survey of smallholder farming. In planning the book, it was not my initial intention to put such emphasis on this class of developing country farmers, large though it is, and I had intended to give comparable weight to diversity in the practices of farmers cultivating all their land continuously under high population density. Yet there is a substantial continuum of practices and adaptations across the range of land rotation farmers, and in the end not an enormous amount is missed.

The balance is to some degree restored in part III, which turns much more specifically to trajectories of change, with a strong emphasis on the sort of substantial change that many would describe as revolutionary. First this question is treated historically, then I turn to people of whose history very little is known but who have changed their farming in modern times as much as those of the Green Revolution regions. Third, I discuss the modern so-called Green Revolution itself, thus preparing the way for the contrasted and sometimes conflicting approaches of modernist farming and alternative agriculture, which occupy the two chapters of part IV. The emphasis in part III is on the role played by the farmers themselves in initiating and guiding the pattern of change.

Part IV moves away from the main focus on developing countries to discuss modern trends and forces in agriculture and in the world around agriculture. In two chapters and an epilogue, I set out to present the gene revolution that has succeeded the Green Revolution and the modern alternative agriculture movements, which relate so closely to agrodiversity as it is practiced in the developing countries. I address the question of sustainability or resilience of agrodiversity and its importance for the future. There is now a substantial literature on this topic, most of it written since the 1980s, when it came to be widely realized that monocultural systems can be very vulnerable to both biophysical and economic variations as well as to pests and diseases. Nonetheless, there must be a balance, and I try to present it. Commerce and politics, only in the background in earlier chapters, become of central importance in this final review. The short conclusion, presented in the epilogue, is guardedly optimistic. I will not anticipate it in this introduction.

EMPHASIS ON CASE STUDIES

Except in two of the chapters of part I and in all of part IV, I rely heavily on a modest number of case studies selected from among the literature. Their locations, either as places or as areas, are shown in figures 0.1 and 0.2, which precede this introduction, and in figure 6.1. Those presented are far from being the only published examples that illustrate the topics to which I relate them, but they seem particularly suitable. I have presented case studies in a fair degree of depth, though selectively in terms of the material extracted from them, rather than refer in a more shallow manner to a larger number of examples. The emphasis in all instances is on the management practices of the farmers, with less discussion of the social and economic context than might seem desirable. This is a book about agrodiverse farmers, not about the farming populations as social systems. Although external stresses are given weight, it is not a book about the political ecology of farming systems. The cases might overemphasize successful adaptations to some degree because so much writing discusses the failures and my purpose is to offset this tendency. But for this selectivity, the book easily could have been twice as long. A great deal of material drafted in the course of writing has been discarded or relegated to endnotes. Had I let it develop into a longer book, it might never have been finished.

■ Acknowledgments

My large debt to Helen Parsons and Muriel Brookfield was set out earlier. They are the only people who read all of the book before it was submitted for publication. One other, who has read an entire penultimate draft, is Dr. Karl Zimmerer of the University of Wisconsin, who read the draft manuscript for the publishers. I am very grateful for his comments and suggestions. Christine Padoch, with whom I started the book, has seen about a sixth of what is in the final text (as well as a good deal more that has since been discarded). Her pithy comments were of great value. The friendship of Christine, her husband, Miguel Pinedo-Vásquez, and Michael Stocking, all active members of PLEC, has led to many insights. My regard for their work is reflected in the frequency with which it is cited in the text. Drafts of particular chapters have been read and commented on by Jim Fox, Geoff Humphreys, Malcolm Cairns, Michael Stocking, and Lesley Potter. Michael Stocking and I wrote a paper that drew on a draft of part of what is now chapter 3, and Michael's contribution to that paper has influenced that chapter. A few others have been shown bits of the book in draft but have kindly, or alarmingly, refrained from comment. Copyeditor Carol Anne Peschke has proposed a large number of stylistic changes, tautening the argument.

My institutional acknowledgments are important. I wrote the book while working as a visiting fellow in the department of anthropology (Research School of Pacific and Asian Studies) at the Australian National University (ANU), Canberra, and most of it in the comfortable office that department provided. The technical support of Ria van de Zandt, Merv Commons, and Fay Castles has been indispensable. The financial support of the ARC, through 5 years, gave me a good deal of independence and research assistance.

I have already acknowledged my debt to the UNU and its PLEC project, which has provided financial support, especially for the overseas travel that was of major importance in creating the book. My colleagues there, especially Juha Uitto, Audrey Yuse, and Liang Luohui, and Timo Maukonen in UNEP, have given vital background support and tolerated periods in which I put their work aside to concentrate on this book. PLEC's part-time Canberra administrator, Ann Howarth, has been centrally involved in the final stages of manuscript preparation, and her skillful work is greatly appreciated. Acknowledgment is made to the editor of the journal *Human Ecology* for permission to reproduce figure 1.1 and to Edward Arnold Ltd for permission to republish figure 1.2. The other new maps drawn for this book are the work of Kay Dancey in the cartography unit of the Research School of Pacific and Asian Studies, ANU. Tess McCarthy also helped.

My PLEC colleagues are too numerous to acknowledge individually, but it would be churlish not to mention the groups headed by Guo Huijun, Edwin Gyasi, Ibrahima Boiro, and Romano Kiome, in particular. They have seen none of the book, but their contribution in information and guidance in the field has taught me much. Also, I should acknowledge colleagues and students from earlier years, who have been with me in the field (and I with them) and whose drafts and final writings have contributed to the formation of ideas and collection of information on agrodiversity. Some are mentioned in the text or referred to among the list of references; I am grateful to them all.

I need only add, but importantly, that no one shares any of the blame for errors, misstatements, false interpretations, and plain ignorance. All responsibility for these failings is my own.

Harold Brookfield
Canberra, May 2000

Notes

1. At least it has been academic tourism from the point of view for which I use it here. In some areas in which I have resided as a researcher for long periods, such as Fiji and the West Indies, my main concern was not with farmers' management of their land and biotic resources. Observations on these practices were gained only incidentally, in the context of other work. In the end, I have drawn on my 18 months' experience in Fiji only at two places in the whole book.

2. I once suffered a painfully sprained wrist from a fall on a slippery path in Indonesian Borneo. The local herbal medicine specialist treated it, some 28 hours later, using a poultice that he said contained herbs of 44 plants, some gathered in the forest, some in the fallow, and some in or around cultivated land. The poultice relieved the pain within a single night.

3. In its north English area of origin, the term *swidden* probably meant a clearing made from heath more often than forest, in which fire was not the only management tool. It was used in a distinctive manner involving burning of the pared-off surface soil as well as its cover, thrown up in ridges or heaps (Kerridge 1967).

Exploring Agrodiversity

Part I

Presenting Agrodiversity

Chapter 1

Presenting Diversity by Example:
Mintima and Bayninan

■ Mintima, Chimbu, Papua New Guinea

MINTIMA IN 1958

I begin the book where I began this sort of work myself, and I will set out what I did there in more detail than I will provide for other areas. The view from the old government rest house at Mintima was spectacular. In the heart of New Guinea lies a range of mountains larger than any other between the Himalaya and the Andes. Rising to more than 4500 m in the east and more than 5000 m in the west, the range contains a series of large intermontane valleys that were not penetrated by westerners until the 1930s but housed more than two million people.[1] Mintima, in the western part of the territory of the Naregu tribe of Chimbu, is situated high along the northern slope of one of these valleys. From the spur just below the rest house one could look 100 km east to the triangular limestone mass of Mt. Elimbari, 200 km west to the extinct volcano of Mt. Hagen, and 50 km south across the Wahgi valley to the southern range of Mt. Kubor. Climbing the steep limestone escarpment behind Mintima on a fine day, one could see the ice-worn summit of Mt. Wilhelm, the highest point in the northern chain and the highest mountain in Papua New Guinea, rising above the intervening ridges. The Wahgi valley itself usually was filled with fog in the early mornings, and only as the mist cleared could the river be seen in the middle distance.

Closer at hand, the view often was obscured by clumps of graceful *Casuarina oligodon* trees, which were planted as seedlings in field plots and allowed to grow for 5 to 15 years. It was also obscured by dense patches of tall grass, some of it wild *Miscanthus,* and, on wetter ground, semicultivated pit-pit *(Saccharum sponta-*

neum). In between these patches were large open sweet potato fields, all dug in a square ditch pattern, and smaller blocks of mixed crops, dominated by sugar cane and bananas. There were also patches of limited-term short grass fallow. In some of the fields, maize, yams, taro, groundnuts, and vegetables were also planted, usually as a first crop after fallow. Further out, beyond fences of sharp casuarina stakes, was an unmanaged land dominated by short grass on the ridges and tall grass in the valleys. Here the family pigs foraged by day, returning at night to houses built along the fences where the women fed them sweet potatoes. The patchwork was at first sight bewildering, and it soon became obvious that the only way to understand it was to map its diversity. This I began to do only a few days after arrival, initially by the most primitive methods, using a compass and counting paces, improving my methods over time. After 1964 I used an accurate map based on low-altitude air photographs and triangulation. The evolution of mapping methods was described in Brookfield (1973).

WORK AT AND AROUND MINTIMA

It was at Mintima that I began the exploration of diverse farming practices that ultimately led to this book. I first went there in 1958 and last visited in 1991. From 1958 until 1970 I was there part of almost every year, but after 1970 I went back only twice. From 1958 to 1965 my work was largely in collaboration with anthropologist Paula Brown, who continued specific work on the Chimbu people right through into the 1990s. We went back there together in 1984, and I went back again alone in 1991. My first interest was in Chimbu farming, Paula's in Chimbu society. At first, neither of us knew much about either because we had not done this kind of fieldwork before. We chose to go to Chimbu because it had the highest population density of any part of the highlands, but the main models we carried with us were African. Mine were from a limited experience in the field a few years earlier, when I held my first teaching job in South Africa, hers from interpretations of segmentary societies in the anthropological literature. We were looking for shifting cultivation and a tightly structured society, but we found something different.

After a few months we were able to produce a map of land use and land holding over a few square kilometers, understand the organization of group territories and their distribution over a larger area, and write a monograph in which we questioned many of our initial assumptions (Brown and Brookfield 1959). With another year of fieldwork in 1959–60 we were able to enlarge these early interpretations and write a book (Brookfield and Brown 1963) and numerous papers, some of them individual, some of them joint.

My part of the job included walking over fields, recording what was being grown and how; being shown their boundaries and being told about who used them; and mapping this information repeatedly, year after year. My principal information came from a mainly self-selected group of companions of about my own age who enjoyed these hilly days and were not burdened with responsibilities, which then fell to their elders. Two of them accompanied me over the same range of country in 1984, but in 1991 I had to find younger guides, little older than my own son, who was a very small boy when I began in 1958.

By the time the last fieldwork had been done, in 1984, we had good data that spanned more than 25 years on a fast-changing society and its agriculture. These data can be bracketed within a historical and ethnohistorical record that covers most of a century. Change, and rapid change at that, began to command attention as early as 1959, so we were not often tempted to use that deadly ethnographic present to which I referred in the introduction. We knew from a very early stage that the real problem was interpreting the trajectory of events.[2] It was uncharted territory and we made mistakes; some of mine are discussed in the last of my papers on Chimbu (Brookfield 1996b). Supposed trends turned out to be false leads. In this chapter I present mainly my conclusions. It may seem as though I knew them all from the outset, but this was not the case.

A LONG AND TURBULENT HISTORY

The 70,000 western Chimbu (or Kuman) occupy land that falls from 2800 m at the upper limit of cultivation to about 1300 m near the edge of the steepening gorge where the south-flowing Chimbu River joins the larger Wahgi River, coming from the west. Today, more than half of the western Chimbu live on land sloping south toward the Wahgi from a major limestone escarpment now called the Porol Range. The others live in the deeply incised upper valley of the Chimbu River north of that range. When the story told here began, and through the first half of the twentieth century, these proportions were reversed.

The area has a long human history; there have been farmers in the central highlands for more than 7000 years. The broad history of their changing use of and impact on the land and its biota has been studied since 1960 by paleobotanists and prehistorians, synthesized by Walker and Flenley (1979), Walker and Hope (1982) and Golson (1982, 1991). The whole reconstruction was succinctly brought together by Haberle (1994). In contrast with Andean farmers, highland New Guinea farmers have domesticated few crops and trees. Almost all their crops are also found in the lowlands, and their farming is essentially lowland agriculture at its altitudinal margin. All the present Chimbu area, except the highest land now cultivated, seems to have been occupied long before the seventeenth century, when the suite of crops used was augmented dramatically by the South American sweet potato *(Ipomoea batatas),* introduced through Indonesia and along trade routes into the mountains. Yielding better than taro and other former staples in the cool mountain climate and more tolerant of poor soils, sweet potato became the dominant crop over very large areas. Because of its tolerance of lower temperatures, sweet potato permitted occupation of land up to 200 m higher than before its introduction. In Chimbu, where there is some of the highest cultivated land in Papua New Guinea, upslope extension of farming continued until early in the twentieth century.

Since the seventeenth century, a range of agricultural practices suited to cultivating the sweet potato evolved in different parts of the highlands. A set of labor-intensive agricultural practices was developed, all including tillage. Those in the western areas involved building large mounds, composted before closure and capable of sustaining cultivation indefinitely on the same land. Everywhere, an explosive growth took place in a possibly older system of indemnifying social relations between individuals and groups mainly by prestations of pigs: live pigs in some areas, killed, butchered, and cooked in others (Feil 1987). By the time rough measurements by professional observers began to be made in the 1950s, up to half of the cultivated sweet potato crop was fed to pigs. This system of competitive exchanges, highly ritualized in different ways in different parts of the highlands, was the outcome of competition between ambitious "big men" who could gain power by manipulating wealth, forcing others to participate. It was also the necessary means of formalizing periods of peace between groups that fought readily. Whatever its longer history, warfare became endemic in the sweet potato period and probably contributed to the association of quasiagnatic subclans and clans into larger tribal groups for common action and defense of territory, pigs, and women[3] (Brookfield and Brown 1963; Brookfield 1964; Waddell 1972; Modjeska 1982; Allen and Crittenden 1987; Feil 1987; Brown 1972, 1995).

In legend and possibly also in fact, the mountainous upper Chimbu valley is the homeland of all Chimbu speakers. By the nineteenth century, tribal groups were large and powerful, and there was fierce warfare between them, usually sparked by some minor dispute. Defeated groups were driven off their land, almost all down the Chimbu valley and some ultimately out of it. The tribe that Paula Brown and I began to work among in 1958, the Naregu, was one of the latter. They had been driven into their present habitat, mainly south of the massive limestone escarpment of the Porol Range, by the 1920s. The ethnohistorical record suggests that their forced migration and resettlement, with many fights along the way (within the tribe as well as against others), took place over several decades.

FARMING AND SOCIETY BEFORE THE 1950S

Colonial government and an enforced peace were established in Chimbu between 1933 and 1936. Precolonial skirmishes, battles, and even campaigns, and the resulting forced migrations, still formed the core of most remembered ethnohistory in the 1950s. Diseases, especially the malaria that arises each year in the lower parts of the main valleys, are less remembered than war, but they caused much death. Malaria reduced the ranks of those who settled in the lower-lying areas toward the Wahgi river. There were devastating new outbreaks of epidemic disease in the twentieth century, including a serious epidemic in the 1940s that killed many people. Yet in the upper Chimbu valley, population densities of 150 to 300 per square kilometer were recorded in the 1950s even after a period of sharp population decline.

The high-altitude area of the upper Chimbu valley, almost all above 1800 m, has been agriculturally occupied right through the sweet potato period, since the seventeenth century. Here, sweet potato grows well but slowly, and some of the tropical crops cultivated at lower altitudes are excluded. At least by the twentieth century, land was enclosed permanently within blocks marked by live fences of *Cordyline fruticosa*. Blocks were cultivated for 6 or 7 to more than 20 years before being planted with fast-growing *Casuarina oligodon* trees and then opened to the pigs. As soon as land was tilled, small erosion control fences were built across the slope. In some areas the live cordyline hedges were also planted across the slope, building up an accumulation terrace behind them over time (Humphreys and Brookfield 1991). In 1989, coring into the root zone of successive generations of these live hedges on one such terrace in the upper part of the valley, Humphreys (personal communication, 1991) obtained carbonized material dated by radiocarbon methods at around 300 years old.

South of the Porol Range, the environment is more spacious, sloping down from the crest of the escarpment at 2200 m to about 1400 m close to the gorge of the Wahgi River. Most of the area occupied by Naregu after their enforced migration had been only sparsely occupied for at least some time before the late nineteenth century. It was under grassland, mainly cane grass (*Miscanthus* spp.) on the upper slopes and short grass (*Themeda* spp. and *Imperata cylindrica*) on the lower. Bringing their pigs with them, Naregu separated cultivation from the large foraging areas by building impermanent fences of sharp-pointed casuarina stakes. Casuarina trees were planted as a wood-producing fallow crop to supplement the limited natural stands growing in wet areas along the streams descending from the escarpment. *Casuarina oligodon* grows fast on all but shallow soils and produces a readily splittable, durable wood, used in house and fence construction and as firewood. In addition, it carries root nodules that fix nitrogen. Chimbu is still the main concentration of casuarina planting in the highlands but not the only one. Farmers are well aware that planting this tree improves the soil. In the pollen record it emerges strongly about 1200 years ago, with a major increase since the seventeenth century.

Across the western part of Naregu territory a huge ancient landslip, many thousands of years old, brought down from the escarpment a large quantity of limestone rubble flow that now spreads across the underlying mudstone, sandstone, and shale (figure 1.1). All the upper slopes are steep, and sharp valleys are cut into the mudstone. This mudstone belt is still moving in a series of frequent small landslips, creating areas of wetter land with deep soils around the heads of the streams between the firm, clayey ground on the mudstone ridges and the very friable blocky soils on the rubble flow. These mobile areas often are separated sharply from the undisturbed mudstone ridges by cliffs several meters high. Closer to the Wahgi river, slopes become gentler and landslips fewer. The lowest slopes have only thin soils, and they suffer mineral deficiencies.

There are four distinct types of land below the escarpment: the ancient rubble

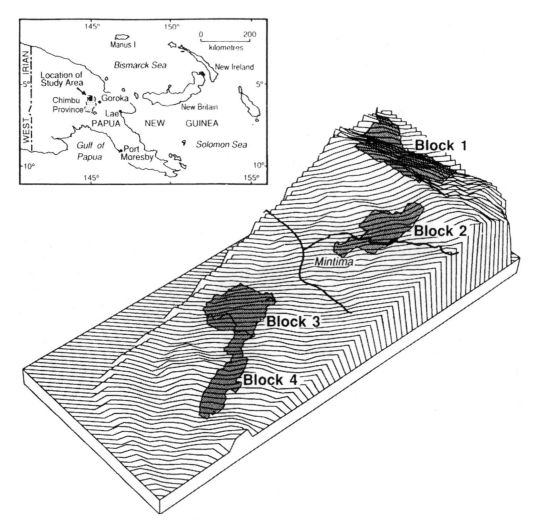

Figure 1.1 Block diagram of the Chimbu area showing the location of the four blocks discussed in chapter 1. (Reproduced from Brown, Brookfield, and Grau 1990:26, with permission from the editor of *Human Ecology.*)

flows below the escarpment, the higher mudstone ridges that emerge from beneath the flows, the large landslip areas around the headwaters of the south-flowing streams, and the lower southern slopes and valleys. When Naregu first migrated into this area, they settled on the escarpment and on the higher slopes of the rubble flows and mudstone ridges, fencing and cultivating irregularly shaped areas, each around a communal men's house. The houses of the women and pigs were along the fences some distance away. Later, incorporating an additional clan from among people who had migrated earlier, they spread onto the lower slopes. Ten or 20 years before our work began, all but one of the four constituent clans of modern Naregu had abandoned the escarpment. Only a few men and women have continued to work there, climbing the steep and rocky face of the range. By the late 1920s, before colonial rule brought political stability, the migration was complete, and the area was being filled in. Almost all men's house sites occupied in 1958 had already been occupied at the time of European irruption in 1933, with the houses rebuilt many times.

FARMING SPACES IN 1958–60

The three western exogamous clans of Naregu occupied some 15 km^2 within which we concentrated our work, mapping land use and land holding over 9 km^2. The farming, pig-keeping, and residential system of the precolonial period was already undergoing change in 1958, following recent construction of a four-wheel-drive motor road across Naregu territory to the large Wahgi valley to the west. A number of separate enclosed blocks had been joined together to form a continuously enclosed belt along the road, protecting the road from damage by pigs. The precolonial system was still visible, and we caught it just in time. There were two main types of arable farming space: the open fields *(kongun)*, dominated by sweet potatoes, and the smaller mixed gardens *(pongi)*, in which a wide range of crops were grown. They included sugar cane, bananas, taro, yams, manioc, maize, beans and winged beans, cucumbers, squash, ginger, and *Saccharum edule*, the edible pit-pit. There were also a range of indigenous vegetables. About 20 different crops were planted in these mixed gardens. Unlike in the comparable field types in Enga to the west (Waddell 1972), land was prepared by the same method in both farming spaces, marked out into near squares of 2–3 m and ditched, with the spoil from the ditches thrown onto the intervening beds. *Pongi* usually preceded *kongun*. Before the late 1950s, steel spades had replaced hardwood spades, and stone axes were no longer used to cut trees and make pointed stakes for houses and fences. The sweet potato fields already included a few garden squares of introduced groundnuts *(Arachis hypogaea)*, and maize *(Zea mays)*, also an introduction but from late precolonial times, commonly was dibbled around the margins of ditched squares.

More restricted spaces were occupied by the two main tree crops. The long-lived nut pandanus *(Pandanus julianettii)* grew only above 1900 m and was found mainly in a dense grove all the way along the foot of the escarpment and in small patches beyond it. The shorter-lived oil pandanus *(Pandanus conoideus)* was grown on wetter ground close to streams at lower altitude. There were also many other nut-bearing and fruit-bearing trees and trees with edible leaves. Among these the most widespread was the highland breadfruit *(Ficus dammaropsis)*. Emerging in the shaded environment of the mixed gardens were newly planted seedlings of coffee, introduced in 1955 and already being planted outside the original small blocks. By the end of the 1950s, some farmers had more than a hundred trees (Brown 1999).

Over time, successive seasons in the field revealed the manner in which these farming spaces were adapted to ecological conditions (Brown and Brookfield 1959;

Brookfield and Brown 1963; Brookfield 1973). Most mixed gardens were located on the moister soils of the landslip areas, and there were few on the limestone rubble flows. The wettest and most unstable soils beside the south-flowing streams were not cultivated at all because slipping was such a frequent hazard and fences would not stand. They were excluded from the enclosed areas even though they contained some of the best soil. They are particularly well suited to growing taro *(Colocasia esculenta),* only a minor crop nowadays but supposedly far more important before the introduction of the sweet potato. We can use modern knowledge of soils to help speculate on the pattern of earlier farming systems, but that is beyond the scope of this chapter.

On the well-drained rubble flow lands, ditching was used much less than on the mudstones, and the whole surface of the land was lightly tilled with digging sticks or spades. On all steeper sites, erosion protection fencing was used, with the dual role of stabilizing the soil and marking the plots planted by each wife or other related woman allocated land by the owner after he had tilled it. Similar allocations were made on other land but were marked only by selected ditches, known to the women but not often to my male informants. These allocations are transitory, lasting for the life of only one crop. Many plots are cultivated again, sometimes many times. Planting areas are then reallocated. Although sweet potatoes were grown on all types of land, there were obvious differences. On interfluves in the mudstone belt some sweet potatoes were planted in a soil consisting only of the weathered shale, recognized and named by a term literally meaning shale soil. In 1958 and 1959 I supposed that such inferior land would not continue to bear crops for long. I was wrong.

COFFEE AND OTHER INNOVATIONS

Coffee was first introduced to Papua New Guinea as a cash crop for Australian and other planters. It soon "escaped" onto indigenous land, and in 1955 it became official policy to develop small coffee blocks in selected areas. The Mintima area was one. Coffee initially was planted under introduced shade trees, as advised. It soon was planted in the mixed gardens, then much more widely, with casuarina used to provide the necessary shade. In a mapped study area of 882 ha, coffee occupied only 7 ha in 1958. By 1963 it had increased to 45 ha, and to 64 ha by 1967, when it was last mapped over the whole original area. Coffee largely took the place of the mixed gardens, which declined from 58 ha to 13 ha over the same period. It also very quickly became concentrated on the moister soils of the landslip areas flanking each small valley and in a specific altitudinal belt, between 1600 and 1800 m. A new farming space, most of it occupying a clear ecological niche, came into existence mainly between 1959 and 1965, while the open field sweet potato space expanded downslope into lower southern areas. All this quick adaptation happened spontaneously, without any prompting by the single agricultural extension officer in the region.

Strong local political leadership emerged in this period, based not on warfare and ceremony but on the will for development, however this was perceived. Paula Brown concentrated her work on this leadership over several years, summarizing her results in two books (Brown 1972, 1995). Under this leadership, the enclosed land was extended. Most immediately important were the innovations associated with coffee. Initially pulped only by hand, from 1963 the beans were pulped mainly by machine. Drying was concentrated in community sites, methods were standardized to improve quality, and Naregu joined in a regional cooperative for sale to a new factory close to the district center. Field management of coffee was improved greatly. Centralized processing and sale failed, and after the cooperative collapsed in the early 1970s all sales were made to private buyers. During the life of these innovations, a number of associated changes emerged that were more enduring.

WIDER ENCLOSURE AND A NEW PATTERN OF SETTLEMENT

Even in 1958, settlements newly made in the old way, by a group of men building a men's house in the open area beyond the fences and enclosing the area around it, were quickly incorporated into the general enclosure. At the same time, and without any specific direction, the pattern in which men's wives and daughters lived in houses shared with pigs along the fences began to fade away. Already in 1958 there were some small house clusters around a few men's houses; by 1965 these were common and the clusters were growing larger. In the 1970s the clusters became villages—formless, but villages nonetheless. Except for those of bachelors and widowers, new men's houses rarely were built after the late 1960s, and by the end of the 1970s most married couples lived together in one house.

The new villages grew up on the interfluves, where men's houses had always been located, but in the same altitudinal belt as the coffee and close to it. By the 1980s only a few families still lived away from the centers of activity. The main road was improved, later relocated and paved, and the villages above and below reached toward it. Several small shops were opened. By the 1980s vegetables were being taken 8 km down the road for marketing at the small district center, which after independence in 1975 became the provincial capital. Even in the 1960s, after the end of the contracted migrant labor scheme in which many men had participated in the 1950s, some people began to leave the area to find opportunity in and around the more distant larger towns. This continued, so that by the 1990s some 10 percent of all Chimbu speakers lived elsewhere in the country. By the 1980s many children had been to high school, but up to 1984 none who had completed high school, or gone beyond it, had ever returned to live around Mintima.

THE CHANGING LIVESTOCK ECONOMY

Naregu had participated in a large regional ceremonial pig killing and massive prestation in 1956 involving thousands of people from many tribes. It took a few years between ceremonies to build up pig numbers, so the ceremonies moved between tribes, and those who did not kill pigs in 1956 did so in 1959. This was the pattern of these prestations in Chimbu. They entailed the coordinated activity of several thousand people over 2 years or more. At the major ceremonies held in neighboring group territories in 1959, we saw as many as 900 pigs killed, butchered, and cooked, and the meat distributed to several thousand recipients, all in a single day. Naregu should have held a new ceremony in the mid-1960s but delayed, and when they did hold it, in 1972, and then another in 1983, there was already an emergent emphasis on attracting paying tourists and less emphasis on the killing and distribution of pigs.

The livestock economy had been greatly disrupted by all the changes. Although the pig-based system had always imposed heavy demands on female labor, the coffee period increased these demands significantly.[4] Concentration of settlement, especially of women, and enlargement of the enclosed area increased the work needed to meet the demand for cultivated sweet potatoes for pigs. For a time in the mid-1960s the enclosed area was again reduced, and some long reentrants of foraging land were made into the enclosed block to bring the pigs closer to the women. Chimbu fences last only a few years, and pigs can break through them when they weaken. Sustaining the long fences proved too labor-intensive. In the late 1960s and early 1970s there were major problems as pigs broke into large areas of cropland. The fencing pattern of this period, and the settlement pattern in a stage of transition, are shown in an extract from the 1965 land use map in figure 1.2.

There were significant changes in animal husbandry. A practice, rare in the 1950s and 1960s, became common by 1984. Pigs were no longer fed at the fences but were brought into the cultivated land and tethered, usually to find their own food in sweet potato gardens from which the larger tubers had already been harvested for human use. By the mid-1980s and 1991, men handled the pigs in this way as often as the women. Partly in the 1970s, then more completely after the 1983 pig killing, fences were pushed further out so that enclosed land took up much more of the total territory than before. Pigs had become less important. Their place in cementing interpersonal and intergroup relationships was taken in large measure by money.

DIVERSITY FROM 1958 TO 1984: PUTTING IT TOGETHER

When Paula Brown and I went back to Naregu in 1984, we decided to resurvey land use and land holding in one part of our former survey area. We chose the land of one subclan that held just under 2 km² of land in four blocks extending from beyond the escarpment most of the way down to the bottom of tribal territory (figure 1.1). We already had land use, land holding, and genealogical data for this group going back to 1958, so we brought the series up to date. Data for 1958–59, 1965, and 1984 were then put into a geographic information system (GIS), yield-

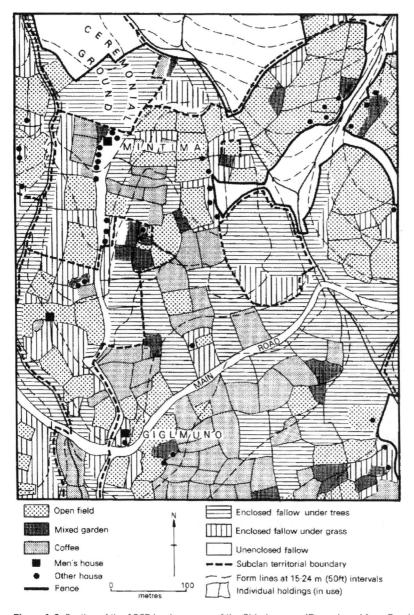

Legend:
- Open field
- Mixed garden
- Coffee
- Men's house
- Other house
- Fence
- Enclosed fallow under trees
- Enclosed fallow under grass
- Unenclosed fallow
- Subclan territorial boundary
- Form lines at 15·24 m (50ft) intervals
- Individual holdings (in use)

N

0   100
    metres

Figure 1.2 Section of the 1965 land use map of the Chimbu area. (Reproduced from Brookfield 1973:132, with permission from Edward Arnold Ltd.)

ing both statistics and maps. The results were illuminating (Brown, Brookfield, and Grau 1990).

The distribution of coffee was not very different in 1984 from that in the late 1960s because the innovation had reached a plateau by 1970 and very few had begun to replant. Nut pandanus distribution also was unchanged, but most of the

shorter-lived oil pandanus had gone and had not been replaced. Of principal interest were changes in arable farming, in the mixed gardens, smaller and more scattered than they had been earlier, and in the remaining unenclosed area. When the
data had been analyzed, it seemed we had been dealing with three, even four, distinctive farming systems. Two might properly be described as shifting systems,
those on the high northern land beyond the escarpment and on the lower southern
block (1 and 4, figure 1.1). In both areas, there was no continuity in the siting of
fields between dates and although some of the same people were involved, there
were also newcomers. Moreover, a good deal of land had been transferred tem-
porarily between members of different subclans of Naregu and, in the northern
case, other tribes from across the boundary. There was often, but not always, an affinal or matrilateral relationship between the givers and receivers.

The two main blocks showed a very different story. In both there was remarkable continuity. In the principal block 3, most members of the subclan lived in a
large village of more than 60 houses by 1984. The village and the central field area
lie along a mudstone spur, with lower, wetter, and unstable ground, mainly under
coffee, on its western side. Not only was much land that had been in use in 1958
and in the mid-1960s still in use in 1984, but it had never been uncultivated for
more than short periods in between. As I later saw, it was still in cultivation in
1991. Some of this was the shale soil land that in 1959 I had expected to go out of
cultivation soon. The mudstones are shales of marine origin that, weathering readily, release abundant mineral nutrients. Even in the absence of significant organic
matter they yield crops for many years. Humphreys (personal communication,
1991) has measured high yields on these soils, a few kilometers from Mintima.

Although there had been substantial redistribution of land between individuals,
those involved in the transfers had a close relationship. They were either agnates or
co-resident affines of those who held the land earlier. The plot boundaries were little changed, but this was not surprising in an area close to where people lived; any
attempt to dig a spade into a neighbor's land would quickly arouse heated protest.

The second block (block 2) lay almost entirely on the limestone-charged rubble
flow. It had a different land use history. Intensively cultivated in 1958, all but a
part of it gradually fell into disuse in the 1960s. For a period, most of it lay outside
the fences and was churned over by foraging pigs. In 1984 it was all in cultivation,
mainly with sweet potatoes. There had been more transfer of land than in block 3,
but the 1984 users included many of the same men as in the 1960s or their sons.
Where this was not so, most of the new users were members of the same subclan or
coresident affines; except for the affines, almost all were members of the same
exogamous clan. The boundaries in block 2 ran between prominent limestone
boulders of memorable shape. The 1984 boundaries ran between exactly the same
boulders as in 1958. Both these blocks, at least since the 1950s, have had quasiper-
manent systems of land use, but on the mudstone the land has never been left under grass or casuarinas for more than short periods. On the rubble flow it has been
cultivated continuously for a few years, then left under casuarinas for a similar
period.

We are here uncovering an important dimension of agrodiversity. Individual small farmers' land in Papua New Guinea has no registered title, and even tribal boundaries remain unmarked. In recent years they are again contested. Among the ordinary rural people, there is no legal impediment to change and adaptation in tenurial and transfer arrangements on any type of land. This is not so in all small farming societies, but it is quite common, and what we see in Naregu is an example of the infinitely variable and flexible arrangements that the rules of land tenure often constitute. We compared all transfers within the subclan, or to coresident affines, with all transfers between our subclan and other groups. There was a great contrast between the four blocks over the whole 1958–84 period. In the outlying northern block 1, beyond the escarpment, 89 percent of land had been transferred outside the subclan, and in the peripheral but increasingly colonized southern block 4, the figure was 38 percent. In the two central blocks the picture was totally different. On the rubble flow block 2, with its history of going in and out of cultivation, 12 percent of land had been transferred outside the group. On the mudstone block 3, where most people lived, the comparable figure was only 4 percent (Brown, Brookfield, and Grau 1990:42, table VI).

Within a territory smaller than 2 km^2, and in a system with a narrow range of main crops and only one basic method of land preparation, data over a long period showed considerable diversity in the frequency and manner in which the land was managed. Importantly, this is reflected in major diversity in the history of land tenure and land transfer. Until the end of our data series, Naregu and other Chimbu had continued to treat land as their most basic resource in organizing livelihood and status. Its valuation differed greatly according to the frequency with which it had been used. This in turn reflected a range of variables: land quality and ease of cultivation, suitability for the crop of highest value (which was coffee almost from the outset in this data run), accessibility, and proximity to the places where the interests of people lay. This last variable changed greatly over time, increasing the weight given to central areas close to the road and downgrading the value of remoter areas that earlier were close to the main foraging lands of the pigs. It had taken us a long time to tease out the full meaning of the trajectory of change.

DIVERSITY AMONG INDIVIDUALS

In the late 1950s the larger landholders were a distinctive group of men known as ground fathers who were the leading men, the organizers of pig and vegetable distributions, and, formerly, leaders in war. They sponsored young families of the subclan and sometimes their own affines who came to live with them. By the 1980s these men and their immediate male descendants no longer held prominent social roles; they held much less land, and there was more equitable distribution according to family size. Even so, some families still held little land, remaining dependent on others for access to resources. There were no tenants or sharecroppers, and the land poor were not prominent among those who had colonized new areas on the margins. These continuing inequalities are similar to those reported in other New Guinea highland societies; they are much less pronounced than they once were.

There was another dimension of diversity among the farmers and their families during this period of transition. Whereas almost all landholders had plots in each main area of subclan territory, in the 1960s and even in 1984 a few specialized in the use of particular areas where they made concentrated private farms. With resumption of intertribal warfare in the late 1980s, such outlying areas again became insecure. The private farms had vanished from the scene in 1991, and the inference is that they belonged only to the period of colonial peace. Such areas often were not close to the modern villages, and farmers in this category were among the more innovative people who, over the years, have farmed the remoter areas in the higher and lower parts of the several subclan territories. Members of a distinctive group have pioneered the occupation of new land, sometimes successively in different areas. The same farmers were prominent among innovators in cash cropping and in other ways. One once drained a pond by digging a deep ditch to obtain land for a mixed garden in which to grow a range of crops. Most of them were early adopters of family coresidence.

Among these innovators were two or three of my frequent companions between 1958 and 1970. Their offer to accompany me was one facet of their curiosity and willingness to try new ideas. In this area of limited economic opportunity, a new generation of keener spirits is more likely to seek employment in the towns than to experiment at home. Among these people we perhaps find reflected the type of New Guinea highland farmer who innovated so successfully in land management in the period between the introduction of sweet potatoes and colonial rule. In all farming societies there is a minority of keener spirits, among whom some become leaders and others provide example. Diversity on the land is better understood if diversity among individuals is recognized.

■ Bayninan, Ifugao, Philippines

PLACE, PEOPLE, AND THEIR FARMING

By 1963 I was ready to widen my explorations into farming systems, and I went to the Philippines. As a central part of that visit, I stayed several days with anthropologist Hal Conklin, then in the early stages of his fieldwork on the cultural ecology of the Ifugao of the mountains of northern Luzon in the Philippines. He spent a total of 38 months there in several spells between 1961 and 1973, based in the village of Bayninan, and ultimately produced the spectacularly comprehensive *Ethnographic Atlas of Ifugao* (Conklin 1980).

The car I hired from Bontoc, in northern Luzon, approached Bayninan along a deeply muddy road. Finally, the driver stopped and showed me where I had to get out and walk. I half-slid down a precipitous slope through cane grass and a sweet potato field, forded a small river, and climbed up and over another hill. Then I walked down a better path along the edge of a basin filled from side to side with young wet rice grown in steeply descending terraces. Relieved of my bag by a kind Ifugao farmer, I could better enjoy the beauty of the basin, and the deeper valley beyond it, and see the village ahead on a spur among a cluster of fruit trees. As soon as I arrived, Hal sent me out on a long walk with a young man who spoke no English and therefore could not answer my questions. On my return, Hal sat me down to write answers to a series of questions about what I had seen. Over the next days we walked and climbed all over a landscape of wet rice terraces, through temporary hillside fields mainly under sweet potatoes, into agroforests (or woodlots), and through the cane grass in between.[5]

I was there at the start of the main crop season, which in this wet and cloudy region begins in the dry season and extends into the start of the wet season. Rice planting mostly had been done by the time I came, and the heavy maintenance work on the terraces of the previous few months was over except for necessary repair of small failures in terrace walls. It was not yet far enough into the drier season for new temporary fields on the unterraced slopes to be slashed and tilled. Nor was there yet a serious need to maintain the irrigation systems that fed almost all the terraces, keeping them inundated during the driest period of the year before the rice harvest in July. The Cordillera gets its heaviest rainfall between July and October but, being on the eastern face, Ifugao gets lighter rain from the Pacific in most months, and only a short period in March though May is reliably dry. The western side of the Cordillera, at Bontoc and Sagada, was in the middle of a long dry period in February. There was heavy work on irrigation and, at Bontoc, on crops grown on seasonally dry fields. Cool and high Sagada, where reliable irrigation depended on steady flow from limestone with some intakes made many meters back into the caves, was a green oasis in that season.

Conklin (1980) provided abundant evidence on diversity, but he did not stress it. In an early paper he outlined a wide regional diversity in Ifugao farming, responsive to differences in altitude, slope, temperature, and rainfall patterns over an altitudinal range of more than 1000 m, but his atlas did not offer any review of diversity at this level (Conklin 1957, 1980). He placed his main emphasis on the one crop uppermost in the life of the people. This was rice, even though from one-third to two-thirds of Ifugao food was made up of root crops. At Bayninan, around 900–1500 m altitude, only traditional *tinawen* types of rice were grown. The varieties among them were not described by either Conklin or later writers (e.g., Andam 1995). However, Chang (1976; Chang, Adair, and Johnston 1984) noted that Ifugao was one of only three main centers in which the distinctive *javanica* rice was concentrated.[6] About a third of each family's plots was in glutinous rice, used for rice wine and the making of rice cakes rather than eaten as grain. Because inundation was permanent, or intendedly so, harvested rice was not threshed in the fields but was carried away to the village areas in bundles to be dried and threshed there. This made it hard to determine the yields. On the basis of his measurements on the product of 10 fields studied over 7 years, Conklin (1980:35) found a mean yield of 2.5 tons per hectare. Other estimates for the Ifugao region have ranged between 0.6 and 3.1 tons per hectare (Andam 1995).

The sweet potato, contrasted in its ecological needs, was grown both on temporary hillside fields (called swiddens by Conklin) and in drained and mounded terraces not found to have sufficient water for the continuous inundation needed by rice. Also important was taro, some of which was grown among the sweet potatoes and some along with rice in the inundated terrace fields. The pond fields provided important quantities of fish and shellfish. Conklin reckoned that each family household of about five people needed half a hectare of pond fields, one and one-third hectares of agroforest, and a quarter-hectare of temporary dry field, cultivated about 2 or 3 years, then fallowed under cane grass 6 to 10 years. A family also needed its livestock, mainly pigs and chickens.

When drought occurred, there was a shortage of irrigation water. More terraced fields were drained and used for root crops and vegetables, and more hillside dry fields were made. However the work time was distributed between field types, it absorbed about 400 person-days of family labor each year (Conklin 1980:37). Most of this labor went into the pond fields, including the heavy work of maintenance and new construction. When there was major work to be done the demand was much higher; because land was individually owned and operated, many farmers found it hard to mobilize labor. Irrigation systems usually were managed by small groups of farmers. Some larger systems were constructed in the colonial period when a class of wealthy farmers grew up, able to control the labor of their poorer neighbors (Eder 1982; Klock 1995a). Coordinated maintenance of long canals later became hard to achieve, resulting in loss of irrigation from some areas.

AGROFORESTS AND DIVERSITY

What Conklin called woodlots, combination tree farms and slope gardens, often were made on land that had been cultivated as temporary dry fields for a few years, then marked out by border plants as the continuing claim of the farmer. Volunteer growth yielded mainly firewood, the clearing of which permitted hardwoods to emerge as dominants. Other fast-growing trees, and rattans, were then planted. The agroforests at Bayninan supplied fruit, timber, firewood, and medicinal herbs; some contained as many as 200 plant types.

At lower altitude than Bayninan, Klock (1995a) later found 300 plant species in one agroforest, of which 77 were considered useful for timber and 121 for firewood. In the early 1990s, some timber was dragged away for sale, but the main uses were local, for house building and increasingly for the handicraft carving that had grown in importance as tourism expanded. Among 36 wild and cultivated fruit species, the most important was a rattan *(Calamus manillensis),* which produces edible and readily marketable fruits. Coffee was introduced by the Spaniards, then more strongly promoted after 1900 by the Americans, who also introduced a number of other fruit trees that have been adopted (Klock 1995b). In the early

1990s, a ficus–rattan–coffee cash crop association had become common, the result of selective cutting and interplanting. The three-story woodlands also were important sources of forage for pigs. Agroforests close to the rice fields had value in conserving supplies of irrigation water and were carefully maintained in the 1990s. Those further afield had been cared for less well. Like the publicly owned natural forests, they were no longer protected by customary sanctions, with adverse consequences for the water supply of rice fields lying downslope (Klock 1995a).

Agroforests were individually owned, commonly inherited jointly with the rice-growing pond fields to which they ranked only second in value. They remained a major element in the total agroecosystem. The temporary dry fields, which grew most of the sweet potatoes in Conklin's time, were on sites selected within community-owned land and had much lower value in Ifugao perception. They were individually owned only while in use, and there was no enduring claim unless the site was subsequently converted into a private agroforest.

Although the system comprised only three main elements and had only two dominant crops, cropping diversity was general. Conklin (1980:25–26) listed a total of 48 cultivated plants and wrote,

> A transect through any valley of swiddens, woodlots and pond field terraces reveals many kinds of crop production within the same field zones. Taro is . . . intercropped in both swiddens and pond fields. . . . Numerous other crops are planted in swiddens, in woodlots (especially near hamlets), around the edges of house terraces, in temporarily dried up pond fields, in mulch-mounds in or alongside of pond fields, along pond field embankments and dikes, and in ridged and temporarily drained terraces.

CHANGE THROUGH TIME

The Ifugao farming system, with all its splendid landscape architecture, was created by the Ifugao farmers and their ancestors. No superior authority more powerful than the wealthy of a small area had ever directed the labor to make the terraces and irrigation systems. They had been sustained through many years, but frequent repair had led to changes in their shape. New terraces, protected by stone walls or earth banks, had been built over old terraces, and comparison of photographs taken at different dates showed that others had been more recently cut out of the upper hillsides. This process may now have ended, but it continued through the time of Conklin's research in the 1960s. The clearing of forest margins in this way was blamed by some observers, especially the authorities, for loss of irrigation water from fields too low to be fed adequately by the mountain streams and springs and too high to receive canal water diverted from the rivers (Eder 1982).

The soil of the terraces is an artificial clay enriched by aquatic plants. Spoil to make new terraces was brought there mainly by directing the flow of water; this is why most of the heavy work on the terrace system was done during the wettest months. Nothing that water could move was moved by hand. All this had been going on for centuries. Although the Ifugao system must have changed greatly after the introduction of sweet potatoes, which the Portuguese and Spaniards brought to southeast Asia from South America in the sixteenth century, archeological evidence takes it back 300 to 500 years earlier. It is still a remarkable system, scenically spectacular, with great flights of more than a hundred hillside terraces flanking the larger valleys. It is also a flexible system, able to make space for new crops, more people, and new demands and to permit survival in climatically exceptional years. And up to Conklin's time, and even later, it has been under construction and modification.

Interpretation through time is made difficult in this case by Conklin's use of the ethnographic present, which he defined as the 1960s. The major post-1945 changes in use and management of land in response to commercial opportunities, described for the more accessible Agno valley in the western Cordillera by Lewis (1992), seem to have had no parallel in Ifugao. Nonetheless, commercialization, tourism, and emigration have all been increasing in Ifugao throughout the period since the American colonial occupation (1898–1942) and have accelerated since the 1950s. Some later accounts, especially that of Klock (1995b), described the Ifugao system in much the same terms as did Conklin, and it is difficult to place a gloomy account by Eder (1982) in this context. Eder wrote of extensive loss of irrigated land, especially in the large valley where the regional capital, Banaue, has grown up. He described this loss as having been progressive since the 1950s and as caused by a mixture of deforestation, social change, and growing dependence on government to take initiatives and provide funds. I can confirm Eder's basic observation. I saw this valley again in 1985, when the continuous flights of verdant pond fields seen in the 1960s were broken by large areas not merely under dry crops but under unproductive grass. But Eder spent only some 2 months in the field in 1977, and his analysis cannot be given equal rank with that of other observers.

If I had thought in terms of agrodiversity in the 1960s, my experience in Ifugao might have been the start of a systematic comparative study. The areas of Papua New Guinea in which I had already worked since 1958 were dominated by a single crop, the sweet potato, and had only a more subtle complexity. I did present a paper comparing agriculture in the mountains of Luzon and New Guinea at a conference but never tried to publish it. It was too soon. The true depth of Ifugao diversity could emerge only from the long observation Conklin was able to sustain. He published its full results only in 1980. Fuller understanding of the complexity in New Guinea could not come until many years of research results had been assembled.

■ Comment: Dimensions of Diversity

Although their landscape architecture is almost unique, Ifugao farmers were not so different from those of Chimbu or other regions. About two-thirds of all the world's farmland is used by small farmers who cultivate the land with simple tools, grow a wide range of crops, and have received only limited benefit from the innovations of the modern Green Revolutions. The farming landscape of these many millions of small farmers is very different from the wide, uniform fields of most developed countries. Theirs is a landscape of great diversity. Small farmers often make detailed use of small local variations in soil, microclimate, and water conditions and often produce a great variety of crops. Commonly described as traditional, this farming landscape is very dynamic; in it the results of learning and experiment are expressed in ways that are constantly new. To try to understand and explain this dynamic diversity and its implications are the purposes of this book.

Elements of the manner in which farmers relate their agriculture to the detail of landscape and often modify that landscape, or transform it in the Ifugao case, are encountered here. Also revealed is crop and plant biodiversity, the genetic diversity of plants that they use and manage to create enduring crop varieties, or landraces.[7] Centrally important is the internal dynamism of so many small-farming systems, yielding a constantly changing patchwork of relationships between people, plants, and the environment.

These elements are widely repeated on the small farms in the developing countries of the South. The three elements are closely linked in practice but rarely in the literature. All depend heavily on what farmers know, how quickly they learn, and how well they use information. Farmers are always coping with new problems, finding new ways. Farmers experiment constantly and have done so over long time, using well-structured methods, as Paul Richards (1985) has cogently argued. Change in small-farming systems arises largely from within and is continuously in progress.

Ability to manage variation in natural conditions is critically important. Not only are drought and flood, varying soil moisture conditions, and degradation involved, but also plant biology, pests, and diseases. However they conceptualize the natural processes with which they have to deal, farmers are adapting to them and seeking to manage them rather than responding to any set of fixed and stable conditions. Most significantly, small farmers must also adapt to changing social, economic, and political conditions over which they have no direct control. Cases of the latter kind are exemplified in chapter 2.

Notes

1. Except for a few places on the fringes, the central highland valleys of Papua New Guinea remained unknown to literate outsiders until the 1920s. Even though a major and ancient trade route from the lowlands entered the central highlands down the Chimbu valley, the Chimbu area was not penetrated by a European party until 1933.

2. Paula Brown used the ethnographic present in some of her writings, specifically to describe "traditional" or precolonial times (e.g., Brown 1995:24–30). I sometimes wrote in the present tense, too, but usually in a context of the immediate past of recent fieldwork.

3. This applies especially to Chimbu, where the larger tribal groups recognized as warring entities in the 1930s already had combined populations of 1000 to 5000. Elsewhere in the highlands the largest political grouping generally was smaller. "Big men" emerged in all areas, in some becoming notably differentiated from the mass by their possession of pigs, shells, and wives and ability to mobilize the labor of their clients (Strathern 1966; Waddell 1972; Feil 1987; Allen and Crittenden 1987).

4. Writing of a more westerly highland people, and following Modjeska (1982), Allen and Crittenden (1987:150) wrote this about the roles of men and women: "In the open fields dominated by sweet potato men do the basic work of land preparation, but in the mounded areas women then break down old mounds and reform them, collect compost and planting material, plant, weed and harvest. They are also wholly or mainly responsible for raising pigs. Men's labour thus creates capital improvements in the land; women's labour, on the other hand, is repetitious and accrues little benefit from cycle to cycle."

5. Conklin's initial term for these important cultivated woodlands was *private forest.* Later he used *woodlot,* and the same term was used by Klock (1995a, 1995b). Nonetheless, they are more generally called agroforests, in no sense natural and in large degree managed.

6. This was the predominant rice in Bali and eastern Java until it was replaced after serious pest infestation in the 1970s (see chapter 12). Northern Luzon was the second area of concentration of this variety, which probably developed late in the history of Asian rices.

7. A basic definition of a landrace is provided by Sauer (1993:275): a "sexually reproducing crop variety developed under local natural and folk selection." Sometimes they are described as local or folk varieties. As such, they are best known to farmers themselves, often are known only by local names, and below the species level are not easily defined in taxonomic terms. Among the 5000 or more landraces of potatoes and 6000 of maize used in Andean farming are many subvarieties distinguished only by color. The potato varieties distinguished by tuber characteristics only are not yet understood in terms of genotypes (Zimmerer 1996:10; Brush 1992:164). In Indonesia alone it has been estimated that there were more than 8000 landraces of rice before the 1960s, and more than 6800 are preserved *ex situ* at the International Rice Research Institute (Fox 1993b; Chang and Vaughan 1991).

Chapter 2

Diversity, Stress, and Opportunity

■ Three Contrasted Examples

This chapter continues the presentation of agrodiversity by example, with an emphasis on externally generated change and the stresses and opportunities that result. The two communities described in chapter 1 had a good deal in common, both being in remote mountainous areas of the southeast Asia–New Guinea region. Most of the developing world's small farmers have experienced stronger external forces than either of these two groups, making it more difficult to maintain their diversity. These other farmers must be introduced by example before the main argument begins. The three cases presented in this chapter are strongly contrasted in their basic ecology. Each has also been affected in a different way by external forces. One is in the Andes of South America, one is in central Africa, and one in the lowlands of Southeast Asia. In the first, there is a long historical experience of enforced change, culminating more recently in major penetration by the market economy. The second has a much shorter history but involves a period of extensive colonial interference and ends in civil war. In the third, rural communities of no great age are being swallowed by an expanding city. Each example raises its own set of themes.

■ Threats to Crop Biodiversity: Paucartambo, Peru

FARMING SPACES ADAPTED TO PLACE AND TIME

The first example concerns a valley in the eastern Andes where cultivated land rises from 2800 to 4100 m, in an area about 50 by 30 km, ranging in terrain climate from superhumid to semiarid. The valley has a history of strong external interference going back around 600 years. First, the Inca empire demanded tribute and labor, then the Spanish demanded more of the same before introducing the hacienda, which swallowed most of the land for more than 300 years, reducing the peasants to a condition of semiservitude. The modern story begins with a land reform in 1969. It brought only limited benefits to the peasants but increased their participation in the market economy. It was described in great detail by geographer Karl Zimmerer (1996, 1998).

Some background is necessary. Ten million small farmers in the Andean countries cultivate a great range of crops; if the many landraces of potatoes and maize are taken into account, they cultivate the greatest crop biodiversity on earth. Most Andean farmers herd livestock in addition to growing crops; this enables them to use land unsuitable for crop production. These include the pre-Hispanic domesticated camelids, alpaca and llama, as well as introduced cattle and sheep. In the Peruvian Andes, a vertically stacked array of land uses from the lowlands to above 4000 m is apparent at a small map scale and seems closely to reflect natural ecological zonation in the mountains (Troll 1968). Secure production entails access to land at an array of elevations and some perceived simplification of natural diversity. Its main common elements have been an upper zone specializing in tuberous crops, the many varieties of potato and ulluco (*Ullucus tuberosus*), and a lower zone specializing in maize and quinoa (*Chenopodium quinoa*). Greater refinements at more local level showed a zonation upward from subtropical fruits in the deeper valleys through maize, introduced wheat and barley, potatoes and ulluco, and finally grazing (Brush 1977). Subsequent work specializing on control of resources has led to the view that production zones did not arise directly in response to natural conditions but were created by people acting as communities (Mayer 1985). In the 1980s, the 20,000 farmers of the Paucartambo valley organized their production within just four farming spaces, corresponding to Mayer's production zones, which overlapped significantly in elevation.

Two of these, the hill farming and valley farming spaces, were ancient. Between them lay a space that came into being only in Hispanic times and has recently expanded; it has been called by several names, of which Zimmerer selected *the oxen area,* or land on which the plow is used. The colonial period in the Peruvian Andes began with the Spanish conquest of the Inca Empire in 1532. Oxen, for work on land seized by the Spaniards, were introduced only a few years later. The fourth and most specialized space was much more recent in creation. Devoted to early-planted potatoes, it emerged only in response to market opportunities. Each had a distinctive assemblage of crops and distinctive methods of management. Their fluidity in modern times reflected a decline in communal management of land in sectors, which were fallowed together to simplify livestock rearing. The land reform accelerated a trend toward individualization of management decisions, reducing the role of community institutions in sectoral land management. The rapid decline of collective management institutions was perhaps indexed by the successively lesser emphasis given to their role in regulating land use by Mayer (1985), Brush (1992), and Zimmerer (1996), based on research in the 1970s and the early and late 1980s, respectively.

Hill farming contained most of what remained of the sectoral fallow system. All crops were tuberous. In other parts of the Andes an anciently domesticated grain plant, cañihua (*Chenopodium pallidicaule*) was also grown in the highest land, as high as 4500 m in southern Peru and Bolivia (Sauer 1993:33–35). The potato was overwhelmingly dominant, with ulluco, oca (*Oxalis tuberosa*), and mashua (*Tropaeolum tuberosum*) in lesser quantity. In the Paucartambo valley alone there were 79 potato landraces and 5 of ulluco. Some landraces were limited to the upper zone. High-altitude potato landraces are frost resistant and bitter and are processed by freezing and drying. Other landraces have a much wider altitudinal range. Hill-farming land was prepared by hoeing the soil into high beds, breaking the sod with the Andean foot-plow, the *chakitaklla,* and pulverizing the clods by hand. Fallowing was normal practice, enduring up to 7 years. Small livestock grazed during the fallow period, the manure being added by folding the livestock on intended field sites at night. Whereas the sectoral fallow formerly separated crops from livestock, and farmers still grouped their fields together to facilitate protection in the 1980s; some fenced their fields, an added labor.

Valley farming was in sharp contrast. A great variety of crops were sown on the steep lower slopes in fields that were mostly walled to protect crops from livestock and define the boundaries against neighbors. The dominant crop was maize, of which Paucartambo farmers cultivated about 10 percent of the 100 or more races that exist in Latin America. Wheat, as important as maize in lower and drier valleys, was not a significant crop here. In the Uchucamarca valley, to the northwest, wheat provided about half the calories and proteins available in the early 1970s (Brush 1977:78). The second most important crop in the valley farming of Paucartambo was quinoa. The nitrogen-fixing legumes tarwi (*Lupinus mutabilis*) and broad beans (*Vicia faba*) were less commonly field dominants, with lower yields. Valley land was scarce, not available to all households, and was used intensively. Intercropping was a common practice and had increased since the 1960s. Formerly, tarwi and broad beans were grown mainly higher on the slopes, but in the Paucartambo valley more than in other areas there has been a reduction in their area as sole crops. They were interplanted in rows with amaranths, kidney beans, squashes, and quinoa, and also in gaps and corners of valley fields dominated by maize. Livestock manure, guano from the coast, and chemical fertilizers were used, together with crop rotation, to reduce fertility exhaustion.

THE GROWTH OF COMMERCE AND ITS CONSEQUENCES

By the 1980s, valley farming had become constrained by growth of the oxen area above it as a major zone of modern commercial farming. During the colonial period, it became common to use oxen-drawn plows on this midaltitude region of somewhat less precipitous slopes for growing wheat. The area has been enlarged in recent years for contract production of barley for a beer factory in Cuzco and for high-yielding modern varieties of potato grown mainly for market sale. The high-yielding modern potato varieties were agronomically superior to the local landraces but were seen as greatly inferior from a culinary point of view. As both Brush (1992) and Zimmerer (1996) showed, by no means all marketing was of the modern varieties, and a range of landrace potatoes were sold in the urban market and exported from the region. They commanded a higher market price. An array of farming habitats were subjected to growing uniformity in both crops and management. Sectoral fallowing continued to be practiced in parts of the oxen area, but with new rotations in which barley ousted ulluco, broad beans, and even wheat from the sequence. Many plots of landrace potatoes remained, and more remained in other valleys where contract production of barley was not developed (Brush 1992). Early-planted potatoes were common in this belt and became particularly characteristic of the fourth farming type, the early-planting land, which was much more ecologically specific in that it depended not only on temperature and slope but also on access to water during the dry season. Much of the early-planting land was naturally watered by springs and seepages so that it had to be mounded and ditched as thoroughly as any of the hill land to sustain a dry habitat for tuberous crops.

The commercialization of farming that brought about this change began before the land reform, but the end of the hacienda system freed peasants to engage much more thoroughly in commerce and seasonal off-farm employment. It imposed a new set of work routines geared to the contract demands of the brewing company and seasonal prices in the urban market at Cuzco. Not only have the commercial farming spaces enlarged at the expense of the older hill and valley spaces, but there has also been significant loss of landraces, the labor demands of which conflicted with commercial imperatives. Zimmerer (1996:148–85) discussed the losses of crop biodiversity that have taken place but also showed that there is significant regional variation and that fewer varieties have vanished than is commonly supposed. Brush (1995) came to a similar conclusion. It is of particular interest that loss of crop diversity was greatest on the land of the middle-ranking farmers. The poor had little access to the land most valued for commercial production and farmed mainly on the hills; the wealthy sustained production of a wide variety of landraces for reasons of dietary preference, for use as gifts, and as payment to workers. Diversity of maize also remained high, for many of the same reasons. As in other parts of the Americas, the utility of a variety of maize landraces for consumption, rather than their advantages in production and for marketing, sustained crop biodiversity.

Zimmerer remained optimistic that a large measure of diversity would survive, but he noted the severity of the stresses, especially where they were combined with severe soil fertility loss caused by neglect or failure of livestock production, neglect of conservation works, and soil erosion. In one part of the valley, many families had reluctantly taken to buying noodles from village stores because these subsidized imports were less expensive than home-grown food. He finally concluded that "tragically ironic, many citizens in the country whose predecessors deeded the world its greatest heritage of diverse farm plants are surviving on floods of the cheapest imported foodstuffs" (Zimmerer 1996:232).

■ A People Resettled Again and Again: The Zande of the Southern Sudan

A REMOTE SOCIETY MADE FAMOUS BY ITS INTERPRETERS

Adaptation and its limits are the central message of the second example. The case of the Zande in central Africa is one in which unwelcome externally generated change has been frequent and extensive.[1] Zande are widely known for other reasons, for they have been subjected to a remarkable concentration of mainly anthropological research. In the period from 1920 to the mid-1950s, successive government officers invited anthropologists and others to work on the Zande. By the early 1950s, the Zande had already become well known through the writings of Edward Evans-Pritchard, who worked there between 1926 and 1930 and had already written one classic. He went on to write much more in his later years (Evans-Pritchard 1937, 1951, 1958, 1960, 1963, 1965, 1971, 1974).

Zande has a huge literature. Evans-Pritchard's earliest paper on Zande appeared in 1928. Several of his papers are reprinted or slightly amended in later books. The first detailed account of an African agricultural system in all its complexity was written about the Zande just before Sudanese independence, by an agronomist, Pierre de Schlippe (1956). He had been hired to assist in monitoring a government-sponsored rural development scheme begun in the 1940s (Wyld 1949). The same program was also studied by another anthropologist, Reining (1966), who arrived in 1952. Including these authors, by about 1970 some 60 people had written on the Zande (Bicknell 1972:41). For reasons that this chapter makes clear, their numbers have not increased significantly since that time.

There is a problem in using this literature. Each principal writer on the Zande drew on a large literature going back to mid-nineteenth-century explorers, but in the days before interdisciplinary or multidisciplinary approaches became popular, they made remarkably little reference to the work of colleagues outside their own fields. De Schlippe wrote that he was advised by Evans-Pritchard not to read the latter's work before doing his own. In his principal book of collected results, Evans-Pritchard (1971) made only passing and technical reference to de Schlippe's work. Reining (1966) made only a single reference to de Schlippe's work, although it was published 10 years before his own book and focused on the same development scheme. The first attempt to put the Zande material as a whole into an ecological context was by Singer (1972).

The Zande live in what are now three countries, Congo (formerly Zaire), the Sudan, and the Central African Republic, before the late 1950s respectively the Belgian Congo, the Anglo-Egyptian Sudan, and French Equatorial Africa. They occupy an undulating region developed on the ancient basement complex of Africa, on the divide between the Nile system to the north and the Congo system to the south and west. This is a large area extending over some 8° of longitude and 3° of latitude, just north of the equator, ranging from the margins of the tropical rainforest to the edge of the subhumid savanna. As a single people, Zande came into existence in the eighteenth century, the product of conquest from a heartland in what is now the Central African Republic. A successful ruling clan built a kingdom that assimilated as many as 30 other peoples in a series of conquering wars that were unchecked until the late nineteenth century. The historical evidence assembled by Evans-Pritchard indicates that hunting was more important than farming until about the mid-nineteenth century. Zande acquired use of eleusine or finger millet (*Eleusine coracana*) as their staple crop from farming people they conquered.

Long before Evans-Pritchard began work in the 1920s, Zande had become a mainly agricultural people. Hunting was of secondary importance, and collecting termites probably was of greater dietary significance. A highly complex system of crop magic, in addition to medicinal plants related to human illness, used a wide range of plants and their juices. The practice of magic, belief in witchcraft as the source of misfortune, and reliance on oracles were of major importance in the 1920s and perhaps only a little less so by the 1950s.[2] Early in the twentieth century, farmsteads were located on the deeper colluvial soils along streams, close to strips of gallery forest, dispersed but in loose groupings that were the fief of a chief or subchief whose power was already significantly less than it had been in precolonial times. Large areas of unused land lay between each group. The thin soil of the higher ground was hardly farmed at all. During the wet season most of the landscape, except the plateau caps, was densely covered in very tall elephant grass (*Pennisetum purpureum*) together with some other grass species, and the paths linking homesteads together were the only way to pass readily through this tangle. Each dry season the grass withered and much of it was fired, creating an open landscape within which new gardens were made, and planted early in the wet season that followed. Agronomist Pierre de Schlippe (1956:45–47) noted that Zande named 700 trees, shrubs, climbers, grasses, and herbs. They also named at least 15 different microenvironments, together with some that were also distinguished by the name of the dominant plant cover. The qualities of each were well known.

A HUNDRED YEARS OF LOST SOLITUDE

The central interest of the Zande story lies in a remarkable amount of external interference with their settlement pattern and society. They were conquered just after the beginning of the twentieth century and subjected to colonial administration from about 1910. In the early 1920s the British, Belgian, and French colonial authorities determined that Zande should move from the valleys onto the ridges to reduce the incidence of sleeping sickness among them and facilitate its control—and also facilitate colonial administration and tax collection. Houses were strung out in loose groupings along newly made roads. Thus farming was shifted from the deeper and more clayey soils of the lower slopes onto thinner and sandier soils on the uplands. Farming had been moved from the lower to the upper parts of a catena, a concept developed in chapter 5. The roadside settlements acquired a look of permanence still present in the French and Belgian territories in the 1950s, but soil fertility exhaustion necessitated more frequent movement of field sites. As distance from compound to field grew onerous, temporary or even principal compounds were set up away from the roads.

Agriculture was based on finger millet, nearly always mixed with other crops including pearl or bulrush millet (*Pennisetum glaucum*), cowpea (*Vigna unguiculata*), and Bambara groundnut (*Vigna subterranea*). These grains and pulses are domesticates of the savanna region between the Niger and Ethiopia, but varieties have been introduced from nearby regions on all sides. Yams were also indigenous, but many other plants that had become widespread in the early twentieth century were introductions. Sorghum (*Sorghum bicolor*) and the green gram (*Vigna mungo*) were introduced from the drier regions to the north. Other crops, including bananas, came from the south and west, though from within Africa. Maize was so firmly established as to be seen as traditional in the 1920s, and this American crop may be as old as any Zande agriculture because it was already in the Congo basin in the seventeenth century (Sauer 1993:233). Groundnuts, by the 1950s the second crop in area on Zande farms, originated in South America and came to Africa from the island Caribbean, very early in the slaving period. They also seem to have been an early crop in Zande farming, although newer varieties have been introduced.

Manioc (*Manihot esculenta*) may not have reached Zande from the south much before 1900. It had replaced eleusine to become the largest single item in the diet by the 1950s, and was invaluable for its tolerance of poor soil conditions and its ability to be kept unharvested in the ground for some time. Sweet potato, grown in gardens around houses, was also a modern introduction. Even more recent was the mango (*Mangifera indica*). Introduced only in the colonial period, mango was first planted along roads and later around all houses. Mango fruit ripens at the end of the dry season in March and April when, in a time of annual food scarcity, it provided invaluable supplements of vitamins A and C. Reining (1966:72) cited a Zande of the 1950s who described it, perhaps very appropriately, as the best thing the Europeans ever brought to his country.

Before the 1950s, Zande farming was wholly for subsistence, and there was little cash cropping in the Sudan part of Zande except for chilies, which, dispersed by birds, were collected rather than cultivated. Toward the end of World War II, the ambitious rural development scheme on which de Schlippe (1956) and Reining (1966) wrote was introduced to the British part of this frontier area. The purpose of the Zande scheme was not only cash cropping, but also the creation of a small industrial complex, making cotton cloth for local and Khartoum use, expressing oil, and making soap. With mixed motives, analyzed both from the records and from Zande experience by Reining, cotton cultivation was enforced on farmers who, only in the Sudan, were again resettled, this time away from the roads.

Beginning in 1946, large areas were divided into parallel rectangular blocks or strips, 10–16 ha in size, and allocated to Zande farmers, who were settled on them in a manner that broke up former groupings. The new distribution almost totally disregarded soil types so that although some resettled farmers got good land, others got mainly poor land. More than 60,000 families had been resettled by 1950. The factory was built and brought into production in 1951. Rewards to the farmers were small, made smaller than they should have been by the pricing policies of the scheme authorities. Production peaked in the early 1950s and then declined. It was in this constrained environment that Sudan Zande farmers set out, once again, to reestablish their evolving farming system. This was the system that de Schlippe investigated from 1948 to 1953, at first with a team of ill-trained assistants, then on his own. He provided impressive evidence of the adaptability of Zande farmers.

ZANDE FARMING IN THE 1950S: SEEMING CHAOS INTERPRETED

In this new context, the Zande homestead may have been a little smaller than in Evans-Pritchard's time, but it retained the same characteristics. Many external observers found the homesteads to be chaotic, and so initially did de Schlippe (1956:101). He went on to show that everything was in fact in its place. The seemingly randomly placed buildings were disposed so that those of each wife formed a group. The crops of each wife lay behind her own house and storage huts, on the outskirts of the compound. De Schlippe drew a distinction between gardens and clearings (or fields), the former close to the compounds, the latter extending over some distance. On these latter, land normally was cultivated for at least 2 years, and for longer on the better soils. Except for cotton, almost all crops were interplanted in associations. At the beginning of the 1950s, the principal association still included eleusine millet associated with maize, manioc, sesame, the local domesticate hyptis (*Hyptis spicigera*), and sorghum, together with fiber crops and vegetables. On better soils, a distinctive eleusine–groundnut association dominated, with groundnuts usually dominant in the first year. Closer to the houses was an association of maize with oil seeds and gourds. Manioc normally was a second-year crop with a life extending into the fallow years, so that manioc in the fallow constituted an additional field type.

Field types were distinguished by the method of land preparation as well, whether cutting and burning, partial burning, or planting among grass or through mulch. Adjacent to the family compound, sometimes spreading along paths, were the more distinctive crop environments of longer-lived gardens, especially on and close to ridges created by moving the topsoil to make the courtyard and paths, on old dump heaps and dead termitaria. Here were mainly grown sweet potatoes with manioc, some of the sorghum, and associated maize. Plants used for magic, vegetables, and tobacco were grown closest to the houses.

Generalizing from the detailed reality, de Schlippe grouped the associations into a small number of repeated field types, each with its own ecological characteristics, its own set of crop associations.[3] Within the fields, however, patches existed in which one crop among a mixed sowing dominated. These reflected microenvironmental differences, or differences in the timing of rain suiting one crop rather than another. The newly introduced cotton, planted in monoculture, interfered with this complex system. The idea had been that farmers would develop their own rotations within their blocks, ceasing to shift homestead sites, and would adopt soil conservation methods at the same time. This system did not work well, and the soil conservation design did not work at all. Zande may not have needed to move frequently for reasons within the farming system, but they still needed to shift home sites when death and misfortune occurred. The manner in which they had been individually settled deprived them of the support of nearby kin and affines at such times. By 1953 the block allocation system was starting to break down, and its enforcement was eased tacitly as movements and exchanges became frequent.

Zande were unfamiliar with land ownership, and in the resettlement system the holding boundaries sometimes were transgressed. Each family, and each woman within a family, owned the crops and the clearings in which their labor was invested. Some fields, on which most work after clearing was also done by women, belonged to men. Field sizes were determined by needs and availability of labor. Labor was shared between adjacent farms. The crops of each woman's fields went into her own granary and were used mainly for feeding her own family. De Schlippe found that each woman had a complete set of field types of her own, extending outward from her own section of the compound. The seeming chaos therefore was order, an order that was related closely to the pattern of natural resources and the created resources of fertility provided by wild vegetation on land formerly cultivated and by household refuse thrown on to areas around the compound.

ADAPTABILITY STRETCHED TO ITS LIMITS

Zande farming has been adaptable from its eighteenth-century outset. Almost all the crops and their varieties were adoptions or introductions. One of the principal established associations described by de Schlippe, that of groundnuts with eleusine, had come into being only after the introduction of a new and more robust variety of groundnut in the 1920s. In the 1948–53 period, de Schlippe and his assistants observed several experiments in crop associations and cultivation method and also imitation of successful practices from neighbors. Sudanese Zande had to adapt to being relocated twice in not much more than a generation. The environment and its elements had to be learned, first in one settlement location, then in another, and then in a third, and field types developed in relation to this knowledge. Colonial prohibition of cultivating land close to the rivers, sustained to the end, deprived Zande farmers of the use of their most fertile soils, and this constraint had to be accommodated. Cotton had to be brought into the system and the strains it created by competing for working time with the eleusine harvest accelerated principal reliance on manioc. The end of colonialism in 1956 did not free Zande from restrictions and initiated a new period of major disturbance.

Independence and appointment of northern Sudanese as government authorities were followed quickly by an uprising that was repressed severely and led to the end of all professional observation by outsiders. Cotton production had been almost halted by discontent with the scheme in the mid-1950s, but it was resumed despite desultory civil war that continued until 1972. There were then efforts to revive the scheme in the mid-1970s, and in 1976 and 1979 the bat-infested but still operating cotton factory was the workplace of index patients in two outbreaks of Ebola virus disease in Zande. Reports by visiting health specialists, seeking other victims in Zande households, suggest that most settlement remained dispersed along the same lines as in the 1950s. Teams of health workers, searching for cases and contacts, visited family compounds "constructed on relatively isolated clearings in the dense grasslands" (Baron, McCormick, and Zubeir 1983:997). Civil war resumed in 1984 and continued for the rest of the century. The more varied cash cropping that had evolved during the 12 years of peace deteriorated, and cotton production "all but disappeared because of marketing problems" (Dickie 1991:284).

There is an irony about the case of the Zande. In most writing about shifting cultivators, de Schlippe's early account has a deservedly prominent place, though one less prominent than that of Conklin (1957) on the Hanunóo of the Philippines, which in this book is introduced in chapter 6. Yet the Zande whom de Schlippe described were adapting to enforced resettlement, and both the Zande and Hanunóo suffered grievously in the subsequent half-century. The point made in the introduction about the need to take account of stressful change is strikingly underlined.

■ The City in the Village: Four Villages Around Kuala Lumpur, Malaysia

FOUR VILLAGES IN A CHANGING SOCIETY

The third example is different again, and this one draws on my own work. It concerns the swallowing up of villages by an expanding city, reducing their agricultural diversity but widening the diversity of livelihood sources. More comprehensively than Paucartambo, it illustrates the role of opportunities as much as that of stressful events. The study I undertook in Malay villages around Kuala Lumpur in the late 1980s, with Abdul Samad Hadi and Zaharah Mahmud, was very different from my earlier work in Chimbu. We could spend only limited periods of time in the field, and we relied heavily on mapping, on the records of land holding and transfer, and on a questionnaire survey administered by students to 731 households. Other than the land records described here, historical sources were scanty, but we were able to map land use accurately from 1966 air photographs in sufficient detail for precise comparison with our 1986 mapping. Fieldwork and documentary research were undertaken at intervals during 1986 and early 1987, and all areas were revisited in 1989 before completion of a single major publication (Brookfield, Abdul Samad Hadi, and Zaharah Mahmud 1991). I last visited the area in 1992.

Kuala Lumpur began life as a tin-mining camp in the 1850s in what had been a sparsely peopled area of small Malay villages among dense forest occupied by aboriginal Orang Asli, who practiced some shifting cultivation but relied mainly on foraging. Further south, from as early as the sixteenth century but mainly in the eighteenth century, there had been substantial settlement by matrilineal Minangkabau from Sumatra. These people brought with them more elaborate wet rice cultivation methods and more complex dry crop and tree crop production systems than those used by the Malays of the peninsula (Abdul Samad Hadi 1981).[4] Minangkabau settled to within 30 km of the site of Kuala Lumpur by the early nineteenth century, and the oldest of our four villages, Beranang, remained predominantly Minangkabau in the 1980s. Later migration from Sumatra led to establishment of two of our other three villages, all of which were set up within about 15 years of 1900 (figure 2.1).

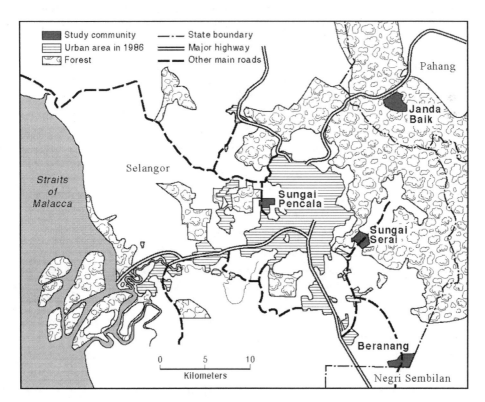

Figure 2.1 Kuala Lumpur and the four villages discussed in chapter 2.

By 1900, there were two important changes in the farming conditions in these villages. The essentially informal land tenure arrangements of the precolonial period were replaced after 1886 by registration and the issue of a title. This happened in all established Malay village areas, and it became the means by which new land rights could be established, especially after Malay reservation areas were first set up after 1913. These titles were an important source for us. They included a lot number, its location and size, the name of the first owner and subsequent transfers, date of registration, annual rent, and the actual or permitted use of the land. The information on transfers included either the price paid or an evaluation by the office staff of what should have been the price. These well-kept but incomplete records were analyzed in the district offices where they are held by Ms. Ho Wai Suet.

Second, commercial agriculture had reached the villages. The first commercial crop taken up widely was coffee, which failed, followed by coconuts. There were many experiments with *Ficus elastica*, planted among fruit trees, before the success of Para rubber (*Hevea brasiliensis*), first planted on estates in the 1890s, was demonstrated early in the new century.

Rubber, not necessarily *Hevea*, was already being planted at Sungai Serai by 1902. General adoption of Para rubber by 1909 led to a sustained demand for title to the hitherto little-used upland areas. Sungai Serai was first settled in the 1870s, when the whole valley was occupied by a more modern wave of migrants from Sumatra. By very early in the twentieth century, its economy was already based on a rice-and-tree-crop complex, but only limited areas were planted to wet rice, which proved vulnerable to drought. Sungai Pencala, an early-twentieth-century offshoot of earlier Sumatran settlement in the nearby lowlands, was set out as a reserve area for rubber and fruit trees only, with the first titles formally registered in 1916. There was never any rice. Its mainly Sumatran people were later joined by migrants from Java who established a distinctive settlement area in the upper reaches of the Sungai Pencala valley.[5] Janda Baik, in an isolated upland basin, was settled by Malay people from the interior of the peninsula early in the twentieth century. In the 1930s it received a government-funded irrigation system, strengthening its rice economy. As land was registered, some of it only after the 1940s, the Orang Asli who had formerly occupied the basin were driven to the edges, where they finally acquired two small reservation blocks.

Before the rubber period, Malay villages had their own areas of wet rice land, and villages stood on firm ground among a range of fruit trees and coconut palms that shaded a large proportion of the houses. Spices and vegetables were also grown in this *dusun* area, around and within the village. The forest behind was lightly used to produce dry crops by shifting cultivation methods. The forest was cut down for conversion into normally oblong blocks of rubber land, usually 1–2 ha, and only the steeper and higher areas remained under forest. Once crowded out by the growing rubber, the dry crops were moved down to the village site, where the *dusun* strip became wider, mingling fruit trees and vegetable crops around and behind the houses. Some villages converted all their forest into rubber. Beranang completed the conversion by about 1930. At this village, an extensive wet rice area occupied half the total village land in 1966, rubber occupied 30 percent, and *dusun* and the village 14 percent. The other three villages retained significant areas of mature or secondary forest.

MODERN ACCELERATION OF CHANGE

Kuala Lumpur and its satellites still had only 450,000 people in 1957. In the late 1990s it had more than three million. Based on reconstructed 1966 land use maps, the four villages at that time not only remained physically separate from the city but also retained a primarily agricultural appearance. This was already deceptive. Data for Sungai Serai in the late 1960s, from the undergraduate thesis of geographer Kamal Salih (1969), showed that 60 percent of aggregate income came from wages and that 25 percent of the workforce already commuted to the city. Although rice cultivation was given up at this village around 1950, in the decade that followed 59 percent of the workforce was still engaged in agriculture. Substantial replanting of rubber with high-yielding varieties was in progress during the 1960s at all four places, and there was new planting in former forest at Janda Baik.

Between 1966 and 1986, Kuala Lumpur and its industrial economy were energetically enlarged under the New Economic Policy, initiated in 1970. Middle-class suburban development reached the edges of Sungai Pencala reservation in 1982, and by 1989 a large low-cost housing estate was almost complete very close to Sungai Serai. A new industrial estate with housing for its workers, one of 16 set up in the urban outskirts between 1952 and 1985, was being developed only 2 km north of Beranang in 1986. It was largely occupied in 1989. There were already plans for a satellite city in the mountains close to Janda Baik, although these plans were not implemented quickly. Sungai Pencala was within the metropolitan planning area, and from 1982 there were proposals for its complete physical transformation, resisted by the resident landowners.

NEW SETTLEMENT AND LAND SPECULATION IN THE 1980S

Two new elements appeared among the village populations after 1970. During a period of steep inflation in land prices in the 1970s, members of the newly affluent Malay middle class began to buy land in the reservations, still cheap because it was separated from the general market by law, which restricted sale to members of "Malay race and religion." They built houses for residence or weekend retreats. Others bought land for intended resale, speculating on a rising market. Inevitably, the price of land in the reservations began to join the general escalation, but the big jump occurred behind that of national inflation. The value of land was redistributed between the communities as its basis shifted from agricultural to urban valuation. At Beranang in 1974–77 median transfer values were still almost twice those at agriculturally marginal Janda Baik, but by 1984–86 they had fallen below the median value at the new mountain retreat, where land values began to rise after road communication was improved dramatically in 1980. By 1984–86, median transfer values at Sungai Pencala had increased by five times since 1974–77.

Although records were incomplete, we calculated that in 1984–86 nonlocal owners held 6.5 percent of registered land at Beranang, 17.8 percent at Sungai Serai, 24.8 percent at Janda Baik, and 31.0 percent at Sungai Pencala. With many missing titles, we believe that the last two figures underestimate the nonlocal ownership of land in 1984–86 by a substantial margin. This was especially so at Janda Baik, where most of the forested land around the edges of the reservation had no more than provisional title. A good deal had been sold nonetheless.

Whether resident or absentee, few of the new landowners invested in agriculture. The majority created urban-style environments on their land or left it either in the care of tenants or unused. Some of the small number of innovations in productive land use lasted only a short time, such as fish pond and market gardening investments at Janda Baik, which were promising in 1986 but dead by 1989. More enduring perhaps, and in line with what many long-resident farmers were also doing, was the replacement of rubber by highly priced durians (*Durio zibethinus*) by a nonresident buyer of several blocks in Beranang.

A widespread alternative use of land was emerging quickly in the 1980s, and this offered greater rewards. This was the construction of residential buildings in a range of styles and standards for rent to tenants in nonagricultural employment. Agricultural tenants are a long-established element in Malay rural society, but they became increasingly hard to find as urban employment opportunities boomed in the 1970s and 1980s. At Beranang, it was remarked in 1986 that the only remaining source of agricultural tenants was among illegal immigrants from Indonesia. The landowners who built houses for nonagricultural tenants were both old residents and new buyers, mainly the former. At Sungai Pencala, where there had been few tenants in 1973 (A. Maulud bin Mohd Yusof 1976), 146 of a total of 657 residential buildings were multihousehold dwellings in 1986. Units in them ranged in number from 2 to 20, and they varied from solid buildings to the most wretched slums. Construction continued despite a general economic slowdown, and we estimated that by 1989 significantly more than half the resident population of Sungai Pencala were tenants.

POVERTY, RUBBER, AND RICE

Few of the resident householders interviewed by our student assistants mentioned rent as a source of income. Most probably had the tax collector in mind. What did emerge from an inquiry into expenditures and incomes among a sample of householders was that the poorest in the four villages were the bottom 25 percent of income-earning households in the two most rural communities, Beranang and Janda Baik. All were farmers, with the rice and rubber combination that for three decades had spelled rural poverty in Malaysia. In 1986, the price of rubber was very low and the wage economy was still booming. It was only in Beranang and Janda Baik that rubber tapping was still widespread. In Sungai Serai and especially Sungai Pencala, rubber was almost totally neglected and large areas were reverting or had reverted to secondary forest. This included blocks newly replanted with high-yielding varieties two decades earlier. In 1989, with an increase in the price of rubber, some of this had been cleaned up and was back in production, but it was apparent that trees were being tapped less than regularly. Replacement of rubber by durians or other fruit trees still continued.

Rice was in trouble in 1986 all through the central west of Malaysia, even in the rice bowl areas of the north. A national self-sufficiency drive had failed, and the political response was to threaten "lazy" farmers with sanctions, including the loss of land. From a peak in the late 1970s the decline was rapid, and it had multiple causes. Poor rewards for effort and expenditure under the Green Revolution packages were common.[6] The rice and rubber combination that had seemed so hopeful to farmers in the early twentieth century now yielded far less than the more abundant opportunities in the manufacturing and service sectors. Away from the cities, rural land prices remained low. Pious farmers offering rice land for sale to finance the *haj* (the pilgrimage to Mecca) faced the dilemma of either abandoning their plans or selling much more than they could afford to give up.

There were other and more local causes of decline. Over the years, soil washed into the valleys from the rubber land on the surrounding uplands had raised the level of large land areas too high to be irrigated from the ditches. Although higher-level plastic flumes were introduced at a number of places, including Beranang, cutting of forest for timber in the uplands made river regimes more irregular. Beranang was a favored area for rice, with yields still averaging 2.5 tons/ha in the early 1980s. Nonetheless, a quarter of the 1966 rice land was uncultivated swamp in 1986, and this area increased by 1989. At Janda Baik, never a high-production area, one of three rice-growing areas was unused by 1986. By 1989 only a few small plots were planted and the irrigation canals were neglected. Then and in 1992 most of the formerly irrigated land remained unused. Some rapidly reverted to secondary bush, like large areas elsewhere in the country.

With local initiative at Beranang and Janda Baik, some former rice land was planted to bananas. With little maintenance, yields quickly declined, and much of the land became unproductive, infested by *Imperata cylindrica* grass. In 1992 Samad and I walked into one of these areas where some attempt was still being made to sustain banana production. We found the farmer, but while we were talking to him we saw four dangerous snakes in the space of a few minutes. We decided we needed to see no more and carefully withdrew. At the end of our short period of observation, the most active agricultural initiative anywhere in the four villages was on the Orang Asli land, and in the forests behind it, at Janda Baik. Chinese farmers from a nearby village rented land from the Orang Asli and employed them as workers on illegal shifting cultivation plots in the adjacent forest. Vegetables were grown in the fields. Ginger, which commanded high prices in the city market, was grown in the forest.

WHAT REMAINS?

Malay agriculture on the peninsula never developed the great diversity encountered in Java or in the agroforests of Sumatra and Kalimantan. Once it acquired a serious commercial dimension, even farmers who continued a mainly subsistence orientation became accustomed to buying an increasing proportion of their goods from the village shops. The widespread success of rubber in the second decade of the twentieth century confirmed the invasion of a cash economy. Although farmers continued to adapt their land use to market conditions and to retain a self-provisioning capacity, they were already conditioned for the major changes that came about after the 1950s and especially in the 1970s. One of the findings of our work was that among even the household heads, whose mean age was more than 40 in all villages, only about one-half of respondents ranked the traditional importance of holding title to land ahead of alternative ways of making a livelihood.

At the time of our study, fewer than 6 percent of household heads at Sungai Pencala and Sungai Serai described themselves as farmers or farm workers; the highest proportion, 55 percent, was at Janda Baik. Yet a good deal of farming was still going on at all four sites, even Sungai Pencala, both for self-provisioning and for commercial sale. By degrees, the still rural people of these villages were becoming urban farmers, linking their part-time land use to wage earning and business. Wholly nonagricultural people among the populations were wealthy builders of palatial homes, some entrepreneurs, and the wage-dependent tenants. In conclusion, it is necessary to add that although we were able to see and measure an important stage in transition, we could only guess where it would lead.

■ Comment Arising from the First Two Chapters

Chapters 1 and 2 have set out five examples of diversity in rural economies and societies and of the trends within each. The adaptability of farmers and their families, both to adversity and to opportunity, is a common feature, even though the adaptation may include abandoning farming for more rewarding pursuits. A number of themes are indicated. They include management of diversity in the biophysical environment, the range of skills and information used in decisions on resource management, and the constantly changing inventory of crop plants. The intimate linkages between changing social institutions and farm practices, both at the local level and at much wider levels, are demonstrated clearly. It is not possible to deal with the structure and organization of a farming system at any time without seeing where it stands in relation to interdependent national and world systems and without doing one's best to gain a historical perspective.

Other perspectives have included land management using field types and creating farming spaces that have common characteristics over an area. The manufacture of soils in wet rice terraces and fields has been touched upon. The importance of introduced crop plants emerges in all five cases. External intervention in these cases includes contract support of a new crop, the imposition of enforced resettlement, and the intervention of speculation in land purchase and resale. The varied nature of agrodiversity and its modern fate has been presented, but even structured description on this semirandom basis can take us no further. It is now time to define what this book is about and to set out how agrodiversity can be analyzed. This is the task to which I turn in chapter 3.

Notes

1. *Zande* is singular or adjectival, and the plural is *Azande*. As is common in the literature, the single term *Zande* is used here.

2. Evans-Pritchard (1937) provided great detail on these plants and on the "oracle poison" used on domestic chickens (which either died from or survived the dosage) to obtain answers to a great range of questions. The poison was derived from the rainforest well to the south, in Congo. De Schlippe became very aware of the importance of these practices and noted that the Christian missions had done little to weaken belief in magic. Reining concentrated on material aspects of the Zande experience and mentioned magic only in passing.

3. Reining (1966:73) was impressed by the "chaos" and doubted the reality of these codified field types. He felt that because de Schlippe got "most of his information at second hand, he tended to oversystematize Zande agricultural practice." From de Schlippe's own account of his observations over several years and the evidently first-hand nature of most of them (he often stated where they were not first hand), Reining's comment does not seem fair.

4. Rice was grown by a variety of methods in precolonial Malaysia, and these had evolved over several centuries (Fukui 1979; Hill 1977). Some rice was dibbled and only casually irrigated, and irrigation technology often was inadequate to cope with dry periods and damaging floods. Outside the Minangkabau region in the south (Beranang), there was little permanently managed agriculture in the modern Kuala Lumpur area before the late nineteenth century (Lim Teck Ghee 1977:36).

5. The meaning of *sungai* is a river, stream, or brook. Villages often are named for the streams by which they stand.

6. The Green Revolution is addressed specifically in chapter 12.

Chapter 3

Defining, Describing, and Writing About Agrodiversity

■ Summarizing the Elements of Agrodiversity

Agrodiversity is a seamless whole in which all aspects are interrelated. It has become a central concept for the People, Land Management and Environmental Change (PLEC) project. As that project advanced, a need evolved to define and debate agrodiversity in greater detail than in the early statements (Brookfield 1993, 1996a; Brookfield and Padoch 1994). In 1998, Michael Stocking and I built on material that had been drafted for this book to develop a guideline paper for our colleagues and wrote a shorter statement for wider publication (Brookfield and Stocking 1998, 1999). After a guidance document on biodiversity method was prepared for PLEC (Zarin, Guo Huijun, and Enu-Kwesi 1999), the agrodiversity document was modified again to match the biodiversity statement with which it was printed in the PLEC periodical (Brookfield, Stocking, and Brookfield 1999). Thus PLEC has fed this book, and this book has fed PLEC. We have found it useful to begin by classifying agrodiversity into four elements; two of them are related to management, and two are explanatory. In text modified from Brookfield, Stocking, and Brookfield (1999), the four elements are set out here and discussed in the first part of this chapter. The two management elements are presented first, then the two explanatory elements.

- *Agrobiodiversity.* This term is much more common than *agrodiversity.* It is commonly used to mean the diversity of useful plants in managed ecosystems. Agrobiodiversity was defined by Guo Huijun, Dao Zhiling, and Brookfield (1996:15) as "management and direct use of biological species, including all crops, semi-domesticates and wild species." It embraces all crops and other plants used by or useful to people, and because it also involves biota having only indirect value to people, it cannot be distinguished sharply from total plant biodiversity. As we will see in chapters 13 and 14, agrobiodiversity evolved into a major field of its own in the 1990s. It has been redefined several times, in more contextual terms than those just used. It has been described both as a major subset of all biodiversity (Aarnink et al. 1999) and as a subset of agrodiversity (Brookfield, Stocking, and Brookfield 1999). Cromwell (1999:11) treats it as "the variety and variability of plants, animals and microorganisms at genetic, species and ecosystem level," involving the part of the whole agroecosystem that is actively managed by farmers, within which many components would not survive without this human interference. Therefore, "indigenous knowledge and culture are integral parts of agricultural biodiversity management" (Cromwell 1999:12).

- *Management diversity.* This includes all methods of managing the land, water, and biota for crop production and maintaining soil fertility and structure. Biological, chemical, and physical methods of management are included, but they overlap. Some biological management, such as reserving forest for watershed protection or planting live hedges, has direct physical consequences. Local knowledge, constantly modified by new information, is the foundation of this management diversity, as with agrobiodiversity.

- *Biophysical diversity.* This includes soil characteristics and their qualities and the biodiversity of natural (or spontaneous) plant life and the faunal and microbial biota. It takes account of both physical and chemical aspects of the soil, surface and near-surface physical and biological processes, and hydrology, together with macroclimate, microclimate, and their variability. This is the diversity to which farmers respond and which they manage. Quite often, their management includes manufacturing soils, and remodeling the agricultural landscape through terracing.

- *Organizational diversity.* Often called the socioeconomic aspects of agriculture, this includes diversity in the manner in which farms are owned and operated and in the use of resource endowments and the farm workforce. Elements include labor, household size, the differing resource endowments of households, and reliance on off-farm employment. Also included are age group and gender relations in farm work, dependence on the farm as compared with external sources of support, the spatial distribution of the farm, and differences between farmers in access to land. Organizational diversity embraces management of all resources, including land, crops, labor, capital, and all other inputs. It underpins and helps explain management diversity and its variation between particular farms, communities, and societies.

The relationship of the farm household and community to the larger society (local, regional, national, and global) is an important element that shapes the relationship of organizational diversity to the larger world. Changes in these external relationships can have a major impact on agrodiversity at the level of the farm and community, as we have already seen in chapter 2. Later in this book, especially in chapter 6 and in the final chapters of the book, we encounter the inability of organizational diversity, at the local level, to protect the farming system from major external forces.

These categories are used throughout this book. Running through all of them is the dynamism of the systems through time. In a very crude manner, we can distinguish three main time scales:

- *Short-term (interseasonal and intraseasonal) sequential diversity in farmers' decisions about the use of land, labor, capital, and other farming resources, and in the security or risk of the harvest.* The time scale is from months to a few years.
- *Longer-term change in cropping and management practices in response to environmental, demographic, social, economic, or political change.* This includes shifts in cropping patterns, land use allocation, and reliance on different income sources. These changes occur as soils and biota are modified by use and natural processes, self-provisioning gives way to commercial production, new crops are adopted and others discarded, and new practices are taken into the system and others abandoned. The time scale is from a few years to many decades.
- *Secular change in farming systems and their environment, taking place over decades to centuries. This type of change calls for a historical interpretation, setting aside preconceived notions derived from a large literature.* Some of the issues involved are discussed later in this chapter.

A definition must include all these elements without being cumbersome. There have been two independent attempts to define agrodiversity, and these must first be reviewed. Neither attempt alone takes the organizational and temporal elements fully into account.

■ Defining Agrodiversity

AGRODIVERSITY AS MANAGEMENT DIVERSITY

Farmers' management of biophysical diversity was central to how I first thought of agrodiversity when I coined the term in 1992. The conceptual link to biodiversity seems to imply an emphasis on agrobiodiversity, but I then saw agrodiversity principally in terms of the diversity of management methods. Two years later, with specific reference to small farmers, Christine Padoch and I defined agrodiversity as "the many ways in which farmers use the natural diversity of the environment for production, including not only their choice of crops but also their management of land, water, and biota as a whole" (Brookfield and Padoch 1994:9).

Viewed in this way, management diversity is related to agrobiodiversity but embraces much more. It includes not only management of the field but also management of what is often called the fallow, better described as successional vegetation with its wide range of plants and other biota, many of which are useful. It specifically includes management of soil fertility, slopes and water, and the great range of agrotechnologies used for these purposes. In addition to the selection or overriding of ecological niches, whether these are stable or highly variable through the year and between years, it takes into consideration crop rotations and combinations. It includes weed, pest, and disease management. It also includes adaptation to resource degradation, both successful and unsuccessful. A temporal dimension is implied but is not specifically stated in the wording of the definition. In any broad definition, it also includes the indigenous or local knowledge on which management is based.[1]

BIOPHYSICAL DIVERSITY AS THE BASIS

Emphasis on management diversity must be balanced by greater attention to diversity that arises more directly in response to variations in the microenvironment and its biota. Conny Almekinders, Louise Fresco, and Paul Struik of the University of Wageningen, The Netherlands, independently developed a separate definition of the term *agrodiversity*. Their writings gave primary emphasis to variation in the biotic and abiotic environments. They did not restrict the term *agrodiversity* to small farms, and they used it in a more general agronomic sense. Almekinders, Fresco, and Struik (1995:128) wrote of agrodiversity in arable systems as resulting from interaction between three factors: plant genetic resources, the abiotic and biotic environments, and management practices. They continued, "We call the variation resulting from the interaction between the factors that determine the agroecosystems, 'agrodiversity.'"

Almekinders and her colleagues wrote of variation within agroecosystems, succinctly defined as all ecological systems with an agricultural component. They argued that productivity variation within agroecosystems is naturally occurring, arising from differences in microenvironment and microclimate, distribution of minerals in the soil, microfauna, and microflora, both pathogens and beneficial organisms. Management is a part of their definition, as are crop biodiversity and biodiversity as a whole, but differences arising from management are put more thoroughly into a biophysical context.

The variations recognized by Almekinders, Fresco, and Struik occur in high-input as well as low-input agriculture. In high-input systems the usual management strategy is to reduce biophysical diversity by uniform management or to override it by applying industrial fertilizers and biocides. This can lead to unsustainability, also revealing sources of variation that were formerly overlooked. In most low-input systems, reducing biophysical diversity is possible only to a limited degree, and the optimal management strategy is to use variability by adapting cultivation methods to take advantage of microenvironmental variation, thereby reducing risks and improving total production and its sustainability. Resource-poor farmers cannot readily override biophysical diversity. Almekinders and her colleagues concluded by saying that the occurrence and importance of variation deserves more attention in all forms of agriculture, not only in small farming, and that its management is of increasing importance as demands placed on the land increase.

Biophysical diversity can be viewed on almost any meaningful scale. On a landscape scale, it is a major element in the widely repeated manner in which farms are structured to allocate land of different intrinsic qualities so that all or most households have access to each. This is one way in which organizational diversity is related directly to biophysical diversity. At a finer degree of resolution, biophysical diversity can arise within a single field, where a crop yields differently in separate parts of the field, whether in all years or in years with drier or wetter climatic conditions, better suited to one or other part of the field. The association of crops in an intercropped field may show subtle differences related to natural conditions. This is one relationship between agrobiodiversity and biophysical diversity.

THE MEANING AND CONTENT OF ORGANIZATIONAL DIVERSITY

No agricultural system can be understood independently from the manner in which farms are organized and the forces that interact to shape this organization. All of the case studies reviewed in chapters 1 and 2 take these elements into account. The Chimbu case details the manner in which farm organization has changed through time, especially in terms of individuals' access to resources and the changes in their priorities. These and other dimensions are developed further throughout this book, but some general considerations are presented here.

Highly important is the manner in which farmers mobilize labor throughout the year to manage seasonal bottlenecks when a great deal of work has to be done at one time. Mobilization of cooperative labor varies widely, and the use of hired labor varies more sporadically. In many farming societies, gender division of labor is one way to mobilize a small pool of household labor effectively. In many systems, only certain tasks are performed by men, leaving most of the work to women, who are sometimes seen as exploited by the men. In others women do little work in the fields. This happens for doctrinal reasons in societies in which either all women or high-caste women are strictly secluded, but no such religious sanctions apply in many other areas where men do most of the farm work. In other societies, men and women perform most farm tasks interchangeably, and in still others the gender allocation of work is flexible, differing even between adjacent farms. But in almost every account, responsibility for the most diverse tasks on any household farm seems to belong to the women. Drawing mainly on his material from Nigeria and Switzerland, Netting (1993) discussed in detail the questions of mobilizing household labor and allocating labor time. Richards (1985) described modern innovations in labor organization in an area of southern Sierra Leone.

Intuitively, agrodiversity seems to belong to resource-poor farmers who depend on the land for a major part of their subsistence and income. We have already seen that this is not so in the Peruvian Andes, and that is not a unique case. Specialized cash cropping occurs on some very small farms, especially in modern times, but the farmer must give up the security offered by a range of crops. Risk avoidance is a question of major importance for small farmers. There is a large theoretical literature on this topic, arising since modern development economists began to advocate the social benefit of market solutions to small farmers' development problems. By maximizing utility, farmers whose goal is optimization can prevent "any major inefficiency in the allocation of traditional factors" (Schultz 1964:39).

In a manner widely demonstrated by empirical realities, first Lipton (1968) and then Weeks (1970) argued against the social benefits of a competitive market. Innovative behavior, especially in agriculture, involves risk and uncertainty. They proposed that people do not become wealthy because they have innovated; those who could best afford to take risks are already the most wealthy. The poor have to avoid disastrous outcomes at all costs, so risk avoidance is a survival strategy for them.

Irrespective of the ability of poorer farmers to bear risk, there has been abundant empirical support in the last three decades for Weeks's (1970:30) conclusion, "The greater is the innovative component of economic growth, the greater will be the tendency for the distribution of income to become more unequal in the agricultural sector." Were all small farmers individual operators, this elegant economic reasoning would suggest that only the more secure farmers would experiment. In the historical literature, there is evidence that this may have been so, with the principal landowners identified as the main channels through which innovations have entered farming systems (chapter 10). However, many societies continue to follow cooperative practices that support those in need and sustain a redistribution of wealth, permitting a wider spread of risk taking. Such "uncertainty sharing" through social support systems at local level is of major importance.

Crop choice often differs between wealthy and poorer farmers. There are many other differences, such as the use of livestock and their manure and of purchased inputs. The type of soil and water conservation practices adopted is influenced strongly by the resources available to different groups of farmers, thus affecting the pattern of management diversity and feeding back to enlarge the differences in natural land quality. All elements of agrodiversity are interrelated, and in any explanatory framework none can be considered without taking each of the others into account.

COBBLING TOGETHER A SHORT DEFINITION

The organizational issues and their theoretical implications must be incorporated, but the basic definition of agrodiversity must remain tied to the land. Adapting an earlier proposal (Brookfield 1996b:4), I now suggest that agrodiversity be defined as "the dynamic variation in cropping systems, output and management practice that occurs within and between agroecosystems. It arises from bio-physical differences, and from the many and changing ways in which farmers manage diverse genetic resources and natural variability, and organize their management in dynamic social and economic contexts."

For the present, that will have to do. There remains a larger and more important question than that of definition: how do we analyze agrodiversity in terms that capture the complexity but are not buried by it?

■ Describing and Classifying Agrodiversity

A TAXONOMY OF AGRODIVERSITY?

When natural scientists deal with diversity, their first step normally is to classify it. Geologists separate rock formations of different types and ages and map them. Climatologists use a variety of different methods to capture the differences in climate, and in variability of climate, between regions. Taxonomists classify plants according to a set of morphological and genetic characteristics into plant families, genera, species, and varieties. The ways in which these classifications are made have grown up with the natural sciences themselves, and in geology and botany their origin goes back to the eighteenth century. Is it possible to do this with the wide range of practices embraced in agrodiversity? For crops, taxonomic methods have been applied widely, although not down to the level of landraces, but the mutual relations and competition between crops plants grown together in the same plot are much less well understood. Indeed, farming practices more often remain misunderstood than they are comprehensively evaluated, and this raises some serious difficulties in writing about agrodiversity.

USING THE WORK OF OTHERS

These are problems enough in interpreting one's own fieldwork and its results. Because no one person's fieldwork can encompass more than a few areas in a lifetime, writing about agrodiversity must draw heavily on the writings of others, and there are several problems in doing this. One of them, concerning the so-called ethnographic present, is discussed in the introduction. A larger problem is that there is no generally accepted checklist of what a fieldworker should observe and record in describing an agricultural system. Conklin (1961) drew up such a checklist for shifting cultivation, but no one has ever really used it. Modern farming system research has produced a number of guideline statements, but they differ substantially from one another.[2]

Most descriptions of agricultural systems have used individual approaches. Even where efforts have been made to develop a comparable set of descriptions, as in monographs associated with the Southeast Asian Universities Agroecosystem Network (SUAN), based on Conway (1985), inconsistencies remain. The most formal and detailed treatment of farmers' practices in the literature is that by Wilken (1987) for central Mexico and highland Guatemala. He deals successively, and in detail, with the management of soil fertility and surface, slopes, water on and within the ground, crop microclimate, and horizontal and vertical space. All are separated in description and analysis. This treatment covers one region only, but the logical form of analysis could have wider application.

MODELS IN THE MIND

Most of the useful literature is in academic books and articles. Some of it is written by agricultural scientists, some by anthropologists, some by ethnobotanists, some by geographers, and some just by keen observers. All tend to see rather different things, so it is often very difficult to compare information yielded by one account with that yielded by another. What is worse, many observers go into the field with models in their minds, derived from other writings that they have read or simply from their background and training. Often they see the model, and no more. It is shifting cultivation, for example, so there is no need to look at it more closely. The main crops may be described, but a large number of minor crops may be overlooked. Their very existence may emerge only where foods in use are described elsewhere in the book or article, or in other publications by the same author or even by other authors. Sometimes certain crops are mentioned only in lists of what is formally presented on ceremonial occasions between individuals and groups. Remarkably, house gardens and mixed gardens surrounding villages often are ignored. If an area has gone into fallow it is often described as abandoned, and surviving crop plants and useful species that grow up within it are not noted.

On the other hand, if there is something remarkable to be seen, such as a skilled use of irrigation on a small part of the land, it may be overemphasized. Just as anthropologists often tend to pay more attention to the ceremonial and dramatic than to daily events, so observers of agricultural systems often pay most attention to what strikes them as of particular interest. Moreover, everyone has or develops their own special interests. Conklin (1980) built his whole account of Ifugao agriculture around pond field rice, writing far less about the dry fields and agroforests. Zimmerer (1996) was interested primarily in small potatoes, as one of his informants remarked to him, and his book is light on some aspects given more stress by other writers on Andean agriculture. Although the quality of observation has improved in recent years, even the best accounts contain omissions.

TWO ATTEMPTS AT CREATING A TAXONOMY

Notwithstanding the difficulties, I did make a substantial attempt to use this varied material to classify farming practices some 30 years ago (Brookfield and

Hart 1971). I compared data on 44 places, ranging from small regions to single localities, in New Guinea and adjacent parts of the southwest Pacific. From several hundred potential sources, I used 55 that offered more complete and seemingly reliable information and extracted data from them. Although most of the sources were written in the 1950s and 1960s, they included some that were several decades old. For about 15 of the 44 places I could supplement the written data from my own fieldwork, in depth in some areas but no more than sketchy academic tourism in others. Data extracted from each source were classified into 49 variables. Fifteen of these related to the environment, 7 to use of wild food sources, 6 to the main staple crops, and 21 to cultivation methods. A presence/absence method was followed, with note taken where a practice was of particular significance. Often, it was not easy to be sure whether the checkmark was put in the right box, and later information made clear that I had made several mistakes.

In 1971 it was not central to my design to evaluate indigenous management skills, and this is reflected in my combining a lot of variety into single categories. Cultivation frequency and the degree to which crops were segregated or mixed (intercropping) were separated, but the classification of cultivation methods themselves was rudimentary. I used only 14 categories to classify a wide range of management practices. They embraced most of what was important, but sparely. There has been progress, largely through growing recognition that indigenous land managers have good reasons for what they do. Indigenous farming is the same as any other farming in that it manages the land to produce crops. That management can be classified into different elements. Spencer (1966) pioneered a logical approach, based on the purposes of management. Adopting an evolutionary perspective, Spencer determined three classes with 18 types, within which an unnumbered range of subtypes was also indicated. Okigbo and Greenland (1976) made a partial attempt, unusual in its effort to cope with intercropping and multiple-cropping systems.

A modern classification of all the agricultural systems of Papua New Guinea in the early 1990s, already different from those in the 1960s, used observation by three fieldworkers as the primary source and used the literature only for supporting information. The resulting computerized database is compatible with geographic information systems (Allen, Bourke, and Hide 1995a, 1995b). It distinguished 37 management variables. They were grouped into classes based on crop components, their arrangement and sequencing in time and space, fallow-clearing practices, soil fertility maintenance techniques, land surface modification (including that on slopes), and other agronomic techniques, including the use of livestock.

There is a clear structure here, but the level of resolution is that of a fairly small map scale (1:500,000). It is not possible to capture the detail of agrodiversity on a scale at which 5 km is represented by 1 cm. Covering all of Papua New Guinea in this way took three people, and a number of local collaborators, more than 4 years. The question of scale is very important. Most descriptions, mapped or not, are at one of two scales: that of the large region and that of the single community that is studied in depth. The map scale of resolution is central to the study of diversity. Conklin's (1980) basic maps of Ifugao are printed at a scale of 1:10,000, but some of his detailed maps attain a resolution of 1:1500. To map diversity in the Chimbu region of Papua New Guinea, I worked on a map scale of 1:3600 but could not capture all detail even on that scale (Brookfield 1973).

FUNDAMENTAL PROBLEMS LIMITING THE VALUE OF TAXONOMY

The pieces of data about particular cultivation practices make sense only in the context of the whole systems of which they are elements. Moss (1969) pointed out that simply establishing land use patterns can never reveal the factors that produced them. Similar units or characteristics may have different functions in different systems, and seemingly different units may have similar functions in the system as a whole. Anticipating more modern thinking discussed in chapters 8, 13, and 14, he went on to argue that even the distinction that is so often made between land use and vegetation must be eliminated. Both land use and vegetation are influenced by human behavior, and the proper approach is to look at the system as a whole and the dynamic interrelationships within it.

This is not the only major problem. Without some clear design to help determine what we are seeking, trying to bring a mass of data together could never be more than a worthless exercise. For example, Harris (1976) remarked that searching for "the facts" without preconceived ideas is neither feasible nor desirable. As in any good and valid scientific research, there must be a structure of questions to which we are seeking answers. Classification systems in all science must serve a useful purpose. My comparison of 44 places in the southwest Pacific was motivated by a search for regularities in the transition from low-intensity to high-intensity systems (Brookfield and Hart 1971). My conditional hypothesis, following Ester Boserup (1965), was that population density might be the underlying cause. I found exceptions to any such rule, leading me into a theoretical debate that has gone on for a long time (Brookfield 1972a, 1984, 1986, 2000). The question is taken further in chapter 11.

THE COMPARATIVE METHOD

It is necessary to use the diverse body of information that is available in other ways. The most formally structured of such ways is the comparative method. In this, one takes data from only some places and only some of the literature and tries to see, systematically, what common ground and what differences can be found between the practices and responses of farmers in different areas. There are many common elements in the ways in which farmers everywhere modify the biophysical system to produce crops. This is especially so in the case of difficult environments such as wetlands and steep and unstable slopes and in the reclamation of grasslands for crops. As Netting (1993) cogently argued, the many means developed by farmers to gain similar ends show remarkable similarities between regions. But the detail is less readily analyzed. I use the comparative method in chapter 6, but find I have to control the comparison within a limited region with a fairly constant set of variables. By itself, the comparative method is not the answer.

■ Following What Farmers Do

AN APPROACH THROUGH CODIFICATION

Notwithstanding all these difficulties, there is still need for a structure within which the elements of agrodiversity can be described and analyzed in a meaningful way. It must be less than a detailed and widely replicable taxonomy, but it must be more than description, which so stresses complexity as to do no more than satisfy intellectual curiosity. One way might be to group descriptive data within the elements, or categories, of agrodiversity described at the beginning of this chapter. In this way, both biophysical and management diversity could be described at scales appropriate to either landscape or local levels, perhaps in the ranges 1:50,000–25,000 and 1:10,000–5000 respectively. Agrobiodiversity can be inventoried within sampled or selected areas, using either randomly or purposively sampled quadrats (Zarin, Guo Huijun, and Enu-Kwesi 1999). Organizational diversity, except for the one aspect of farm layout, cannot be mapped but has to form a context to what is observed or measured on the ground. This is description at a point in time, perhaps enlarged to capture the short-term variations that take place seasonally or within a short span of years.

Description is necessary, but if there are to be meaningful interpretations that can overcome the objections to a taxonomy, they cannot rely only on observation. Descriptions must be related to the whole systems of which the recorded facts of diversity form part. To do this, they must take account of the codification of land use and management that is often, if not always, developed by the farmers themselves. For example, rice farmers in eastern India describe and name a *beusani* system of land, crop, and weed management designed to adapt to uncertainty in hydrological conditions (Fujisaka 1997). One possible way of finding a farmer's structure is suggested by the field types and farming spaces identified in Zande and Paucartambo in chapter 2. Recognizing these farmers' categories is less than the guiding theory called for by some writers, but it would approach the structure Harris (1976) sought. It relates what happens on the ground to the manner in which it is organized by society and to the economic and social context of farming.

SIMPLIFYING DAILY DECISION MAKING

Every day, everywhere, men and women farmers must decide what work is to be done in which fields. Although a program of work already is determined by their choice of crops and by seasonal conditions, a lot of day-by-day decision making is needed. Sometimes they need to work with other farmers or to hire labor of their own, and their helpers need to know what to do. Large tasks such as land preparation, fencing, planting, and harvest often involve mobilizing additional labor or arranging with other farmers to work in the same area at the same time, all performing work that meets local norms and expectations. Coordinating livestock management and protecting crops from birds and animals, are achieved most economically by following a standardized timing of activities. Farmers develop a kind of coding to assist them in all these decisions.

Commonly, farmers recognize different sets of field types, widely repeated between farms and often over a large area. In Conklin's writing on Ifugao, cited in chapter 1, more detailed divisions within them were described as field zones and planting spaces. In each field type, farmers tend to grow the same sets of crops, with limited variation between fields and farms. In a shifting cultivation system, these field types may shift across the landscape from year to year, leaving behind them fallow types from which plants are still taken and which may themselves be planted. They have arisen through a process of experimentation over time and are constantly being modified, a fact that must be taken into account.

In his pioneer study of the Zande system, reviewed in chapter 2, de Schlippe (1956:117–18) speculated as follows on how such field types arose:

> Theoretically, one could think of thousands of different ways in which the great number of crops and varieties and the astonishing mosaic of soil–vegetation types could be combined into field types. In practice one discovers, however, that field types are few and that it is always the same field types that are repeated by all members of the group. One can imagine that through a long process of trial and error many of the theoretically possible combinations have been tried out and all those useless among them and possibly a certain number of useful ones have been discarded, whereas only a few useful combinations have been accepted and perpetuated. Field types, therefore, are the result of a process of simplification and of codification.

In more stable systems, the field types are more permanent and commonly are grouped within areas of broadly similar ecology. They form recognizable farming spaces, known to the farmers who have developed them. One example is the intensively cultivated and manured infield versus the more extensively used outfield. Another is the division of land between seasonally irrigated terraced or ponded fields, dry fields that are alternately cropped and fallowed, planted and managed agroforests, and the very mixed home gardens. This repeated pattern of just four main types is commonly found across southeast Asia. Except that there were no home gardens as such, this classification applies well to the Bayninan example (chapter 1) and to the rice–*dusun*–rubber combination in Malaysia (chapter 2). The four farming spaces of the Paucartambo valley, also described in chapter 2, similarly represent conscious simplification by the farmers. These four systems have developed over a range of ecological niches. Zimmerer argued that this resulted not only from simplification of work routines but also from seed selection practices, especially of potatoes, ulluco, and maize. Farmers rotate their activities over a range of land types and altitudes, and they rotate the seed in the same way, so that over time landraces have evolved that are adapted to wide ecological ranges. Some more complex or more dynamic environments are very finely subdivided, but it is common to find the land used under only three or four basic management systems, even across significantly different ecological zones.

Farming spaces embrace more than crops and technology. Field types and farming spaces adapt not only to environment but also to social and economic conditions and reflect the objectives of the farmers themselves. To use them for description and analysis is to try to seek the hidden order farmers of any region collectively use. Through it, they seek to reduce the range of uncertainty contained in the variables with which they must deal in their daily, monthly, and yearly rounds of decision making. Among the examples reviewed in chapters 1 and 2, field types have been mapped over whole areas by Conklin and Brookfield and in sample areas by de Schlippe and Zimmerer. These examples are unusual in the literature. It has not been common to use land use mapping systematically. This is a possible alternative to further attempts to find a taxonomic basis for the description of diverse farming systems. But although farming spaces are more enduring than the detail of agrobiodiversity or even management diversity, they too can be captured only at a point in time. The issue of longer-term dynamism remains unresolved, but it is central to explaining how the systems, and the diversity they contain, have come about and continue to evolve.

■ Analyzing and Writing About Agrodiversity

THE WOOD AND THE TREES

Writing a book about agrodiversity demands thematic approaches. I have to describe the trees, but I have to see the wood as well. More importantly, I have to capture adaptation and change in such a way as to make possible general as well as particular statements. All of us who write about indigenous farming systems face such problems. We have to discuss variation and flexibility while making large generalizations. For the most part, the context in which people write their generalizations is only that of a single region. Those who try to write about agriculture on the global scale normally have to settle for the generalizations only, and the variation and flexibility are suppressed even in the most comprehensive studies, the outstanding example being that of Ruthenberg (1980). Less detailed studies, such as that by Grigg (1974), offer only generalizations. That will not do for this book, which has to be written around themes generated, in a logical manner, from the facts about agrodiversity, agriculture and ecology in general, and the people whose farming is characterized by diverse production conditions and diverse practices.

INTERPRETING CHANGE

One very basic theme is the interpretation of change. The science of ecology has changed radically in recent years. As conceived from the time of Clements (1916) until the 1970s, ecosystems were seen as seeking to regain equilibrium after each disturbance, so that all postdisturbance forms were stages along a equilibrium-seeking transition to a climax state. Following this approach, equilibrium ecologists concluded that deflection of the natural succession would arise unless systems are protected from further disturbance. In consequence, deterioration would ensue. This set of views paralleled the contemporary notion of traditional farming systems as enduring, basically unchanged, over long periods of time, only to be broken up by modern interference. Although the shadow of these old views still falls across the literature, neither is current. The notion of homeostasis is out in both areas of inquiry. Ecologists now recognize frequent and widespread disturbance caused by natural forces as integral to the structure and content of ecosystems, even as necessary to sustain biodiversity. Disturbance can occur on a huge range of scales in both time and space.

Modern developments in nonequilibrium ecology and environmental history have developed linkages through studies of vegetation transformation, especially in regions of high natural variability (Botkin 1990; Zimmerer and Young 1998). They have also spilled over into a set of major revisions of long-held views about land degradation, overgrazing, and the threat of system collapse, especially in semiarid regions. In their place notions have been advanced of unstable but resilient systems capable of renewal with variations after major disturbance (Scoones 1994; Mortimore 1998). In Mortimore's version of the argument, instability in the biophysical system is responded to by unstable yet highly resilient adaptation in what are here called management diversity, agrobiodiversity, and organizational diversity.

The study of agricultural change itself is only starting to be affected by these newer ideas, and the prevailing notion remains one of unidirectional transformation. Usually, attention is given to an often alarming deterioration in soil–plant dynamics. Sometimes, this is linked to a concern with deterioration in social conditions, also viewed as unidirectional. All this is close to the linear models of vegetation change under human impact that have widely misinformed interpretation (Blumler 1998). In nonequilibrium ecology, human-induced change sorts out the conditions for plant life in new ways; such change is the normal condition, but both destruction and permanent improvement are rare. The same is true of farming communities and populations in all continents. We can profit greatly from taking the notions of unstable equilibrium, the positive aspects of interference, and resilience into the study of agricultural change.

Much of this book is concerned with agricultural change, involving a certain amount of historical reconstruction. It is important to set doctrinal assumptions aside and, in particular, not to assume without evidence the existence of any processes such as degradation or intensification. Change can take many forms, and the adoption of practices that are more intensive in use of capital or labor is only one form. Comparatively, it is useful to describe one system as more intensive than another, but the terms *to intensify* and *intensification* are used loosely in the literature. I introduce the theory of intensification in chapter 4 but discuss it in some depth only in chapter 11. Elsewhere in the book, I avoid these terms as far as possible. Certain other themes are also of major significance.

■ Themes for a Structured Argument

The case studies on which most of this book rests are not drawn at random. They illustrate specific themes. Taking the importance of studying change as paramount, I identify four other themes drawn from within the dimensions of diversity as follows.

ADAPTABILITY THROUGH TIME

The adaptability of farmers, using a diverse range of strategies, is a central issue. It is still greatly undervalued in a wide professional literature on agricultural development. Yet the circumstantial evidence of longer-term adaptability is overwhelming. The purpose of managing agricultural land resources is to enhance crop plant environments. Crop plants differ greatly in their preferred microenvironments, yet we do not find them growing only where natural conditions are optimal; much more widely there is deliberate modification. To this fact has to be added the wide diffusion of crop plants across the world that has taken place throughout the history of farming. It accelerated after linkages were established between Old and New Worlds in the sixteenth century and became more rapid still in the past 100 years. In many parts of the world, a majority of the most commonly grown crop plants are recent imports. Either the preexisting systems contained ecological niches that suited them, or the systems have been modified in their favor. This much ought to be obvious. Nonetheless, it seems not to be obvious to many writers, and warrants careful examination.

INNOVATION AND INVESTMENT

Agrodiversity can be viewed in two ways. One way is to see it as the contribution of skillful but essentially passive adaptation to local diversity in the context of changing conditions. Another is to stress innovations that impose greater control on the farming environment. Both are elements of diversity and sometimes both are present in the same general region, the same community territory, and even on the same farm. Ecological niches are carefully selected, but whereas some are used with little modification of the land surface and biota and others are managed in a uniform manner, still others are used only by the creation of substantial and enduring physical changes. These are the so-called landesque capital, investments with a life expectation well beyond the present crop or crop cycle.

Landesque capital is defined by Blaikie and Brookfield (1987:9) as "any investment in land with an anticipated life beyond that of the present crop, or crop cycle. The creation of landesque capital involves substantial 'saving' of labor and other inputs for future production." The term was developed within agricultural economics, and I cannot recall from where I borrowed it when I first used it in 1984. In the broad field of cultural ecology it is often attributed to me, but unfortunately I cannot claim that credit. Landesque capital includes terraces, irrigation and drainage systems, large mounds, and deliberate changes in the mixture of trees grown on land that has been cultivated. Landesque capital also includes soils that are so modified by their users as to be almost manufactured, as will be discussed in chapter 5.

RESOURCE MANAGEMENT AS A WHOLE

In studying agrodiversity, it is also necessary to take account of diversity in resource management as a whole. The small farm regions of the tropical world contain a high degree of biodiversity, sometimes greater in species numbers than the wild land itself. Only a portion of this biodiversity is directly manipulated for production, although a much higher proportion provides services of various kinds to the farmers. It provides food and economically valuable fruits, nuts, and leaves, restores soil fertility, yields wood and raw materials for artisanal production, offers a range of medicines, shelters wildlife that can be hunted, and so on. Whether or not agroforests are part of the system, the farm does not end where the field meets the wood. We need to look at the continuum between the two and examine the manner in which the managed and unmanaged wild elements relate to the cultivated land and plants.

Inclusion of managed and unmanaged wild elements in the content of agrodiversity or agrobiodiversity is contrary to the view of some modern students of agricultural biodiversity. Wood and Lenné (1999a:2) excluded "wild plants and animals of food value outside the agroecosystem" from the domain of agrobiodiversity because they are of small importance on the global level. On the local level, they can be of major importance. Chapter 4 introduces the integral significance of the wild populations historically. The topic is taken further in chapter 6, and chapter 8 develops the close relationship between farm and forest in depth. Food and useful plant sources beyond the field are indeed part of the farm.

SUSTAINABILITY AND RESILIENCE

In modern practical terms, perhaps the most important theme concerns what is called the sustainability of diverse farming practices. Following the earlier discussion of change, I use the term *resilient* more often than the more popular *sustainable* throughout most of this book, although I justify the preference only in chapter 14. Are systems of resource management that exhibit both local diversity and flexibility more likely to be resilient in the long term than systems that impose more uniform management on land resources? There is strong support for the view that this is so, and there are scientific reasons why it should be. Nonetheless, I need to be careful to avoid any simplistic association of all agrodiversity with resilience or sustainability. To do this would be as wrong as to write, along with an earlier generation, of the destructiveness of all indigenous cultivation methods.

A final question of importance arises. Which is more important in sustaining the resilience of diversity: the diverse practices or the adaptability of farmers to changing conditions? Much of the book provides evidence to help answer this question, but I reach a conclusion only in the epilogue.

■ Two Cautions

In viewing change through time, it must not be forgotten that very large areas that were formerly managed for crop production are now in a degraded condition. The question of adaptation so as to manage degraded land and ameliorate it is therefore of critical importance. So also is the failure of adaptation and the diversity loss that can arise from such failure. This must also be considered in a historical context. Sometimes the reasons for degradation can be established. Sometimes, modern successful management occupies land that has been degraded in the past. I shall examine examples later, but here I note just one.

Wilken's (1987) study in Mexico and Guatemala showed that many of the efficient practices he described are either identical or closely similar to those of preconquest times. He mentioned failures, and the results of past erosion, but stressed the very practical nature of the techniques described. History was not his main purpose. In the 3500 years before the Spanish conquest, three major periods of soil erosion have been recognized in lake sediments in central Mexico, the last of them not long before the short Aztec period, which preceded the Spanish conquest (O'Hara, Street-Perrot, and Bert 1993). Karl Butzer (personal communication, 1999) reported evidence of very widespread land degradation in central Mexico during a period of lower population density than either before or afterwards, around 1000 A.D. There is therefore presumptive evidence that these "sustainable systems" might have emerged in a context in which resource degradation was already a serious problem.

A second caution arises from the modern emphasis on local or rural peoples' knowledge (also called indigenous technical knowledge). From being regarded as something of an ethnographic curiosity before about 1980, rural peoples' knowledge has come to be seen not only as dynamic and well informed but also as crucially important for an understanding of what the farmers actually do and why they do it. In common with others, I treat it as an integral part of agrodiversity, especially of its agrobiodiversity and management diversity components. Often, however, outsiders make the mistake of assuming that local knowledge can be made mutually complementary with formal, scientific knowledge. Although it may be true that farmers and scientists both use observation and classification, they do so in very different contexts. Seeking an understanding, scientists often ask questions in terms of their own categories, not necessarily or even likely to be those of the people. Sometimes, the conscious or unconscious purpose may be to validate the scientists' conclusions by seeking corroboration from farmers' own observations.

A particularly penetrating discussion of these problems is one by James Fairhead (1993), showing how the new research fashion of seeking indigenous knowledge can have serious pitfalls. It is very hard for any outside observer to view resources and their management problems, as the farmers do, in the context of their whole livelihood and system of beliefs. Farmers themselves, seeking to simplify the comprehension problems of visiting observers, often give intendedly helpful but less than complete answers to questions. Efforts to overcome these difficulties meet with varying degrees of success, and even a long-resident field researcher can sometimes get sudden, illuminating surprises.[3]

Both these cautions are serious. Although land degradation may be exaggerated in some of the literature, this is precisely because farmers' efforts to manage it are not often taken into serious account. It has to be one further major theme. Local knowledge is important everywhere, and it is an essential element in the creation of the field types and farming spaces I emphasized earlier. But in this case, the spaces can also be established by observation. This is not so in all cases where some reliance on local knowledge is needed.

■ The Way Forward

Two common strands run through most of these themes. First is the importance of trying to establish the historical basis of dynamism in agrodiversity, its adaptations and innovations. Second is the use and management of the biophysical diversity in the land. I therefore present two further introductory chapters that relate specifically to these two questions. After that, I go on in parts II and III to discuss the evidence around a series of topics, within which the themes identified in this chapter are developed.

Notes

1. Netting and Stone (1996) also invented the term independently but used it without definition. They emphasized agrodiversity as a way in which resource-poor farmers spread risk and support their food security, but they referred to what was grown and used. Their detailed account of management practices, resting on Netting's earlier work (Netting 1968; Netting, Stone, and Stone 1993), was also put in the contexts of labor management and marketing.

2. Farming systems research (FSR) is usefully discussed by Norman and Collinson (1985) and more critically by Chambers, Pacey, and Thrupp (1989). An extensive illustration of the complexity FSR can attain is provided in Shaner, Philpp, and Schmel (1981). Although in principle FSR is intended to determine farmers' goals before introducing improvements on their land, the practice has more often been to explore the conditions for transfer of technology developed on research stations. Revealingly, today it is often described as FSR and extension (FSR/E).

3. One of the anthropological classics on resource management, cited in chapter 7, reports such a surprise. Walking through the "monotonous" African savanna–forest in the 1930s, Audrey Richards (1939:232) realized that her Bemba companions saw a large range of resources in the forest, interpreted in great detail. She wrote, "These people's thoughts and interests are in fact entirely concentrated on objects that the average white man simply does not see."

Chapter 4

Learning About the History of Agrodiversity

■ Two Very Relevant Questions

EVIDENCE AND INFERENCE

Most of this book is about the practices of small farmers in the twentieth century. Few of the cases reviewed have any firm information on what happened in earlier times. Little is documented in any form before the sixteenth and seventeenth centuries. Yet the book is not only about the diverse ways in which modern small farmers manage their resources. It is also about the resilience of agrodiversity, and it becomes important to know how deeply embedded this might be in agricultural history. This chapter looks at that question as a whole. Some initial thinking about the categories of agrodiversity that were presented in chapter 3 may help. All elements are dynamic but in very different ways.

Biophysical diversity, which was present before any agriculture, varies significantly through time and is locally very dynamic in detail. All farmers at all times have had to manage it either passively by responding to its variations, especially in climate, or actively by seeking to establish control. Social organization to manage production has changed comprehensively and has had to respond to a great range of external forces. Agrobiodiversity, or crop diversity, has enlarged as new sources of germplasm have become available; it has also been impoverished as new crops and new varieties have replaced old ones. At first glance, management diversity seems likely to have been elaborated through time. There are both practical and theoretical arguments to suggest why it should have done so, the former including the need for increased production from more varied land as population has grown and expanded its living area.

On the theoretical plane is the influential model of Ester Boserup (1965, 1981), treating population growth as the principal driver of change. Fuller consideration of this important model is deferred to chapter 11. I take only one point out of it in the present chapter: the proposed sequential change from simpler to more complex and labor-demanding agricultural systems. Boserup (1970:100) summarized her model of progress toward greater intensification of production as follows, disregarding land used for nonfood crops and for nonagricultural purposes:

(a) Cultivable land in completely uninhabited regions;
(b) land used only for the collection of vegetable foods and for hunting;
(c) land used only for grazing by domestic animals;
(d) land used in long-fallow agriculture (sometimes called shifting cultivation);
(e) land used in short-fallow rotations with grazing on fallow land;
(f) land used for annual cropping without fallow;
(g) land cropped twice a year, without fallow;
(h) land cropped three times or more annually, without fallow.

Long before Boserup wrote, it was already often assumed that all modern systems were preceded by a phase of shifting cultivation. For students of tropical agriculture, in particular, the assumption was forcefully stated by the authors of what remains a classic on the relations between farming and the soil. Nye and Greenland (1960:1) wrote,

> One of the most remarkable features of shifting cultivation is its universality. Primitive people evolving from the status of simple food gatherers, in which they had little more influence on the soil or vegetation than wild animals, have responded to the challenge of the forest in the same way. In regions so widely separated as the tropics of Africa, America, Oceania and south-east Asia they cut and burned the forest, planted their crops with a simple digging stick, and after taking one or two harvests they abandoned their plot to the invading forest. . . . Later some of them developed the hoe, and with this implement they were able to clear tenacious grass roots and cultivate the savanna, abandoning their plots as in the forest after a few years of cropping.

This assumption has continued to affect interpretation, as we will see at several points in this book. The farmers I discuss in chapters 6 to 8 mostly practice forms of shifting cultivation. Often, they are criticized for doing so and described as primitive. If we follow Nye and Greenland and Boserup, it seems that their practices are only slightly more elaborated than those of the very earliest farmers. In fact, all have greatly changed their crop inventories, and many have made important changes in their practices. It is of more than purely antiquarian interest to know more of the past. Did agriculture begin with clearings in which a small range of crops was grown, or was diversity of resource management present from the beginning?

An alternative view of agricultural origins proposes that the first farmers might have been selectors or creators of favored sites in which a fixed-field or even permanent cultivation of particular trees or field crops evolved early. Often, such sites might also have adjacent resources of fish. The question is important in relation to agrodiversity because burned clearings are not a likely place of origin for diverse cropping and land management. The alternative is that diversity of food sources and resource management, which already characterized foraging systems, was carried from the beginning into a fairly stable form of agriculture. The view that the first agriculture was in favored sites, which has gained much support among archeologists and paleobotanists, was first firmly advanced by Andrew Sherratt (1980, 1981). Writing of the Near East and southern Europe, he proposed that most early grain agriculture in these subhumid environments probably was fixed-plot cultivation on the moist land that was easiest to manage and that differentiation into other forms of both wetland and dryland farming, shifting cultivation included, came later.

This chapter addresses these linked questions. Firm answers are lacking, but a lot of work has been done since Nye and Greenland and Boserup wrote. Some of it helps provide answers to questions, but most of it is only indicative. This chapter, drawing mainly on research done since the 1960s, is as much about the sort of evidence on which we can draw as it is an attempt to reach any firm conclusions. New paleobotanical methods, discussed later in this chapter, are already leading to big improvements, and it will be possible to draw much more positive conclusions in the future, replacing the speculation that is all we can achieve at present. The speculation is useful, however, because it offers a lot that can be carried forward into the rest of the book and into a possible future history of agrodiversity itself.

EVIDENCE FROM WORK ON EARLY AGRICULTURE IN THE OLD WORLD

The vast literature on crop plant domestication seems to be the first place in which to search. It does provide some answers about agrodiversity, but the main work has concerned the principal food crops, so that most information more relevant to the questions of diversity has emerged only incidentally. Most of the discussion has been about the seed-bearing plants, angiosperms, which have been of fundamental importance in human history. Among these, it has been centrally concerned with the seed-bearing grasses.[1] Since the pioneer work of Vavilov (1926) it has been known that the diversity of these food-bearing grasses is not uniform across the world but is concentrated in centers of diversity. Vavilov and many of those who followed him thought these would be the centers of origin of the domesticated landraces that are the breeding source of all the major modern cereals. This is true for wheat and barley and probably also maize, but it has been demonstrated that several other major crops, including the African millets, sorghum, and perhaps even Asian rice, were domesticated at places widely separated over large areas, which J. R. Harlan (1975) described as noncenters.[2] Others, including rye (*Secale cereale*) and oats (*Avena* spp), probably developed by selection from among weeds in fields of already-cultivated crops.

In regard to the grains, the domestication argument contains a revealing circularity. Hillman and Davies (1990, 1992) carried out two decades of largely experimental work on the conditions needed for domestication, which confirmed earlier African observations by Stemler (1980). Most wild seed-bearing grasses have heads that shatter when ripe, and one major element in domestication is the selection of nonshattering mutants that hold their grain until after it has been harvested.[3] But the natural incidence of such mutants is low. They occur in tiny proportions in wild stands, and for domestication they must be selectively advantaged and multiplied. They need disturbed or new ground in which to become established and escape being overwhelmed by nonmutants. Moreover, they need to be either uprooted or cut with a sickle when near ripe; stripping the heads or beating into baskets, common methods of harvesting wild cereals in the modern world, were shown not to lead to rapid selection of domesticated varieties. This seems to say that creating the raw material for fully domesticated cereal crop agriculture entails managing the land surface and its cover that we associate with agriculture. This consideration, together with others, led Harris (1996) to propose that an unusual conjunction of biological and cultural conditions was necessary for domestication to occur on a sufficient scale to persist and generate agriculture. In this argument, domestication and the invention of agriculture did not happen repeatedly in many places, as is often proposed, but only in two main areas in the Eurasian landmass.

The hitherto less regarded role of wild cereals and other wild plants has therefore come to be seen as highly important in agricultural origins. A pattern still common in most of Africa until a short time ago is now viewed as having prevailed in other continents since the late Pleistocene. In the African drylands and savanna, numerous formerly hunting and foraging peoples have adopted agriculture only within the past 100–400 years, so modern comparisons still have relevance. Collection of wild foods, tubers where available and grains more widely, was facilitated by immense areas of grain-bearing grasses that fluctuated in distribution after long-term variations in rainfall in Sahel- and savanna-type regions. Mustard, legumes, capers, and fruit were collected as well, leaving abundant remains at the best-researched sites, some of them now wholly arid. The most important of these sites is Nabta playa in the modern desert of southwestern Egypt, occupied between about 9500 and 5000 BP (before present), with the richest finds dated close to 8000 BP (Wendorf et al. 1992; Wetterstrom 1998). Domestication of crops, as distinct from livestock, came very late in sub-Saharan Africa because for a long time there was no need to plant crops in the presence of large natural stands (Wetterstrom 1998; Marshall 1998). Certain wild crops, used for thousands of years, have never been domesticated and others, often called minor crops, were domesticated only late in prehistory when the necessary agrotechnology was already widely known. They may have been cultivated casually, but this leaves little record.

In the Near East, Anderson (1998) found abundant examples of flint sickles in use from as long ago as 12,000 BP, used in harvesting both wild grains and the later domesticated species. He experimentally established that the wild grains, however closely harvested, reseeded sufficiently to yield another crop. Accepting the necessary separation of selected seed to produce stable domesticated varieties, he saw this as likely to arise only if farmers migrated, carrying seed with them. In sub-Saharan Africa, Wetterstrom (1998) suggested that the retreat of rainfall from more northerly latitudes in and after the mid-Holocene created such conditions. They led to the domestication of pearl millet and sorghum, at successively lower latitudes about two millennia apart, long after crop domestication took place in the Near East.

The earliest clear evidence of domestication anywhere comes from the Jordan valley, in modern Israel and Jordan. Within one millennium it covered a suite of crops including barley, einkorn and emmer wheat, pea, lentil, chickpea, vetch, and flax (Zohary 1989; Harris 1998a, 1998b; Willcox 1998). Within at most a few hundred years, clear evidence of domestic herds of goats and sheep began to emerge. For more than another millennium, a major proportion of subsistence continued to be derived from hunting, fishing, and foraging for a large range of wild plants. This coexistence of systems continued as farming of domesticated crops, plus the rearing of domesticated livestock, evolved and spread east, west, and also south into the Nile valley over a period of some 3000 years. The eastern Mediterranean was one of the few places where temperate zone cereals grew wild in the late Pleistocene, and domestication took place during and after an early Holocene cool pause. Warming conditions then continued to facilitate its spread. In Harris's (1996) interpretation, domestication was thereafter a spreading process of displacement of wild crops by domesticated crops, probably with new domesticates being added as farming systems reached climatically different areas. Cotton and rice in India are examples, both several thousand years later (Meadow 1998). Domestication of rice and then millet (*Setaria italica, Panicum miliaceum*) in China, within at most 1000 years of the emergence of domesticated crop agriculture in the Near East, seems to have been entirely separate (Cohen 1998).

On why it all happened, the commonsense reasoning of J. D. Sauer (1993:207, 268) is helpful: "If people were not satisfied with the yield of natural stands and began planting seed elsewhere, the nonshattering forms would be favored and spread. . . . Almost any member of the [hunter–gatherer] group is familiar enough with the reproductive biology of the plants they harvest to know how to propagate them if it were considered worthwhile." Sauer's last five words highlight what is perhaps the most important consideration. Before the 1960s it was generally thought that agriculture appeared rapidly in the course of a neolithic revolution in the first half of the Holocene in just a few limited areas (Childe 1936). It was explained in a number of ways, including rising population pressure and the adverse consequences of environmental change. Now it is evident that agricultural origins involve a long period of time. J. D. Sauer (1993) and J. R. Harlan (1992) are among writers who have come to reject even the possibility of any single world-wide model for agricultural origins or plant domestication.

THE CONTENT OF PROTOAGRICULTURE AND EARLY AGRICULTURE

The origins of agriculture and of domestication are not necessarily the same. The fact that most of the important food annuals, each one a complex represented by many varieties, were present when agriculture began was already recognized in the 1930s (Ames 1939; Anderson 1952; Harlan 1957). The crop plant botanists who first remarked on this saw the necessity of a very long process of selection and initially involuntary breeding among wild food plants, extending over thousands of years. This is now a widely accepted notion, and the probability is that a very long period of such protoagriculture, characterized by selective management of wild plants, preceded agriculture in all areas of origin.[4]

Protoagriculture may well have involved forms of management of the species that were of greatest value and that were later to be domesticated. It is highly un-likely that such casual cultivation involved any systematic tillage of the soil. In Africa, but not in the Near East, domestication of livestock preceded any domestication of plants, and the first farmers were primarily herders. In areas not well endowed with wild food species, they may have made small holes in which to plant groups of seed that would germinate and grow to ripen when they returned to the same place next season (Wetterstrom 1998). This was done by the west African Tuareg even within the last few decades (Nicolaisen and Nicolaisen 1997).

Fire was certainly used in Africa from an early time to drive animals for hunting and later to encourage new grass growth for herded livestock. Widespread evidence of fires, probably both of natural occurrence and deliberately set to clear woodland and create savanna, emerged in Burkina Faso around the time of pearl millet domestication, around two and a half millennia ago. A shifting cultivation origin therefore appears possible in the African context. Most of the domestication literature, especially from the Near East, is remarkably silent on the question of fire, and the common emphasis is on cultivation in the moister alluvial soils, with grazing on the higher and drier areas once herded livestock were incorporated into the system.

The full content of protoagriculture or early agriculture does not yet clearly emerge from the archeological and paleobotanic literature from the Old World. Research emphasis on a narrow range of crops is partly responsible, but more is known of the content of the hunter–gatherer economy than of early agriculture. Although domesticated tree crops emerged only later, wild tree crops were important even before the farming of cereal crops in the upland northern fringes of the Near East. They included almond and wild pistachio (*Pistacia atlantica*) nuts, which were used abundantly (Hole 1998). There is nothing to suggest that the nut-bearing stands were managed or favored by their users.

Much more on the latter question emerges in Japan, where tree crops, especially chestnut and walnut, together with acorns and water chestnuts, were major sources of food throughout the whole long Neolithic (Jomon) period from around 10,000 to 2300 BP, before cereal crops were introduced to Japan (Nishida 1983; Imamura 1996). Chestnut and walnut occur only sparsely in the natural deciduous forest, and their emergence relied on opening of this forest around settlements that became permanent. In central Japan, the reason for permanence was possibly the invention of installed fishing equipment to catch freshwater fish. The consequence of village establishment was clearance of forest for timber and firewood and a major increase in the sun-loving nut-bearing trees that were preserved. Expansion of arboreal food crops was symbiotic with human activity modifying the natural landscape, principally by creating permanent openings in the forest. There was no domestication. Tree crop domestication relied on changing the reproductive biology from sexual reproduction to vegetative propagation, with the later addition of grafting. Plant generations are long, and changes in morphology or ecological adaptation occur slowly. In fact, walnuts and almonds are two of the few tree crops that can be grown successfully from seed (Zohary and Hopf 1993). Domestication of tree crops in the Near East and China took place after 7000 BP, mostly well after that date.

Absence of domestication notwithstanding, Nishida (1983) argued that there is a real sense in which there was farming in Jomon Japan within an integrated economy based on fishing, hunting, gathering in the wild, and gathering from managed stands of trees. He went on to propose that "gathering and farming should be distinguished by the relationships between man and the plants used. The morphological traits of the plants provide only secondary evidence for such relationship" (Nishida 1983:318). Where the environment was managed to suit wild wheat and other cereals in the Near East, a similar argument might apply. Anderson (1998) found only differences in processing technology (threshing) and none in harvesting methods across the boundary between use of wild and domesticated grains. It must be presumed that agriculture led to a greater concentration of production as management extended to soil management, but there is little information in the literature on this topic.

The question of the place of diversity of crops and practices in early agriculture must be approached obliquely. Early farmers left records only in the soil, and until lately archeologists and paleobotanists have not been seeking evidence of diversity. One area in which they have provided clear evidence is in eastern North America, where understanding of agricultural origins, much more recent than in the Near East, has come together quickly since the late 1970s using methods different from those of earlier archeology. A consistent and highly interesting story has emerged.

INSIGHTS FROM EASTERN NORTH AMERICA

The agriculture of eastern North America at the time of European colonization is well described in modern literature, and the scale of the transformation of the forest by farming people is made very clear (Williams 1989; Denevan 1992; Peacock 1998). One early piece among a large relevant literature is a splendid account by Cronon (1983), analyzing environmental change in colonial New England. It offers one important clue. Cronon's main object was to understand how English settlers had changed the New England environment between 1620 and about 1800. To do this he had to reconstruct, from contemporary accounts, how the Indians used the land before and into the early decades of colonization.

The Indians whom the settlers displaced were cultivators who also derived a large part of their subsistence from hunting, fishing, and collecting wild produce. In southern New England, maize was planted annually in hoed fields worked, as the crop grew, into ridges or small mounds. Adjacent to the maize, in the same mounds, Indian farmers also planted beans, squashes, pumpkins, and tobacco, so that the entire surface of the field became a dense tangle of plants, discouraging weeds and preserving soil moisture. Methods included only a modest use of fire.

Most significantly from the present point of view, this system lay within a wider managed context. The Indians set fire to large areas around their villages each spring and fall, making a more open landscape under the trees, driving animals for hunting, and creating the conditions favorable to strawberries, raspberries, blackberries, and other gatherable foods. The southern New England Indians thereby manufactured an environment that attracted herbivorous animals and birds and their predators. Cronon (1983:51) writes of this extensively used practice:

> Indian burning promoted the increase of exactly those species whose abundance so impressed English colonists: elk, deer, beaver, hare, porcupine, turkey, quail, ruffled grouse, and so on. When these populations increased, so did the carnivorous eagles, hawks, lynxes, foxes and wolves. In short, Indians who hunted game animals were not just taking the "unplanted bounties of nature"; in an important sense, they were harvesting a foodstuff which they had consciously been instrumental in creating.

A people who had no domesticated animals were therefore modifying the environment to build up the population of species that were important to them for meat and skins. There is a real sense in which all this has to be seen as part of the agroecosystem. The practice was identical with that of many hunter–gatherers, in the Americas as elsewhere, but here it is set in agricultural context. The argument says that resource management is not linked only to deliberate planting of crops and that different sorts of farming spaces can be created by other means. This view has been strongly developed since 1980 in work on resource management in Latin America, and I will follow it more closely in chapter 8. It is a familiar view in the Australian context, where no agriculture evolved before European settlement but where there was a great deal of deliberate environmental management over much of the continent to encourage production or attract prey. Sometimes this was close management. Until about 1880, when they were overwhelmed by advancing European settlement, a dense Aboriginal population along the lower reaches of the Murray–Darling river system subsisted mainly on uprooted wild cereals and fish.

They repeatedly and maybe continuously used the same set of resources (Allen 1974; Pfeiffer 1976).

SOME REAL EVIDENCE ON EARLY AGRODIVERSITY

To return to the eastern woodlands of North America, the maize–beans–squash complex described by Cronon probably was only 200 years old and maybe only 100 years old in cool-climate New England when the English colonists arrived (Galinat 1985; McBride and Dewar 1987). Within eastern North America, maize did not spread north as a significant crop until varieties suitable for shorter growing seasons had evolved. Maize entered the midwestern archeological record around 200 A.D., but it became important only after 800 A.D. and did not become the dominant food source until after 1100 A.D. The common bean (*Phaseolus vulgaris*) did not arrive in the Midwest until between 1000 and 1200 A.D. (Galinat 1985; Smith 1987, 1992). Archeological and paleobotanical research in the upper Mississippi basin of the Midwest and in the Southeast has established that a much earlier phase of agriculture had a very different basis (Asch and Asch 1985; Smith 1987; Butzer 1992). A group of six grain-bearing grasses and a vegetable supported an ancient gathering economy, increasingly with patchy semicultivation. Six of the seven domesticates were derived from native wild seed-bearing plants foraged, nurtured, and then domesticated (by the evidence of their changing morphology) from before 2000 B.C. to 1000 B.C. They included two starchy-seeded autumn-maturing annuals, chenopod or goosefoot (*Chenopodium berlandieri*) and knotweed (*Polygonum erectum*); two starchy-seeded spring-maturing annuals, maygrass (*Phalaris caroliniana*) and little barley (*Hordeum pusillum*); and two oily-seeded autumn-maturing annuals, marsh elder (*Iva annua*) and sunflower (*Helianthus annuus*). Chenopod was still cultivated by Indians in the lower Mississippi valley in the eighteenth century. The seventh and earliest domesticate was the squash (*Cucurbita pepo*).

These crops became the principal basis of a developed agriculture. This system has an extensive literature since around 1980, most of it brought together by Smith (1992). The discoveries depended heavily on the analysis of plant remains identified in samples taken from the soil and habitation sites, using the modern paleobotanical methods discussed later in this chapter. What are described as farming economies developed during the last millennium B.C., and field agriculture became general between 250 and 1150 A.D. (Smith 1992).

Reasoning from the archeological and paleobotanical evidence in the woodlands of eastern North America, Smith described a sequence that is similar to that also suggested for the Near East of a few thousand years earlier by Zohary and Hopf (1993) and more explicitly by Harris (1996). Smith (1992:272) wrote,

> A gradual co-evolutionary progression of increasing human intervention would have gone from initial toleration through first inadvertent then active, intentional encouragement of these food plants, to the deliberate storage and planting of seed stock. . . . This co-evolutionary progression, which occurred over a

span of two or three millennia, transformed stands of colonizing weeds first into inadvertent or incidental gardens and finally . . . to intentionally managed and maintained gardens of domesticated crop plants.

By the time clear paleobotanical evidence of domesticates could be identified, a "host of other quasi-cultigens and cultigens" also became archeologically abundant (Smith 1992:60). Work using paleobotanical methods has found similar early diversity in India. By the end of the stone-using period, 3500 to 2500 years ago, cultivated or semicultivated species already included rice, sorghum, wheat, barley, green gram, black gram, hyacinth bean, horse gram, lentil, grass pea, pigeon pea, Deccan hemp, Indian gum arabic, and Brassica species (Kajale 1994). This is an impressive diversity at early dates, possibly already implying differences in farming methods. In the drier Near East, on the other hand, the literature has less to say about plants other than the range of basic grains and pulses.

Worldwide, protoagriculture explored a very large part of the total range in the plant kingdom. A selected list of some 400 food, forage, and fiber plants that have been domesticated, most of them a long time ago, is drawn from among 55 botanical families (Harlan 1992). Among these, only 12 or 15 species today provide most of the world's food supply, but almost all the others are still used. Drawing on several thousand more species, protoagriculture also provided the world with a great number of other plants used for beverages, relishes, medicines, and decoration, while the present range of domesticated livestock became available early in the agricultural period.

■ Selection of Favored Sites

FLOOD PLAINS AND HOME GARDENS

There is another reason to stress the recent work in eastern North America. It has emphasized the importance of riverbanks in early agriculture and better brings this question together than does most of the literature on the topic from the Near East. Riverbanks are one of two places identified in the plant domestication literature as likely locations for first deliberate cultivation of a range of crops. The other was in the dump heaps around houses, where seeds and other germplasm material would be included in the household refuse.

I deal with the latter first. Dump heaps, or the disturbed areas around houses in which food remains are discarded, often are associated with house gardens. Not many have suggested these areas as places of origin for cereal crops, but they could have been more likely places of origin for some domesticated vegetable crops, fruits, and tubers. They were sites where women, who threw out the household wastes, saw useful plants growing and might nurture them. They continue to do this today in many parts of the world. What can be said both from modern analogy and from logical reasoning is that disturbed and enriched ground near an enduring home site would be a particularly suitable place for the curious—most likely women—to experiment with plants and for plants themselves to be separated from wild relatives and evolve new characteristics.[5] The dump heap theory enjoyed a vogue for a time but has been increasingly disregarded by field crop specialists. It may have a more significant role to play in the fuller history of agrodiversity, when horizons are widened beyond the cereal grains to include root crops and vegetables that must be put into the ground but without extensive land preparation.

People carry food plants from where they grow to where they are consumed and deposit germplasm along with other fertilizing refuse near their houses. Animals and especially birds add to the range of germplasm transport but in a rather random manner. Rivers can transport seed material from place to place naturally, and deposit it in a specific environment. On floodplains regular natural flooding replaces the nutrients taken from the land by crops, and wild foods are both abundant and readily supplemented by aquatic life in the rivers. Recession of the flood not only exposes new alluvial material in which planting is easy but also leaves it with residual soil moisture that would last some time even in a dry climate. Flood-recessional farming certainly has been practiced for millennia and, though without tillage, probably was an important element in protoagriculture. It is still practiced in many areas today and has been adopted even by people who are new to farming.[6]

There is now a lot of support for the view that well-watered lowland soils were the sites of the first sustained agriculture in many parts of the world. The earliest sites in the Jordan valley were of this nature. Harris (1998b) proposed that the first use of drier uplands in the Levant was for early pastoralism, always around cultivation on the better-watered lowland soils. In South America, Roosevelt (1991, 1999) assembled strong evidence of agriculture from around 4000 BP along the floodplains of the Orinoco and Amazon rivers, supporting prehistoric populations already stabilized in preceding millennia by specialized and intensive use of fish and shellfish resources. They developed the earliest pottery in South America. Nutrient-enriched ground around permanent village sites may have facilitated manioc cultivation (Roosevelt 1999). It is possible that farming here was the oldest on the South American continent, ahead of full agriculture in the Andean center of diversity, long supposed to be the first agricultural region in South America.

■ Diversity in Early Management: Evidence from the Ground Surface

THE ROLE OF LAND MANAGEMENT

The evolution of land management has gone hand in hand with the selection and breeding of crops and has been closely related to the soil and water needs of crops that farmers have selected for use. Both fire and digging to uproot rhizomes and tubers were as well known in protoagriculture as they are by more modern foragers. It is a reasonable presumption that evolution of site preparation would have been an important stage in the development of management, thus creating conditions in which domestication could take place. Site preparation must have begun casually, clearing and preparing small patches before such methods were extended over wider areas. Unfortunately, it is only in rather specialized areas that the archeological evidence can say much about purposive early management of the land surface. Tools should be important, but there is doubt about the identification of early hoe blades, which could have served other purposes.[7] The earliest tools, probably of wood, have perished. The literature on wetland systems in prehistory is much larger than that on dryland systems, but both are worth a glance within the framework of this exploration into the past.

Wetlands occupy a much smaller area than drylands, but they offer a distinctive set of farming environments. The most important variable is the level of water in or above the soil, and it varies not only from place to place but also from season to season and year to year. The value of alluvial soils has ensured that major efforts have been made to manage water levels since very ancient times by means of earthworks, irrigation, and drainage ditches. These leave durable marks and have attracted a lot of attention from archeologists.

EVOLUTION OF IRRIGATED FIELDS

Irrigation is a very ancient and widespread practice. When the Near Eastern complex of crops was introduced into the Nile valley between 8000 and 7000 BP, these winter-growing crops initially were planted behind the receding flood (Wetterstrom 1998). A more purposive agriculture evolved between 6500 and 6000 BP, and in this dry environment it was immediately associated with irrigation. A basin system to hold the overflow of the Nile flood, identified as the earliest major system in the valley, continued to be used until the nineteenth century (Wendorf and Schild 1976; Butzer 1976). There is also evidence of water management coincident with early agriculture in the Jordan valley (Prausnitz 1970). Whether irrigation is of the simplest flood-spreading forms or embraces the construction of diversionary channels feeding carefully leveled and maintained terraces, it has many obvious purposes. There are areas in which no cultivation would be possible without it. There are many others in which agriculture without irrigation would be at the mercy of seasonal and annual variations in rainfall, so that a greater measure of security is provided by adopting such practices. There are still other cases in which irrigation is supplied only to certain crops that have greater need of water so that diversity of production and diet is achieved. Very widely, irrigation carries mineral nutrients from areas where they are less useful to sites where they can add most significantly to productivity.

There is a great range of skill levels. Carefully designed indigenous irrigation systems exist or have existed in many parts of Asia and tropical America and are or were scattered across large parts of intertropical Africa. Complex systems, called *qanats,* of tapping groundwater by means of lateral tunnels are described from southwest Asia and the Mediterranean. They are also found in Mexico but were probably introduced there from Spain rather than being an indigenous invention.[8] At the other end of the scale, a great many small systems use only a limited range of technological skills (Sutton 1984, 1989; Kirch 1985). Water often is simply lifted from wells in pots or buckets or scooped or splashed onto fields from ditches. Human and animal labor are used, and there were mechanical devices in several ancient irrigation systems. There are simple diversionary systems made of bamboo tubes, fitted into one another, that may carry water a kilometer or more. Systems that serve as irrigation suppliers at one time of year may serve as field drainage at another time. There is so much about water management in the literature that it would be easy to give it undue emphasis and become overawed by the technology.

DRAINING OF WETLANDS

Systematic swamp drainage and the building of ridged or mounded fields in shallow swamps is the converse of irrigation. It has the purpose of creating artificial dry environments for field crops, very few of which will tolerate, still less prefer, continuous or near-continuous inundation. Rice and Asian–Pacific taro (*Colocasia* spp.) are the outstanding exceptions. Although seasonal wetlands can be used for flood-recessional cultivation, drainage provides great advantages. Flood-recessional cultivation is a risky business, always carrying a chance of crop loss. Where the flood range is limited, creating ridges would greatly increase security. It would also have security advantages over dryland cultivation, despite the much higher initial labor needed to create the landesque capital of the ridges. Mud, plants, and water from the ditches can be thrown up to add mineral nutrients, especially dissolved phosphorus, to the dry soil.

Other benefits from swamp drainage include microclimate modification, important wherever there is risk of frost. Like irrigation, drainage also leaves durable traces on the ground, but in only a few areas can it be established that the main wetland crops differed from those grown on nearby dryland sites. Draining is fundamentally a form of management diversity, making possible more secure use of a particular set of environments. There is a good deal of speculation in the archeological literature about the relation of the wetland evidence to cultivation on the nearby drylands. The power of the shifting-cultivation-first view has been sufficient to make most writers suppose that wetland cultivation would have been adopted of necessity as shifting cultivation proved incapable of supporting larger populations, but there is little or no direct evidence.

A New Guinea example is important because of its great age. Discovery in the 1960s of a succession of ditch systems at the Kuk swamp in the highlands of Papua New Guinea, by Jack Golson and his colleagues, attracted a great deal of attention (Golson et al. 1967; Golson 1977). When fully explored through a series of phases after years of digging, successive systems of field drainage were exposed, each used for a period and each successively more sophisticated in its construction. At Kuk, the earliest, primitive phase seems to have been created very close in time to the origins of agriculture anywhere, around 9000 BP. Ditch-drained, raised-bed systems are still in use in the highlands of Papua (formerly Irian Jaya, the Indonesian half of the island of New Guinea) where, in the Baliem Valley, pollen evidence gives them a suggested age of 7000 to 6500 BP (Haberle 1994).

The New Guinea systems are tiny by comparison with the vast extent of ridged fields in northern South America and in parts of Central America. Massive and widespread evidence remains of raised fields throughout large areas of lowland northwestern South America and in the swamps around Lake Titicaca in the Andes. They cover some 500,000 ha in Colombia alone (Denevan 1992). Although there are suggestions of earlier origin, these extensive earthworks belong mainly within the past 2000 years, very much later than in New Guinea. There may have been older ridged fields in the deltas of the Amazon and Orinoco rivers, but any evidence of fields accompanying large mounded settlement sites is buried under later alluvium. Roosevelt (1991:404) went no further than to suggest that "raised and drained field systems" might "perhaps" have been constructed to support the large prehistoric settlements on Marajó island in the mouth of the Amazon. Later, although it still seems highly probable that maize cultivation was implicated, the management of tree crops may have been more important (Roosevelt 1999). In some parts of South America, raised field systems were still in use at the time of the Spanish conquest, but only the *chinampas* of the highland Valley of Mexico have continued in use to the present. The record of excavation in the *chinampas* indicates elaboration in construction method over a period of more than a thousand years (Parsons 1985; Turner and Denevan 1985). At least in modern times, they include remarkably intensive use of canal muck in seedbed construction (Wilken 1987).

EVOLUTION OF DRYLAND FIELDS

The much larger areas of dryland fields preserve, on the surface, a much more limited archeological record than wetland fields. Often there is no record at all except that of small differences in vegetation and soil, the durability of which is uncertain. Sutton (1984), who remarked on the lack of knowledge of ancient field systems and their technology in Africa, asserted that these seem almost to have been forgotten except where they left large and very visible systems of landscape modification. Even where there is interest in the details of agricultural history, information is scant. In Mauritania, in northern west Africa, there is just one remarkably early set of evidence concerning large stone-bounded fields whose bounding walls may have had the purpose of ponding rainwater. They belong to a period around 3000 BP and may have been used to grow pearl millet close to the time of its domestication (Amblard 1996). Evidence from other areas has not been explored adequately.

In describing prehistoric field farming in the woodlands of eastern North America, Smith and others wrote of fields adjacent to both large and small settlements. By the last centuries of the prehistoric period, systems of hoe-built ridged fields were widely developed (Riley 1987), but because the fields were on land that has since been selected for cultivation by modern farmers, much surface evidence has been obliterated. Denevan (1992) mapped the available evidence. When Europeans first visited the valleys of the American Midwest and Southeast, intercropped infields could be distinguished from large single-crop outfields, and community field systems were linked over extensive areas (Smith 1992). Line drawings show the pattern of these fields as they were, near the Atlantic coastlands, by the seventeenth century (Williams 1989).

Surface marks can remain as indicators of disturbance on dry ground, especially where stones were taken out of the soil to make it arable. Farmers found it simplest to put these neatly at the side of the field or in small cairns, where they served as boundary markers (Grove and Sutton 1989). Much of the evidence of ancient field systems in Africa, in the uplands of southwestern Britain, and in the highlands of South America consists of stone lines. Sometimes stone lines were built up into terrace risers and whole systems of hillside terraces; more often they were not, although they might have been aligned along the contour to form barriers against soil creep and erosion. Where there are no stones, ancient ridging and mounding can leave durable traces that last a very long time, even for thousands of years. If not still visible on the surface, they may still remain in soil moisture patterns and hence can be seen on air photographs and especially on infrared images. The clearest example of an agricultural landform that is still widely visible or can still be traced on images is the plow-built ridge-and-furrow systems of large parts of premodern England and other parts of Europe.

By themselves, field systems say more about the organization of production than about its content. In the British Isles, where the most detailed work has been done, remaining banks and walls reflect land-holding patterns that can be shown to have endured from before the period of Roman occupation in the first four centuries A.D. In the mainly arable regions, in Britain and across Europe, they were overlain from before 1000 A.D. by the planned countryside of nucleated villages and their open strip-pattern fields. This reorganization represented a collectivization of agriculture as drastic as the twentieth-century collectivizations in socialist countries and more enduring (Rackham 1986).[9]

■ Evidence from Within the Soil

A SERIES OF MODERN APPROACHES

A great change in the methods of archeology and paleobotany since the 1970s makes this speculative chapter necessarily only an interim statement. Archeological sites are no longer only those where ancient structures are found. They now include the fields, valleys, and hillsides. Pollen analysis (palynology) has been an important tool since the late 1940s, but its utility is limited to wet situations in which anaerobic conditions have permitted pollen preservation. Although pollen analysis is improving, it is still limited by the site conditions. This is less a problem with other and newer methods of exploring the vegetation, cultivars, and agricultural systems of the past.

A growing body of modern interpretation, including most of the reconstruction in eastern North America discussed earlier, relies on sampling soil and settlement sites to yield botanical macroremains: seeds, fruits, and other plant remains that, where they have been charred, are preserved. Soil stratigraphy can be a useful indicator of the history of land management and erosion, but plant remains within it can be relocated by the bioturbation activities of soil fauna, further discussed in chapter 5. The oldest samples can be found above younger samples, even at or close to the surface. The utility of botanical macroremains depended on the development of modern methods for dating small samples independently of their stratigraphic location. Current practice is to take large numbers of soil samples to yield carbonized plant remains generally 0.5–4 mm in diameter, poorly visible to the naked eye, and separated from the soil by fine-sieving and, when the remains would not thereby be destroyed, by flotation.

Where soil and subsoil conditions are continuously wet, creating anaerobic conditions, uncarbonized material is more abundant than the carbonized material, and this can be separated from the soil by sieving. Also of growing importance since the 1970s is the study of phytoliths, plant opal silica bodies that occur in stems, leaves, and inflorescences of many plants and bear distinctive signatures. Because they are inorganic, they are preserved in environments in which pollen and organic macroremains are not.[10] The quantity of plant remains surviving in the soil is very great; its use in paleobotanical research entails complex sampling strategies and the use of multivariate analysis to analyze large amounts of data. Methods are reviewed by Greig (1989). Also increasingly important is the chemical analysis of skeletal remains, providing important information on the nutrient sources of people and animals.

These newer sources become more meaningful when used in association with older methods, now being applied in new ways, all in the context of constantly improving dating techniques. Plow marks can survive below the surface horizons, and so can planting holes, post holes, and ditches. Reconstructing past patterns of erosion and deposition on hillsides greatly illuminates the effects of ancient agriculture and can also illustrate the history of land management. With Geoff Humphreys, I have been involved in a small piece of landscape history in Vanuatu, a history that shows what can be done even by very simple methods (Brookfield and Humphreys 1998).

THE CONVERGENCE OF EVIDENCE: A SMALL CASE IN VANUATU

In 1992, we had the opportunity to examine what had been described to us as an old pattern of slope management on loose, friable soils derived from volcanic ash on the Vanuatu island of Pentecost. We soon saw that these apparent terraces were simply the cross-slope members of land plot boundaries marked by live hedges that, over time, had checked the downslope movement of soil to build up accumulation terraces (*taluds*) behind them. Local farmers, very willing to provide information once we had identified the origin of the benches to their satisfaction, said they had been present all their lives, and long before. They understood the value of the accumulated soil on the benches, in which they planted their yams much more closely than on the eroded slopes. Trenching through one of these *taluds,* Humphreys established a soil stratigraphy that included old holes, now completely masked by the top layer of deposition. They could have been post holes, but they resembled holes we saw being made nearby for planting taro, a subsidiary crop to these yam-growing people. A few pieces of charcoal were found and later dated (Humphreys 1996). One of these, within one of the holes, returned an age of around 400 years, and its stratigraphic position clearly indicated the continued use of the *talud* both since that time and before it. Another date, almost 1000 years, seemed to represent the unmodified topsoil at the base of the system.

What also seemed to be established was the very long duration of the system of plot boundaries. The people of this part of Pentecost have a matrilineal system of social organization. The heart of this lies in the definition of plots of land, the male owners of which constantly change through the system of inheritance but that themselves remain permanent (Bonnemaison 1986). An implication of our discovery of the enduring nature of the land management system is that the land allocation system might also have great age, surviving all the shocks of the colonial period, with its partial depopulation caused by devastating epidemics in the nineteenth century. It has also survived changes in the crop inventory. Today, planting live hedges is neglected because of demands of cash cropping, but the banks are well stabilized and they still hold the soil.

This is a very small example. The much larger examples of reconstruction in North America demonstrate the power of modern paleobotany in association with archeology. In the Netherlands, Kooistra (1996) provided striking evidence of the remarkable detail obtainable on old farming practices and on their crop and livestock content. Using different methods in conjunction, he reconstructed the whole content of agriculture in the Roman period and succeeding centuries for two areas, one alluvial and one in a loess region. Diversity in both crops and management is clearly exemplified in this work. The future of any history of agrodiversity lies in further exploration of this type. What has already been revealed is a great advance on earlier speculations.

■ Toward Answers to the Questions

SHIFTING CULTIVATION FIRST, OR A LATER DEVELOPMENT?

We can now answer some of the questions posed at the beginning of this chapter. The case that real farming began in favored sites is strong. In addition to the evidence presented from North and South America and the Near East, there is evidence that wild rice and then domesticated rices were cultivated initially in river-bank locations in Asia and Africa (Chang 1976, 1989). This evidence notwith-standing, because the avoidance strategy of shifting cultivation is so effective in site clearing and fertility restoration, it remains likely that the practice was adopted early in the spread of agriculture. It might even have been the initial method, where only single domesticates were used in a context of abundant wild food, as perhaps in sub-Saharan Africa. Even in riverside areas, the first domesticated crops may well have been planted in clearings made with fire, although in such locations shifting fields often would not have been necessary. What is more probable is that when upland forest areas came to be occupied, land rotation methods using fire were the initial means of cultivating a range of plants that had already been domesticated elsewhere.

There would have been important differences between types of plants. In the West African context, Harris (1989) drew a speculative but important contrast between the development of agriculture in the yam-based vegecultural regions and in the drier cereal-based regions. He relied on the contrasting ecology of root crops and cereals, the differing need for partially closed and open canopies, and the more rapid soil fertility depletion under protein-rich cereal crops. He therefore found it easy to propose an extension of vegeculture away from the home site into the surrounding forest, with a fairly low rate of garden resting and relocation. With cereal crops, conditions are different. Once cereal cultivation moved from its putative origin areas in seasonal wetlands onto the dry lands, following selection, storage, and planting of suitable varieties, it would soon have become unstable, and frequent clearing of new fields would have been necessary.

The evidence does not permit replacement of one model by another, but what is clearly proven is that matters are not as simple as they seemed to be in the 1960s. The products of modern research are important for understanding modern as well as ancient trends, but they have made little penetration into a more general literature on the history of agriculture. This is also true of the origins of agrodiversity, about which it is possible to be much more positive.

THE PROBABLE ORIGIN OF AGRODIVERSITY

Convergence of evidence suggests very clearly that a diverse agriculture grew out of a diverse foraging economy. The essence of foraging, together with hunting and fishing, is to provide a varied diet and a supply of fibers, medicines, relishes, and other needs. Within this complex, some species are managed simply by being protected from damage or choking weeds and being given space and light. Once a small number of plants was purposively cultivated and field systems emerged, the managed use of a wider range of semidomesticated plants probably would have followed. Work since the 1980s has shown that ethnobotanical knowledge often is more extensive among cultivators than among hunter–gatherers. This counterintuitive finding, tested and again demonstrated in Borneo by Voeks (1998), is perhaps explained by the wider range of species useful to permanently settled farming people than to migratory foragers. Wherever agriculture evolved out of foraging through forms of protoagriculture rather than being established de novo in new areas, plant and management diversity evolved both before and together with agriculture. Speculating, deductively rather than from evidence, on the history of intercropping, Plucknett and Smith (1986) long ago concluded that the very earliest agriculture involved a range of semidomesticated and domesticated crops, all in mixtures.

Once durable patches of cleared ground had been created and maintained, new crop plants and trees probably would have been introduced from nearby areas and later from a distance. Within particular crop species, selecting seed from a distance is an ancient practice. The establishment of farming economies created the basis for further diversification as changes in the plant environment favored one species or variety over others. Selection from natural mutation and crosses, aided by deliberate farmer selection, created some crop plants that interbreed freely and others that do not or that can be propagated only clonally (Cox and Wood 1999). Selection may well have been helped along by breeding experiments carried out by women with plants found growing in their dump heaps and by both men and women with edible plants found abundantly on riverbanks. It therefore seems most likely that diversity in agriculture first merely supplemented wild diversity, later took its place, and then elaborated as germplasm was carried to other areas.

Notes

1. It was argued early that vegeculturally reproduced root crops were easiest to domesticate and therefore perhaps were the earliest to be brought under cultivation (Sauer 1952). In the Near East, edible wild rhizomes were important sources of carbohydrate before cereals dominated (Hillman and Davies 1990). Domestication ahead of cereals has also been argued for pulses, but the evidence is unclear.

2. Harlan (1975, 1992). Among others, Fowler and Mooney (1990) provided a readable and well-referenced account of the evolution of diversity within crop species and discussed the 1920s work of Vavilov at length. Earlier, C. O. Sauer (1952:21) expanded these notions, insisting that "the hearths of domestication are to be sought in areas of marked diversity of plants or animals."

3. Also important is selecting mutants with seeds that lack the dormancy characteristic of wild grass seeds, enabling them to survive through successive seasons. Domestication for annual production requires that seeds will germinate when they are planted. This requirement applies to all the grain crops.

4. Readers who want to follow the argument might read J. R. Harlan (1975, 1992) and chapters in Harris and Hillman (1989) and Harris (1996). They provided important detail on both specific crop plants and general principles. Zohary and Hopf (1993) presented an important summary for the Old World. J. R. Harlan (1992) provided summaries for all major regions. An invaluable crop-by-crop survey, rich in historical detail and covering a wide selection of crops, is that by J. D. Sauer (1993).

5. Although proposed earlier, this theory was first fully developed by botanist Edgar Anderson (1952). He noted that hybrids often cannot flourish among established plant complexes but that where freed from competition on open sites, and especially on open sites with disturbed ground, they can both flourish and cross-breed freely to generate new hybrids. Species that have never intermingled might find their way to such sites. This finding led him to suggest that natural hybridization of plants taking root there would breed "strange new mongrels," among which some might be useful plants that would take the fancy of people sufficiently to encourage deliberate planting (Anderson 1952:149). This theory has been taken up mainly by those whose interests embrace trees and vegecultural plants (e.g., Harris 1972, 1973, 1976, 1989). Like most cereal crop specialists, J. R. Harlan (1992) had little time for it.

6. A modern example in the lowlands of southwestern Ethiopia indicates how readily flood-recessional farming can be adopted and elaborated. The adoption of an agricultural way of life by a group of tribes in southern Ethiopia probably is quite recent. Highly seasonal small rivers flood the plains twice a year, and people who were formerly pastoralists now cultivate along the river banks or in areas flooded and supplied with groundwater from the rivers (Matsudo 1996; Kurimoto 1996; Miyawaki 1996). Methods range from seasonal cultivation in shallow rain-fed depressions in areas with strong groundwater flow, and more laboriously in marshes when the rains are insufficient, to works that include limited irrigation and embanking to protect crops from late flood after they have been planted. Secular changes in the location of rivers and groundwater flow take place. Rules of access must be adapted to this variability, and the most reliable depressions are the most highly valued.

7. In principle, one might expect that the tools used would be critical in establishing the history of field agriculture. In fact, they yield only secondary information. The literature makes clear that tools were developed according to needs, and although there are many specialized tools, there is much similarity between the tools of different regions. Although shifting cultivation sometimes is loosely described as hoe culture, the hoe is used far more often in fixed-plot systems in which the land must be broken up continually than in land rotation systems in which dibbling is often all that is necessary. The major contrast in tool design lies between the areas in which animals were domesticated to provide traction power and those in which no such animals existed or from which they were excluded by disease. The whole of the New World and large parts of Africa developed only hand- and foot-powered tools. The evolution of animal-drawn tillage instruments, from the surface-scratching ard to the plow in its various forms, took place entirely in Asia and adjacent regions of Europe and northern Africa. In the whole of intertropical Africa, only in the highlands of Ethiopia was the plow used in ancient times (McCann 1995).

8. The Mexican *galerias filtrantes* are in all essentials identical in construction to the *qanats* most strongly developed in Iran and other parts of southwest Asia and northern Africa (Cressey 1958; Wilken 1987). A number of *qanats* were constructed in Spain during the early Muslim period (Watson 1983). Other irrigation technology of Near Eastern origin, including the *noria* in both its Persian wheel and chain of buckets forms, was introduced by the Spaniards.

9. Much less is known of the history of the largely stone-built walls and terraces of Mediterranean Europe, but because of greater continuity in land use, a good deal can be inferred; it suggests that a very specialized use of the landscape goes back a long way. Rackham and Moody (1992:123) remarked that "if as much were known about the history and functioning of terracing in Greece as has been discovered about ridge-and-furrow in England . . . we should be well on our way to understanding the development of the Greek landscape."

10. "Silica that forms the phytoliths is carried up from groundwater as monosilicic acid and is

deposited in epidermal and other cells of the growing plant, eventually forming bodies composed of opaline silica. In many plant taxa phytoliths take on distinctive shapes which may be diagnostic (i.e. permit the identification of the plant) at varying taxonomic levels" (Pearsall 1994:115–16). Phytoliths have been used in Central and South America, especially in reconstructing the history of maize as a crop (Piperno and Pearsall 1993).

Chapter 5

Understanding Soils and Soil–Plant Dynamics

■ Introducing Soils

AN EXPLANATION

Soil diversity is central to the definition of agrodiversity proposed by Almekinders, Fresco, and Struik (1995), and its importance and that of plant nutrients have already been mentioned at several points in this book. It is a major element in biophysical diversity, and soil–plant dynamics are the most important of the sets of processes contained within that fundamental element. Farmers have considerable knowledge about the physical and biological foundation they manage. For an understanding of agrodiversity, some basic scientific knowledge of that foundation is as important as knowledge of the crops, agronomic practices, social, political, and land tenure conditions that shape farmers' activities. What follows includes material familiar to most natural scientists concerned with the land and to agronomists, but not all of it is necessarily familiar to all of them because the field is very compartmentalized. It is certainly unfamiliar to many social and development scientists who write about small farmers or seek to learn about them and who are likely to be interested in agrodiversity. I concentrate mainly on tropical soils because most of the case material in this book comes from the tropics. To ease the path, I place the more technical matter in boxes, and most of the more general references are also in those boxes. But the content of this chapter also has to do with people, and I begin with the soil as one group of farmers sees it.[1]

SOIL AND THE PEOPLE

Many farmers not only understand differences in the soil but also have ideas about the processes that create these differences. One well-described example, idiosyncratic though it is, is valuable in showing what is most relevant to farmers. The Burunge of central Tanzania occupy an unpromising area of considerable variety, so that "as one walks through fields and talks to farmers about their experiences one comes to think of the area as a mosaic of soils and micro-environments" (Östberg 1995:33). Burunge have distinctive ideas about the formation of soil and its qualities, examined in depth by Östberg, and it will be useful as well as interesting to begin with these notions.

Land, the Burunge insist, has life, arising from the warmth within it and the rain falling onto it: "Land is what brings life to plants. If you cut a branch from a tree, it is dead. But put it back in the soil and it will grow again. It draws its life from the land. . . . Two things come together to produce life, water and land. Life is not in the seed that is planted, but in the land that receives the seed" (Östberg 1995:96, 98). The Burunge do not stop there. The heat of the land pushes soil up to the surface. The steam coming from the surface after rain shows that the land is breathing to rise. The lower levels of the soil come up toward the surface, like seeds germinating. Soil layers are formed as penetrating water dries, their color reflecting what was on the land at the time. When old graves, originally deep, are found near the surface, this demonstrates that the soil has risen up. Old utensils and the black soil marking old kitchen fires sometimes are found in the ground, which has risen around them. People now live higher than they did in the past. On the surface new soil is formed from plant residues and household waste. New soil is also added by the rain that brings it from the breaking waves in the distant sea, and it is carried by the wind. "Do you not get soil stains on your clothes when you are out in the rain? The rain contains soil and sand" (Östberg 1995:94).

As the Burunge view the world, the sky, or heaven, is male and the earth is female. It is the strength of raindrops that enables them to infiltrate. Water also transports soil from place to place, and the patchwork of soil colors on the land results both from this transportation and from the rising of soil layers from below. For proper crop development, the warmth of the ground must mix with the coolness of the rain, and the fertile topsoil must mix with the infertile mineral soil below. It is the balance that is important.

This construction has logic, strange though some of it may seem to those schooled in conventional or even unconventional ways of thinking about the soil. Yet not all of it should be strange to scientific observers. First, the importance of organic matter in the soil is common to both systems of knowledge. Second, the formation of earth over fragments of hard matter left on the surface is an old observation. It was first discussed scientifically by Charles Darwin in 1837; he attributed it to the activity of earthworms. Later, Darwin devoted a whole book to earthworms in which he calculated an accumulation rate about 0.5 cm per year for a smoothed depth of measured castings at a number of sites (Darwin 1881). Until lately, the topic of bioturbation has been taken up only sporadically, mainly in the context of soil fertility, but it is important enough in modern paleobotany to warrant the use of highly specialized methods for extracting, dating, and analyzing botanical macroremains, as discussed in chapter 4. It was comprehensively reviewed in the wider context of soil formation in general by Paton, Humphreys, and Mitchell (1995), whose mobile topsoil view is discussed later in this chapter.

Third, formal science has long recognized the contribution of wind to soil building, especially in the case of loess, which is also found and is still formed near the margins of deserts. Along the southern fringes of the Sahel in west Africa, dust plumes have carried large quantities of fine particles from desert basins and deposited these in wetter areas leeward, enriching the soils of northern Nigeria. In smaller quantity, the Harmattan dust continues this process today (McTainsh 1980, 1984, 1987). The role of the wind is also recognized at a field level in the Sahel in the rippling microhighs and microlows of large tracts. The soil is moved from the lows to the highs so that the topsoil becomes concentrated on the latter. Indeed, as the Burunge would remark, it "comes up." Some Sahelian farmers not only recognize the phenomenon but also understand the process (Lamers and Feil 1995).

Fourth, the contribution of rain to the soil is not wholly discounted by formal science, although it is viewed differently. Modern soil scientists agree that rain deposits dust and smoke, but the specific contribution of rain itself is seen more as of dissolved nutrients than of fine particles; the nutrient contribution can be significant. For example, in a savanna in Central America with very poor, deeply weathered soils that could contribute almost no nutrients, slow forest development was starting to take place after a generation of protection from fire. The nutrients needed were not imported by surface flows and were being supplied almost wholly from the capture and retention by the soil of diluted mineral inputs from the rain. The woody transitional plants able to profit from this accession must be very slow growers, but the nutrient capital sufficient to sustain a whole forest community could be captured in about 500 years by this means (Kellman 1989).

This much can be drawn from a comparison between what one east African people believe about soil formation and some information available in the literature of formal science. Burunge base their knowledge on observation, reasoning from a slender understanding of natural processes. Modern scientists observe more systematically and reason from a deeper understanding of process. Yet their methods, applied only to one aspect such as soil physics or soil chemistry, can lead to reductionism. One area in which such reductionism has been particularly damaging to the advance of useful knowledge is soil formation. I return to the Burunge understanding later in this chapter, but I first plunge more deeply into what formal science has to say about soils, with the specific purpose of seeking what we can learn about diversity in the soil.

■ Soil Taxonomy and Its Problems

GREAT SOIL ORDERS

Those who write about the agriculture of any area normally include a description of the soil, and the usual starting point is with some form of taxonomic classification. Unless dealing with small areas that have been mapped in detail, it is common to refer only to the higher taxonomic orders, listed today in the U.S. Soil Taxonomy. These orders are presented for ready reference in box 5.1. Among these, two great soil orders are regarded, for different reasons, as particularly poor in available nutrients: the Oxisols and Spodosols. The Andisols and Mollisols are particularly well supplied with nutrients. How much does this tell us about the soils farmers have to use?

Box 5.1
Brief Definitions of Soil Taxonomy "Soil Orders"

These categories include all the earth's soils, with approximate mapping equivalents in the Food and Agriculture Organization and United Nations Environmental, Scientific and Cultural Organization system (in italics).

 Five soil orders in order of pedogenic time from unweathered parent material to maximum weathering:

Entisols: young soils with little or no profile development (*Regosols,* some *Yermosols, Xerosols, Rankers*)

Inceptisols: soils with incipient profile development (*Cambisols*)

Alfisols: soils with clay-enriched subsoils, moderate to high pH, and base saturation (*Luvisols*)

Ultisols: soils with clay-enriched subsoils, low pH, and base saturation (*Acrisols*)

Oxisols: soils with little surface to subsoil texture differentiation; subsoils with low cation exchange capacity and high iron and aluminum oxide content (*Ferralsols*)

The remaining six soil orders:

Andisols: soils derived from volcanic ash, with organic-rich surface horizon (*Andosols*)

Aridisols: soils dry more than 50 percent of most years, lacking organic-rich surface horizon (some *Yermisols, Solonchaks*)

Histosols: hydromorphic soils with more than 30 percent organic matter to more than 40 cm depth (*Histosols*)

Mollisols: soils with deep surface horizon, high organic matter content, moderate to high pH, and base saturation (most *Kastanozem* [Chernozem], *Rendzina*)

Spodosols: soils with subsoils having high accumulations of organic matter, iron, and aluminum (*Podzols*)

Vertisols: soils with more than 30 percent clay in all horizons, with large shrink–swell properties upon drying and wetting, respectively (*Vertisols*)

Sources: United States Soil Conservation Service (1981); Sanchez (1976); basic typology and some detail from Richter and Babbar (1991:322, 337–39, 344, 379, and discussion). Richter and Babbar (1991) also provided an excellent review of the principles of the U.S. Soil Taxonomy in a tropical context.

Modern classification relies on a set of diagnostic characteristics throughout the soil profile. This is a proper way to classify whole soils, but there is poor correlation between indicators used in soil classification, which are drawn largely from the more stable lower horizons, and the surface soil characteristics important for agriculture (Buol and Couto 1978). Following the same approach in detailed mapping, work carried out at large map scales produces much more complex classifications, but even these cannot take account of the intricacies of growing crops, balancing crop requirements, and all other physical and ecological aspects of the environment. Even adjacent sites differ in slope, soil moisture, and other characteristics not necessarily taken into account by the soil scientist but important to the farmer. The conventional approach to the study of soil formation has little to say about what happens on and near the surface, except in the matter of plant nutrients. A lot of discussion about tropical soils fails to take local and even regional variation into account. This conventional approach is briefly set out in box 5.2.

Box 5.2
Formation and Properties of Soil

In the classic view, soil formation is governed by climate, the nature of the parent material, the physical characteristics of the site and region, the activity of soil flora and fauna, and time. Within and above weathered or newly deposited material, surface soils form through the downward or leaching movement of water, carrying nutrients that are created mainly by organic matter and biotic activity at and near the surface, thus forming the eluviated (A) and illuviated (B) soil horizons, which overlie the weathered C horizon. This view has been challenged, at least in part, by the modern mobile topsoil school.

Ultimately, all soil everywhere originates from the weathered products of the earth's crust, with additions from decaying plant matter and from the activity of soil flora and fauna. Additions from the atmosphere come in the form of dust, diluted minerals in rain and, in some areas, volcanic ashfalls. The first stage in soil development is epimorphism of the eight principal and several minor chemical constituents of the earth's crust into the weathered base, or saprolite. It involves weathering, leaching of rock minerals, and the formation of new minerals, especially clay and various oxides and hydroxides. Weathering involves the breakdown of almost all minerals in the earth's crust except quartz.

Weathering takes place mainly through hydrolysis, in which hydrogen ions are formed through the reaction of atmospheric carbon dioxide with water, forming carbonic acid that then ionizes to form bicarbonate and hydrogen ions. An ion is an atom or molecule holding an electrical charge resulting from the gain or loss of electrons. Ions are electrically charged particles having either a positive electrical charge (**cations**) or a negative charge (**anions**). The tiny, positively charged hydrogen ions penetrate rock minerals, release other ions from them, and modify or disintegrate the mineral structure. The result is a set of cations together with a negatively charged silicate framework, the clay fraction. Cations are then exchanged with anions, and cations with a stronger charge can displace more weakly charged cations from the lattice. Ions on clay colloids are in a mineral form available to plant roots. Ions differ widely in their stability and solubility. Some quickly rebond with others, and some are carried away, or leached, in solution. When ions are added or withdrawn between the solid clay particle phase and the soil solution, exchange takes place. The cation exchange capacity (CEC) is one of the most important attributes of a soil, controlling nutrient availability to plants.

Cation exchange is not a uniform process; it is greatly affected by the acidity or alkalinity of the soil and subsoil. The usual measure of this latter property, the most easily measured if not necessarily the best, is the **pH** (the *pouvoir hydrogène,* or hydrogen power). Properly, the pH of a solution or suspension of a soil is the negative logarithm of its hydrogen ion activity or concentration. It varies through the day, at different levels within the soil, between different parts even of the same field, at a micro level between soil solution close to a clay aggregate and further from it, and with a number of other conditions. However, a pH of 7.0 is that of a neutral soil, with lower values in the range 0–7 representing greater acidity and higher values (greater than 7) greater alkalinity. Within a spread of 0–14, the usual range in agriculture is between 4.5 and 8.5. The pH has a great influence on the mobilization and leaching of ions because their mobility varies greatly. With high acidity, a number of ions, including the very important potassium and phosphorus ions, are depleted or are very difficult to take up in plants.

Sources: Paton (1978) and Foth (1984) both provide accessible discussions of soils in general. A comprehensive summary treatment of soil formation, characteristics, and management, is given in Gliessman (1998). Management of soils in temperate lands is well covered by Donahue, Miller, and Schickluna (1983). On tropical soils and their relationship to plants, Nye and Greenland (1960) is still invaluable. Sanchez (1976) is more comprehensive but has not replaced Nye and Greenland's classic. A source of comprehensive value on the subject matter of this chapter is Young (1976), and useful updating can be found in Syers and Craswell (1995).

UNSOUND GENERALIZATIONS ABOUT TROPICAL SOILS

Detailed soil mapping is the exception rather than the rule in the tropics, and where detailed information is lacking it has been common to use climate and vegetation to define the soils supposedly associated with them. This is especially so in regard to the rainforest soils. A large proportion of these soils is classified in the poorest and oldest of all soil orders, the Oxisols, and in the next oldest, the Ultisols. These acid soils are described in recent literature as occupying more than 60 percent of the total land area of the humid tropics (National Research Council 1993; Lal 1995).

Oxisols are seen as the old product of tropical climatic conditions, deeply leached and very acid, with low capacity to retain nutrients, and very homogenous in their physical and chemical properties both over the landscape and in the soil profile. Weathering of parent rock beneath Oxisols sometimes extends from 10 to

30 m or even deeper (Burnham 1989). Mineral-bearing silts have been washed out of the surface layers so that most nutrients are far beyond the reach of plant roots or inextricably bonded to the clays. The near-surface soils are deficient in most nutrients, particularly phosphorus and potassium. The resulting problems were well described by ecologist Julie Denslow in an elegant simplification of the rainforest environment in a semipopular book. She used certain experiments in Amazonian Peru to illustrate the general infertility of acid soils under the tropical rainforest. Some crops grown in burned clearings exhibited deficiencies of nitrogen and potassium in as little as 8 months, of phosphorus and magnesium within 2 years, and of calcium, zinc, and manganese within the next several years (Denslow 1988). The significance of these deficiencies is made clear in box 5.3. However, the old view that tropical soils are almost universally poor in soil organic matter has been firmly debunked since the 1970s. Despite the higher rate at which plant litter decomposes in the tropics than in cooler regions, the organic matter content of tropical soils often is comparable to that of temperate-zone soils and by some estimates is greater (Richter and Babbar 1991; National Research Council 1993).

Richter and Babbar (1991) offered a careful resurvey of knowledge about tropical soils, aided by a 1980s soil survey of Brazilian Amazonia. They downgraded the importance of Oxisols as the dominant soil order. The actual area occupied by Oxisols may be not more than about 12 percent of the tropics as a whole, less than half of the area given in several still-current estimates. They emphasized that the time needed for evolution of soils through the sequence from very young Entisols to Oxisols has been greatly underestimated, leading to the misclassification of large soil regions (see box 5.1). They concluded that the extent of the most deeply weathered soils is far less than was thought.

There is a major contrast in age between very old soils on the stable surfaces of the great tectonic plates and the old sediments derived from them, and soils formed on areas that have experienced active mountain building and associated erosion and deposition during the past million years. Soils in these uplands not only are younger but also are being rejuvenated by ongoing erosion, landslipping, and deposition, exposing new material to weathering. In volcanic areas and in the alluvial valleys along rivers draining from mountains, soils often are renewed by deposition. Alluvial soils differ greatly according to the regions from which the rivers derive. Not all are fertile, but all are youthful. Soils of the older and younger environments should be separated for purposes of even the most general discussion (Whitmore 1989).

DIVERSITY OF TROPICAL SOILS

Oversimplification of the variety in tropical soils has long been criticized. Writing their classic in midcentury, Nye and Greenland (1960) described many aspects of tropical soil diversity. In 1977 a symposium was held on diversity of soils in the tropics, but it made limited impact (American Society of Agronomy 1978). Much later, in 1989, a soil scientist with extensive experience in west Africa wrote that "a major problem with tropical soils is their diversity, which is probably why there are so many half-truths about them. Generalizations describe tropical soils as 'acid,' 'nutrient poor' and 'high phosphate fixing,' all of which are often wrong. Such misconceptions can be avoided if we accept soil diversity as a fact" (Wild 1989:453). But the generalizations continue.[2] The long-lived failure to appreciate tropical soil diversity adequately has had a major impact on the literature about agriculture and rural development. It has adversely affected the manner in which authorities and experts deal with farmers, whose knowledge of their soils is quite different from that of conventional soil science.

Box 5.3
The Main Plant Nutrients and Their Functions

Nitrogen (N). After hydrogen, carbon, and oxygen, this is the most important element for plants, and although it is abundant in the free atmosphere, plants derive it almost wholly through the soil. It is fundamental in determining the mass of the plant. Soils lose nitrogen rapidly to water bodies, and both soil and water lose nitrogen to the atmosphere. Most is held in organic matter. Because of the large demand, it is often deficient.

Phosphorus (P). In the soil, phosphorus usually is compounded with minerals that are of low solubility and are inaccessible to plants. Symbiotic mycorrhizal fungi (box 5.4) are very important in acquiring it for plants. Second only to nitrogen, it is most often the limiting element in soils.

Potassium (K). This mineral activates many enzymes essential for photosynthesis and respiration and for forming starch and protein. Though less abundant than nitrogen in plants, it is more abundant than phosphorus and all other minerals listed below. Highly soluble, it is readily leached in heavy rain.

Calcium (Ca). About half as abundant in plants as potassium, calcium is seriously deficient under conditions of high acidity. Accumulating in leaves and bark, it is important in plant development.

Magnesium (Mg). Obtained largely from soil solution, magnesium accumulates in leaves, where it is essential for chlorophyll formation. It is rarely a limiting factor. High concentrations of both calcium and magnesium reduce acidity in the soil (as they do in the human digestive tract).

Sulfur (S). The only remaining element needed in large quantity, sulfur is readily soluble and some is obtained in rain through the leaves. It is a principal building block of proteins.

Iron (Fe). Although it is needed in much smaller quantities than other macronutrients, iron is of low solubility and is obtained largely by chemicals secreted from plant roots. It accumulates in leaves.

> Other essential elements are called micronutrients or **trace elements.** They include chlorine, manganese, zinc, copper, molybdenum, boron, and nickel. These minor elements have various but often critical functions in plant growth.
>
> *Sources:* A comprehensive, up-to-date, well-illustrated, and (even for those with limited knowledge of biochemistry) fairly accessible source is Larcher (1995). A useful text is by Salisbury and Ross (1992).

Richter and Babbar (1991) showed that all the world's soil orders are present in the tropics, and variation in tropical soils is as great as in the temperate zone. Even among the soils of the tectonically stable areas there may be far less uniformity than is often supposed. The intensity of cropping that may be achieved without noticeable soil deterioration is determined largely by the parent material. Soils with very different capabilities may lie closely adjacent. In hilly and mountainous areas the soil pattern is complex. In the southeast Asian tropics, true Oxisols are almost absent, and the influence of parent material is very strong. For example, there are great contrasts between both soil and the landscape in areas developed on granites and on shales. Soils derived from granite are commonly very deep, with rounded slopes prone to landslides, whereas soils derived from shales, especially where metamorphosed, are much shallower and more finely textured (Whitmore 1984). In the tropics as elsewhere, limestone soils always are distinctive. Soils developed on shales of marine origin retain their capability well under cultivation. This was demonstrated by example in the Chimbu region of Papua New Guinea in chapter 1.

To continue within the same geologically disturbed region, large upland areas of southeast Asia have clayey red–yellow soils, even brown on slopes. Good structure, a moderate level of inherent fertility, and the constant release of new subsoil minerals for weathering by slope processes make some of these soils surprisingly good for agriculture (Ho 1964; Brookfield, Potter, and Byron 1995). In contrast, soils on loose sands and coarse sandstones, from which almost all clay and silt particles have been removed by leaching, can be of very low value. Below a thin layer of organic matter, almost all components have gone from the surface soil except for large quantities of quartz sand. Such Spodosols are well described in Borneo and in parts of Amazonia, where they occupy several percent of the total area. They carry a distinctive vegetation (Whitmore 1984; Jordan 1985; Moran 1995). Padoch (1986) discussed the avoidance of these soils in agricultural site selection practiced by the Lun Dayeh of north central Borneo. Similar avoidance has been described elsewhere.

■ Soil-Forming Processes

TAKING ACCOUNT OF SURFACE AND NEAR-SURFACE PROCESSES

An approach based less on soil taxonomy and more on the physical and biological processes involved in creating the environment for farming can bring us closer to the type of diversity interpreted by the Burunge and many other small farmers. The classic approach to understanding soil formation, briefly presented in box 5.2, gives little weight to processes near the surface, yet these processes produce the natural units of the landscape. Wherever there is even the slightest slope of land, there is movement of water within and over the soil surface. Water carries both soil and nutrient particles, and gravity assists downslope movement of soil itself, and of the weathered regolith below it, in which structures sometimes can be bent downhill. A widely repeated sequence of soils, from thin and pebbly at the top of a slope to deep and with clay accumulation at the bottom, develops over time and is called a catena. Soil catenas were first described in Africa in the 1930s and today are much more widely recognized as a common feature of soils in sloping lands in temperate as well as tropical regions (Milne 1936, 1947). Soil contrasts developed along catenas are within the daily experience of a great many farmers, whether or not they appreciate that surface erosion and deposition are integral to soil forming. Catenas are of major importance in explaining land use even in gently hilly country (Young 1976).

Added to the now widely recognized catena process is modern understanding of bioturbation by soil fauna, including the earthworms discussed in 1881 by Darwin, and by a range of physical near-surface processes. The significant and rapid soil-moving activities of earthworms, termites, ants, and other mesofauna have major effects on the surface soil. Some of these, especially termites, bring minerals to the surface from below the depth of root action, even from the weathered subsoil. Soil fauna can be present in huge numbers in all but arid and waterlogged soils. Jones (1989, 1990) found densities of up to 400 macrotermites per square meter in soils of dry tropical Africa. They forage a high proportion of all plant litter from an area extending 35–50 m from each mound. Together with other species, they are responsible for sequestering into the mounds most of the end-products of tree litter disposal that survive fire (Trapnell et al. 1976). The effect of bioturbation by all species can be to mix the surface soil completely. In one example, bioturbation has blended together four or five ash layers and an organic carbon layer to a depth of more than a meter; the ashfall horizons remain clearly visible in nearby peat where soil fauna are absent (Burnham 1989).

THE MOBILE TOPSOIL AND ITS IMPLICATIONS

Together with the catena concept, recognition of these processes leads to an alternative view of soil formation in which the topsoil is seen evolving independently of the more inert layers below, not simply as the weathered upper horizons. In this view of a mobile topsoil, water flow within the soil and gravitational movement augmented by biological activity are constantly bringing material to the surface, where it is in a location and state amenable to movement by the elements. Surface and near-surface physical and biological processes create a soil that is distinct even in its mode of formation from the weathered subsoil (Paton 1978; Paton, Humphreys, and Mitchell 1995). The processes envisaged are rapid, achieving soil formation in decades rather than the millennia that weathering takes. The very slow rates of soil formation by weathering that are often cited become largely irrelevant.

The classic vertical evolution approach, with its emphasis on the soil horizons that are so important to taxonomy, and the modern catena and mobile topsoil approaches have not been brought together in a way that will identify the units most significant for farming. Importantly for agrodiversity, the rapidity of topsoil formation has not yet been fully recognized by those who write doom-filled scenarios about soil erosion. To write this is not to downplay the utility of classic soil science in understanding the basic diversity that lies in the land but to express a caution, to which I return in chapter 9.

■ Introducing Nutrients and Soil–Plant Relationships

NUTRIENTS: NITROGEN AND LEGUMES

Plant nutrients have been mentioned several times in chapters 1–4 without much explanation. Every crop takes nutrients away from the soil. Some nutrients are needed in large quantity, others in much smaller quantity, and different plants have different needs and different tolerances. On soils without natural means of nutrient renewal, the available minerals are depleted unless steps are taken to replace them. This may take a long time if the soil is rich in mineral nutrients, but a short supply of one vital nutrient will lead to reduction in yields or enforce change to a more tolerant crop. Some plants can survive with very small supplies of nutrients that others need in abundance. If one basic element is scarce, plants that demand it in quantity cannot thrive even though they have everything else that they need. Nutrients must be supplied in a balanced manner to ensure healthy growth. Where several nutrients are depleted but not exhausted, a long-term equilibrium at a very low yield may be the outcome. The nutrients plants need are summarized in box 5.3.

So central is one plant nutrient, nitrogen, that steps to enhance its supply arose in each of the ancient farming regions. Nitrogen, the element needed in greatest amounts, is acquired by plants in several ways and is the only nutrient of which the main source is in the atmosphere. The most important means by which it becomes available to plants is via bacteria that live in large numbers in the soil, where they absorb inert atmospheric nitrogen and make it available for plant roots in the form of ammonium or soluble nitrate. Young (1976) provided a good summary of the complex process, and Larcher (1995) wrote a comprehensive description. Nitrogen is taken up from the soil as either nitrate or ammonium ions involving chemical (enzymal) processes through at least 25 symbiotic and free-living bacteria. Among these, *Rhizobium* spp. are the most prominent in nitrogen acquisition by plants, where it is incorporated in carbon compounds. Mycorrhizal fungi (box 5.4) also obtain nitrogen for plants. Although inorganic nitrogen is now a principal source of supply in the agriculture of developed countries, natural fixation remains of major importance. There are many associations between plant roots and a number of free-living bacteria and, on wet soil, blue–green algae, or cyanobacteria. Flooded crops, especially wet rice, get nitrogen from cyanobacteria, which live symbiotically among the water fern *Azolla*. A large group of plants permit *Rhizobium* bacteria to live symbiotically with them; the bacteria obtain supplies of carbon and enzymes from the plant, and the plant gets nitrogen. Generally soil biota (bacteria, fungi, protozoa) are limited by the supply of carbon, whereas plants are limited by the supply of nitrogen.

Box 5.4
Mycorrhiza

There are two principal types of mycorrhiza. **Vesicular–arbuscular mycorrhiza** (VAM, or endomycorrhiza) penetrate root systems and live largely within them, communicating with external hyphae through gaps in the root wall. They are widespread in all environments throughout the world. Specific mainly to certain tree and shrub species are **ectomycorrhiza** (ECM), which form a complete sheath around the root. In addition, they spread in the soil to make short-lived fruiting bodies, some of which are edible mushrooms and truffles, providing important ingredients in human and animal diets.

Plant species can be broadly divided into those that are **obligately mycorrhizal,** unable to make growth without infection; those that are **facultatively mycorrhizal,** able to benefit from mycorrhizal associations on less fertile soils and also able to refuse infection beyond their needs; and those that are wholly **nonmycorrhizal,** able to refuse infection while young and healthy. The latter include most early successional plants. Plants also differ greatly in the efficiency with which they use mycorrhiza to gain mineral elements, and within any one plant species there are differences in the efficiency of mycorrhizal use that seem to depend significantly on the mineral content of the soil.

Active mycorrhiza are heavy consumers of carbohydrate, which they obtain from the infected plants. Although the woody root mass of a number of trees amounts to as much as 30 percent of the total biomass, the presence of mycorrhiza adds to the below-ground consumption of carbohydrate, especially where the bulkier and more extensive ectomycorrhiza are involved because these generally are thought to have higher cost. Janos (1996:133) remarked, "Those who would advocate planting pines and eucalyptus indiscriminately throughout the tropics should consider that vegetation conversion to [ECM]-dominated stands ultimately may sacrifice harvestable aboveground production to ectomycorrhizal fungi."

A caution is needed. Most of the literature on the ecology of the plant–mycorrhiza symbiosis depends on the results of pot experiments. Mycorrhiza are microscopic and difficult to observe in the field, and what field evidence there is suggests that they may be less nutrient-effective for some plants than the experimental literature suggests. A number of field studies have shown responses to mycorrhizal infection considerably less than those in pot experiments. These latter are performed mostly under ideal conditions, with soil sterilized before mycorrhizal introduction, and with single plants rather than groups of plants. It has been noted that "several factors probably contribute to the low effectiveness of VA[M] fungi as [phosphorus] uptake systems in field conditions, including incompetence of strains and species of fungi, grazing of hyphae by soil animals in field conditions, [and] the known ability of mycorrhizal hyphae to transfer metabolites between plants and variations in soil water status. . . . The dominant factor may well be the interaction between phosphate supply and soil moisture, and hence the improved survival of infected plants in dry soils" (Fritter 1985:263).

Sources: Mycorrhiza now have a very large literature and even specific journals. Although they still find only limited place in more general texts, clear accounts by Janos (1980), Alexander (1989), Högberg (1989), and Janos (1996), together with a comprehensive (if somewhat impenetrable) review by Brundrett (1991), provide background for most of this box and the short discussion in the text.

Rhizobium bacteria may form nodules on plant roots, most commonly on leguminous plants, members of the Leguminosae or Fabaceae family, among the largest of flowering plant families. Among domesticated crop plants their number exceeds that of any other plant group. The agronomic benefit was fortunate but secondary; their popularity resulted first from the edible and often tasty seeds or fruit, which are valuable sources of protein and carbohydrate. Beans of various kinds remain the main protein source for a large number of the world's people. Other legumes, including lucerne (alfalfa), clover, and vetches, are excellent fodder for animals. It seems very likely that in several parts of the world, legumes were domesticated at about the same time as the starch-seeded grasses. Farmers soon would have seen that all their crops grew better when planted after legumes or interplanted with them.[3]

The use of legumes is a very old and widespread adaptation to two of the greatest needs of agriculture: dietary diversity and soil restoration. Zohary and Hopf (1993:86) described how the agronomic and dietary value of legumes was appreciated even in the early days of agriculture, in all regions:

Wheat and barley agriculture in west Asia and Europe had pea, lentil, broad bean, and chickpea. Maize in Meso-America was accompanied by *Phaseolus* beans; and in South America also by the ground nut. Pearl millet and sorghum cultivation in the African savanna belt were associated with cowpea and Bambara groundnut. Soybean was added to cereal cultivation in China, and hyacinth bean, black gram, and green gram in India.

CLOSED AND OPEN NUTRIENT CYCLES

Another myth must be disposed of, although in this case it is only a partial myth. It was thought for many years that all mixed tropical forests lived largely by recycling nutrients between the trees and the ground, in effect looking after themselves. The belief was that nutrients were always in the biomass, with few in the soil. A tight, closed nutrient cycle was postulated as normal in tropical forests. In closed cycling, leaves, branches, and other litter, including whole trees when dead, would fall to the ground and provide food for a wide range of decomposers. The latter convert them into mineral form, in which they can then be taken up again by the plants. Ants, termites, and earthworms move the nutrients around and often bring them back close to or onto the surface. In this model, plants draw almost nothing from the deeper soil, which, it was generally supposed, had little to offer. Thus was explained how, after a vigorous and luxuriant forest is cleared, the soil is often found to be of very poor quality and will not permit sustained cultivation.[4]

Work since the 1970s has shown closed cycling to be more complex than this. In carefully measured sample plots in Sarawak, Sri Lanka, and Thailand, chosen to represent different climatic regimes, Andriesse and Schelhaas (1987a) demonstrated that in all cases more of the five principal nutrients are in the soil than in the biomass. Studies in three areas of Amazonia showed that even under old forest, the soil quantity of calcium, potassium, manganese, and available nitrogen made up about one-third of the total in biomass and soil combined. This was despite the fact that one of the three sites had much greater abundance of both soil and biomass nutrients than the other two (Jordan 1986; Saldarriaga 1986; Scott 1986). Recycling depends on the ability of tree roots to capture nutrients either before they are leached or from greater depth while they are being leached. The nearest approximation of a truly closed cycle is where root systems form a sievelike mat to recover nutrients newly released from mineralized litter and do not penetrate deeply into intrinsically poor soils.

Soil depth is an important variable. Nutrient exchange between the soil and plants takes place in the top 2 m, and where soils are shallower than 2 m, open nutrient cycles are always found. In open cycles the weathered soil parent materials contribute importantly to plant growth. These conditions arise wherever there are active slope erosional processes including landslipping, in frequently renewed valley alluvium, and where soils are enriched periodically by volcanic ash. They are also important wherever a mobile topsoil overlies the inert subsoil. Even on deeply weathered plateau land, catena formation on all but the gentlest slopes creates shallower, rejuvenated soil. Most tropical soils under most kinds of forest have a large store of nutrients, although some are not in a form readily accessible to plants (Whitmore 1989).

ROOTS AND THEIR COLLABORATORS: MYCORRHIZA

Plant root systems vary enormously. The dense root mats close to the surface developed by some trees and grasses intercept the mineralized products of organic litter and rain before they can be leached away. Roots that penetrate deeply capture leached nutrients and tap water during droughts. Some root systems have extensive root hairs that increase the range through which they can tap nutrients. By their own energy-consuming respiration, plants can release organic compounds that form soluble complexes with minerals that can be taken up into the root. All create a depleted zone around the limits of their reach.

This is where symbiotic mycorrhizal fungi become important in plants' acquisition of nitrogen and, more critically, phosphorus, as well as other nutrients and even water (Janos 1980, 1996). Spreading around the roots and linking to hyphal networks in the soil, mycorrhiza extend the reach of a plant's root system hundreds to thousands of times. Tracer experiments with pot plants have shown that nutrients can travel to plants through the mycorrhizal systems for distances of several centimeters rather than the millimeters that are the reach of root exudates. Mycorrhiza have been known throughout the twentieth century, but their role is not much discussed in most of the literature on tropical farming systems. Mycorrhiza are a fungal infection of plant roots, the inoculum being derived from the hyphal mycelium and spores in the soil. Spores can travel in water, on the wind, and both in and on animals and birds. They are everywhere except in saline, waterlogged, arid, and very cold soils. In nutrient-poor soils their contribution is central to both natural ecology and agriculture. More information is provided in box 5.4. Dependence on mycorrhiza probably is greatest among the canopy trees of the tropical rainforest; some trees (certain palms, for example) seem to have lost the ability to take up mineral nutrients without them (Janos 1996). Among crops and fruits, wheat, barley, maize, manioc, sugar cane, citrus, mango, and avocado all make substantial use of these fungi. On the other hand, many crops and trees can thrive with or without them. The whole Brassica family is an example.

Many other soil microorganisms are important and perform a great range of services. Others are pathogenic and create problems for agriculture, and still others can reduce damage from pathogens and pests. The complexity of soil biodiversity is now an important topic for research.

SCAVENGING AND STORING NUTRIENTS

All plants scavenge the nutrients they need from the soil within their reach, some in a more selective manner than others. Trees and shrubs accumulate far more of the principal nutrients than herbs, but the proportions vary greatly between species and between different stages of their growth. One example is the high concentration of potassium in a bamboo (*Dendrocalamus hamiltonii*) common in northeast India; in the early stages of postcultivation growth, more potassium is held by the bamboo than in all other species combined (Ramakrishnan 1992). Litter falling from trees, shrubs, and bamboo returns minerals to the soil surface, and in some instances this may be more important to soil nutrient content than the nitrogen fixed by bacteria associated with leguminous and other species. After a field has been cleared for cultivation and then left to regrow under forest, a vigorous secondary growth accumulates nutrients in the biomass but continues to deplete soil nutrients until litter fall becomes substantial.

In the soil itself, the major store of nutrients is in the soil organic matter. This includes the soil biota and their dead residues but consists mainly of decaying plant remains that feed the soil biota, which therefore thrive more abundantly in their presence. Everywhere, soil organic matter has a direct effect on nutrient supply, of which it provides a very large part. It is the source of 90–95 percent of soil nitrogen and also of available phosphorus and sulfur. It facilitates water penetration and retention. Soil organic matter is highly variable between sites, in the tropics as elsewhere.

After decomposition, the stable residues that are not converted into mineral forms usable by plants and microbes, or converted into carbon dioxide and lost to the atmosphere, become humus. These dark brown carbon compounds are nearly insoluble and very enduring, rich in nutrients that can be extracted by the large microbial and fungus population they support. The cation exchange capacity of humus (box 5.2) is many times greater than that of clay colloids, so long-enduring humus in the soil is a valuable property. In some soils humus makes up 50 percent of the organic matter content. It forms most readily in clay soils, where it combines with clay particles, and least readily in sandy soils. Humus is mentioned in many sources. An accessible account of its nature and importance is provided for farmers and other nonspecialists by Sachs (n.d.). A more detailed discussion is that presented by Hüttermann and Majcherczyk (1998). An excellent and accessible review of the role of organic matter and its management is provided, but only in a North American context, by Magdoff (1992).

The function of humus, and organic matter in general, in the soil is strongly emphasized in the modern biological management literature, written in criticism of the extensive use of chemicals since the 1940s. Because they lack organic carbon, mineral fertilizers do not contribute to humus formation, and processes that aerate the soil, such as deep plowing, facilitate decomposition of organic matter before it can become humus. I come back to this issue in chapters 13 and 14 because it has major relevance not only to modern practices of alternative agriculture but also to the resilience of the set of soil management practices contained within agrodiversity.

SOIL STRUCTURE AND THE QUALITIES OF THE LAND

When account is taken of the soil biota and the store of nutrients held in the soil, the Burunge are indeed partly right: the soil does have life. The organic element has a great effect on the structure of the soil, a vital property for arable use and, together with pore size, of prime significance in its ability to permit water percolation and storage, as opposed to overland flow. It thus strongly influences resistance to erosion. A compacted soil, poor in organic matter, lacks the pore space to allow water to penetrate. A good structure facilitates root penetration and the spread of soil biota; it also promotes gas exchange within the soil. The available soil water capacity, which closely reflects structure and hence the organic matter content, sometimes can be used as a valuable proxy measure for the intrinsic capability of soils (Lu and Stocking 1998).

The term *soil structure* refers most simply to what soil is like when taken up in a hand sample; does it crumble, or does it hold firm in blocks? Technically, it means the presence or absence of fairly stable aggregates into which soil breaks down when exposed. Soil structure is distinct from soil texture, although the two are related. *Soil texture* refers to the particle size or mixture of particle sizes that is dominant in the soil. From among a variety of measurements used, we may generalize a range for sand at around 1–2 mm to silt and clay at less than 0.002 mm in diameter. Raw sands, and some dense clays, are almost structureless.

In a strongly structured soil, the aggregates can be handled fairly easily without breaking. For a long time, structural units of soil were thought to result mainly from combinations of swelling and shrinkage, depending primarily on the nature of their clay content. Geoff Humphreys (personal communication, 1995) advises that it is now realized that in many topsoils the structural units are influenced by the bioturbation of soil fauna, with shrinking and swelling playing only a subordinate role. Soil organisms also provide adhesive substances such as long chains of sugar molecules that are produced mainly from organic matter. At greater depth, the structure of the soil reflects patterns inherited from the parent rock.

All aspects of the soil—physical, biological, and chemical—are interrelated. The quality of soil as a resource cannot be expressed in its taxonomic classification or in any of its properties taken separately from one another. In finally approaching how soil diversity affects farming and farming affects soil diversity, a classic discussion of a small part of southwestern Nigeria by the late Richard Moss is useful because it emphasizes what is more variable or less variable. In its day, the approach was revolutionary (Moss 1969). Moss first distinguished morphological soil properties, including depth, particle size distribution (texture), slope, and soil water table levels, which are all fairly immutable. Then he noted that the ecological soil properties of nutrient status, organic matter, and the activity of microorganisms, and the resultant structure, by contrast, are highly variable. They change with any alterations in the plant–animal component, but particularly with land use.

■ The Human Factor

MANUFACTURED SOILS

The final diversity does not come from nature; it comes from the impact of those who use the soil. They can override even "immutable" morphological determinants and radically change the ecological properties of the land. Although little account is taken of this fact in most soil science writing, any soil that has been used and worked over for a long period of time has been changed in structure and other properties. In the context of a literature that emphasizes degradation caused by human use, especially where this use is heavy, it is valuable to recognize the widespread incidence of an opposite trend. In northwestern Europe, long continued manuring has created thickened topsoil horizons in what are called *plaggen* soils. Many other soils have been improved by marling where light, and by applications of sand, lime, and loam where heavy.

Everywhere in the world, farmland close to urban areas has received large inputs of manure, domestic ash, street sweepings, and usually, night soil from the town, carried out to a distance as far as is economically worthwhile by vehicle, animal carriage, or human porterage. On a smaller scale, this is equally true around many villages, where it is a part of the basis of distinction between infield and outfield. Several examples are discussed in the chapters that follow. Market gardeners have been particularly active users of this material, and as long ago as the seventeenth century those near London "could almost manufacture their own soils" with these large inputs (Thick 1985:510). As late as the 1970s it was still possible to see market gardeners riskily cultivating the sea beach on inlets around Hong Kong, with soil laid on the sand, fertilized from concrete pits in which human wastes were fermented. Soil creation and reclamation are widespread activities, still ongoing. They are laborious and it takes time to achieve results, but large areas of the world's farmed soils, in their present form, have been created by the people who use them or who developed them in the past. In the terminology of chapter 3, they are manufactured soils.

In the areas described in this book, the most complete transformation has taken place where land has been leveled, stepped, or terraced for cultivating wet rice. The manner in which soils have been created on the terraces of Ifugao, Luzon, is described in chapter 1. Soils modified by people can retain their characteristics for a long time. All along the uplands close to the Amazon and its tributaries, and to a lesser degree elsewhere in Amazonia, are patches of black soil, rich in carbonized organic matter and in nutrients that are scarce in the surrounding region, especially phosphorus. These patches of *terra preta do Indios* range in extent from 3 to 80 ha, and are from a few centimeters to 2 m deep. They are the product of centuries of low-temperature cooking fires and the legacy of large precolonial populations that were wiped out 400 years ago. The soils are highly valued by modern farmers (Smith 1980; Eden et al. 1984).

THE *TAPADES* OF THE FOUTA DJALLON

Many examples of a modern manufactured soil could be discussed, but one will suffice for this chapter, where enduring good farmland is made from some of Africa's worst soil. The Fouta Djallon plateau of Guinea, in west Africa, is very ancient land. Its fine-grained sandstones have been subjected to weathering under a succession of climates for millions of years. Below the surface lies a hardpan or *bowal*, very ferruginous and almost sterile. It is sometimes as much as 5 or 10 m thick. Quite widely, the crust is exposed at the surface or has only a thin, gravelly soil above. Farmers and pastoralists have occupied this land for at least a thousand years. Its dominant Fulani people, who conquered it from others, settled to farming mainly in the eighteenth century. The account that follows relies principally on Barry et al.(1996) and on sources cited by them. Additional information rests on personal communications by the same Guinean authors and on my field observations in their company in 1996 and 1999.

Leaving aside large areas with too little soil for any cultivation and some wetland areas along the rivers and streams, the greater part of the landscape is either in fallow or under cultivation for fonio millet (*Digitaria exilis*). This west African domesticate is often miscalled hungry rice and is able to yield on shallow and rocky soils where most crops would fail. Annually cultivated land, plowed or hoed, is left under short fallow with livestock pasture each dry season. After 5–7 years of cultivation the land is left fallow for about the same period, under grass or a scrubby low bush according to soil type. The subsequent plowing is shallow so as not to bring up too much of the gravel. Yields of fonio are highly variable, from as little as 100 kg to more than 1000 kg per hectare.

Within this landscape are tracts, less than a fifth of the whole, on which a totally different type of farming is practiced. These tracts, or *tapades,* within which the villages are located, consist of closely linked sets of plots, each smaller than 1 ha, enclosed by live fences, and permanently cultivated. The core areas of some *tapades* are more than 100 years old. Reclamation of new areas continues, and there is not much loss. One enters these areas over stiles or through gates and, if the latter, finds a small cattle corral immediately inside where livestock are brought in each night. Goats enter through narrow fenced pathways, leading to pens within the *tapades.* The manure provided by these animals is the basis of soil improvement, together with domestic refuse, heavily applied mulch, carefully made compost, and nutrient recycling by trees.

On the edges of the *tapades* and on the newest *tapades,* the main crops are fonio and groundnuts, the latter said to be planted for soil improvement. Further within, close to the houses, is much denser growth in a complex of mixed crops with trees; the land is thus kept permanently covered from raindrop impact. The soil is tilled by hand and the inputs are thrown from criss-crossing small paths onto the beds; all land is accessible from the paths. The crop complexes are dominated by maize in the wet season and by root crops (manioc, taro, and sweet potatoes) in the dry season. Also grown are a wide range of vegetables and condiments. Principal tree crops include avocado, mango, oranges, mandarins, grapefruit, bananas, and several trees grown for medicinal purposes or to provide fibers and other household needs and material for new live fences. Almost all work except initial preparation, and most work in the outfields also, is done by women.

Depending on the soil and the work put into the land, it can take 10–20 years to bring a *tapade* into full sustainable production, with soil inputs replacing large and annually repeated harvest losses. All soil nutrient indicators are higher in the *tapades* than on the external fields, and there is a rising gradient toward the houses within the *tapades*. Whereas the external pH may be as low as 4.5, it can rise above 6.0 within the enclosures. If a *tapade* is abandoned by its owner, the site is not taken over freely by others. A *tapade* is a major capital investment, and control of its use remains with the (male) owner and his descendants and assigns.

Although a vitally important element in this case is the use of livestock manure, these are soils created by human effort and management and sustained by the same means. Land made good by human effort is of high value. As we go further forward in exploring agrodiversity, we shall see increasingly how one central product of the diverse efforts of small farmers lies, and has for centuries lain, in the improvement of land and its biota.

Notes

1. I am grateful to Geoff Humphreys for invaluable comment on two earlier drafts of this chapter.

2. Writing about the state of knowledge, Richter and Babbar (1991:333) discussed the different purposes of soil scientists, agronomists and ecologists, saying, "At least some of the problems in developing realistic ideas about soils in the tropics are associated with poor communication among scientists of different disciplines that have a purview over natural resources of the tropics. Historically, interdisciplinary communication has been notably deficient between soil fertility specialists and soil taxonomists, and even worse between general soil scientists and ecologists. This latter deficiency is especially significant considering that soil science is fundamentally a biological and ecological science. . . . Lack of communication between soil scientists and ecologists has allowed overgeneralizations about soils in the tropics to persist within scientific communities that have had little contact with the latest developments in the soil sciences."

3. Not all leguminosae form nodules or fix nitrogen, and there is a range of other plant species that do. Species differ greatly in the efficiency with which they take up nitrogen. Plants also obtain nitrogen from soil solution by root exudates and through the aid of mycorrhiza, which are discussed later in the chapter and in box 5.4. Some other bacteria, particularly certain *Actinomycetes,* also form nodules and fix nitrogen while living symbiotically with woody plants, but these have little importance in agriculture. The range of known nitrogen-fixing symbioses and associations continues to widen. An excellent, if dated, review of legumes as food crops and as soil improvers, and of other forms of nitrogen fixation, is that by Isely (1982).

4. There is a modern, very different use of the terms *closed* and *open nutrient cycling* in which the former is taken to exclude inorganic inputs purchased from off the farm and the latter includes their use. This contrast, often used in the alternative agriculture literature, is discussed in chapter 13.

Part II

Diversity Within Land Rotational Systems

Chapter 6

Analyzing Shifting Cultivation

■ Introducing Part II

SHIFTING CULTIVATORS

From this point forward, the use of case study material in this book becomes systematic and no longer purely illustrative. The purpose is to analyze agrodiversity in light of the themes outlined in chapter 3, initially without the stress on change through time, which becomes a central theme in part III. I focus in part II on a particular but large class of small farmers who use land rotation methods. In much of the world, they are the most derided of all farmers because of the alleged unsustainability of their practices. My purpose is not specifically to defend them but to demonstrate their adaptability, to show how they react to their own perceptions of unsustainability, and to show how passive adaptation can shade into active innovation. This is also an area in which the principle that the farm does not end where the field meets the wood can most readily be exemplified.

Farmers who use land rotation methods usually are described as shifting cultivators even though they may use rotation methods on only part of their land. Shifting cultivation is very diverse, and a single descriptor is inappropriate for the great number of systems that are practiced. There are major differences between practices in a rainforest environment and in savanna, the latter placing much less reliance on fallow growth for fertility restoration and, in almost all cases, using tillage, which is of minor importance in the forests. In this chapter I concentrate on use of the forest environment, offering a fairly detailed comparison of information from within one small region of southeast Asia that has some of the most valuable literature. Before I begin, there is a problem of definition. Everyone involved with tropical agriculture knows what shifting cultivation is—or do they?

DEFINING SHIFTING CULTIVATION

Conklin (1957:1) proposed that shifting cultivation be minimally defined as any system "in which fields are cleared by firing and are cropped discontinuously (implying periods of fallowing which always average longer than periods of cropping)." Ruthenberg (1980) proposed that the term *shifting cultivation* be restricted to systems in which cultivation occupies not more than a third of total time between one cultivation and the next; where it occupies more than this, the term *fallow system* should be used. This proposal came late: the best regional study of shifting cultivation in the literature, that of Spencer (1966), included a number of systems in which cultivation occupies more of the total period of any cycle than does the fallow. Within systems commonly described as shifting cultivation, appropriately or otherwise, the fallow period may range from a few months to most of a human lifetime; it may be much longer than the cultivation period, about the same length, or shorter.

Conklin (1957) regarded clearance by fire as a diagnostic characteristic, and the pejorative term *slash-and-burn cultivation* requires it. This creates further problems. Use of fire in fallow-based systems ranges from burning for total clearance, through patch burning and burning only of debris, to the complete absence of fire. There are some very "shifting" systems in which crops are planted in the mulch, without any burning. Trees may even be felled only after the crops are in the ground (Schiefflin 1975; Orejuela 1992; Thurston et al. 1994). At the other end of the scale, controlled use of slashing and burning forms part of some of the world's most permanent and intensive farming systems (Michon and Mary 1994). The term *slash-and-burn cultivation* obscures even more than does *shifting cultivation*. Fujisaka, Hurtado, and Uribe (1996) have proposed a typology based on the type of forest used and its subsequent treatment, but its complexity makes general adoption unlikely.

The terms *shifting cultivation, slash-and-burn cultivation,* and sometimes *swidden cultivation* are used in everyday parlance. Because all systems of this type are still widely believed in otherwise well-informed circles to be backward, unsustainable, and destructive, indiscriminate use of these terms can have serious political consequences (Brookfield, Potter, and Byron 1995). Widely popularized photographs of forests "recently destroyed" by the fires of shifting cultivators present a graphic view of the damage supposedly done by this "primitive" system, disregarding the rapid natural regrowth that follows. In a conservationist era, seemingly destructive use of fire is a very sensitive topic, and to concern over biodiversity is now added alarm over the input of carbon dioxide into the atmosphere, and its effect on global warming (Tinker, Ingram, and Struwe 1996). The reasons why fire is used, and the care with which small farmers generally manage it, tend to be ignored.[1]

The effects of burning on soil chemistry are not discussed in this chapter, but are examined in chapter 7, especially in box 7.1.

By contrast, some studies have identified no long-term environmental damage. Watters (1971:8) wrote that the shifting cultivator, if left alone, will be in a "stable state of balanced equilibrium within his ecological environment." An idealistic modern interpretation that the system is in perfect balance with nature has grown up and gained some currency. More realistically, 30 years of accumulated scientific work has established that in the presence of conditions for rapid recovery of a nat-ural plant cover, the shifting system with fire can be the most efficient method available for getting crops from the land and then restoring fertility. One major In-dian study, among the best in the literature (Ramakrishnan 1992:370), concluded that "modern agricultural technology has not been able to develop a more efficient way of maintaining soil fertility in the humid tropics than that possible through natural processes during the secondary successional fallow phase after juhm [shift-ing cultivation]." The conditions are rigid. Ramakrishnan went on to show how shortening the fallow or extending the cultivation period can readily lead to im-poverished successional growth and degradation.

■ Farming in the Forests of Borneo

BEGINNING IN A WET PLACE

Rather than attempting a comparison across the world, and thus trying to take an impossible range of variables into account, I use a group of studies in northern Borneo to exemplify shifting cultivation in the forests. There are three reasons for this choice. First, the studies contain some of the best accounts of shifting systems in the whole international literature and provide considerable information on change. Second, they describe systems that have a lot in common, not least that they are practiced within population densities in the range of 1 to 20/km^2, within which, as everyone agrees, the system should be sustainable. Third, I have traveled in inland northern Borneo, seen its shifting cultivation, and visited students and colleagues at work there.

Among the large Borneo literature I draw most heavily on the most thorough studies. These are those of Freeman (1955), Padoch (1982), Dove (1985), Chin (1985), Lian (1987), and Colfer, Peluso, and Chin (1997). The first three concern Iban or Ibanic people in Sarawak, or just over the border in Indonesian Kaliman-tan.[2] The other three concern Kenyah people, two groups in Sarawak and one in Kalimantan. Freeman's was a pioneering study and is less complete than others. It has become a classic, twice reprinted without significant change (Freeman 1970, 1992). Dove's has also attained classic status. Padoch's and Colfer's were published internationally but with limited distribution. Chin's was published only in Sarawak, and Lian's remains an unpublished dissertation. These are backed up by a large number of others, especially Rousseau (1977) on the Kayan and, for related people in Indonesian Borneo, studies by Colfer and Soedjito (1996), Ngo (1996), and Lahjie (1996), among others. Geddes (1954) wrote of the somewhat different Bidayuh, or "land Dayak," people in western Sarawak.

Some of these studies have a strong historical perspective supported, for the dominant Iban people, by the work of Pringle (1970). The close relationship of religion to cultivation was strongly developed by Freeman (1955), Dove (1985), and especially Jensen (1974) but was almost ignored by those who wrote of Christianized people. Only one of the studies details changes into the 1990s (Colfer, Peluso, and Chin 1997). Politicized sources, written in a period of confrontation between opponents and supporters of the forest farmers, include Hong (1987) and Colfer and Dudley (1993). Brookfield, Potter, and Byron (1995) offered a somewhat biased summary. Titles listed here cover only a part of the literature. There are several other publications by the same authors and others. Figure 6.1 shows the areas discussed.

COMMON ELEMENTS AMONG THE PLACES AND PEOPLE COMPARED

Although the people described belong to different language groups and had been on their present land at the time of study from as few as 2 to more than 300 years, cultural similarities greatly exceed the differences. Padoch's (1982) monograph was specifically comparative between her own three sites and Freeman's, all four in the Iban language area. Most of the more recent reports have made extensive reference to earlier work and have tended to enlarge the treatment of aspects that previous writers neglected. With caution, sources that are stronger than others on certain aspects can therefore be used to identify possible gaps in the latter. There are many elements in common:

- Sarawak and adjacent parts of Kalimantan are close to the equator, and despite large differences in mean rainfall all areas have weak and unreliable dry seasons; drought does occur, but untimely rain is a more common problem.
- Population densities varied within a range below 20/km^2. Rousseau (1977) calculated the density of the Kayan on the Balui river at about 0.5/km^2, comparing this with a figure of 3.5/km^2 for the Baleh Iban, which he derived from Freeman's (1955) description of his research area. In a region where large areas remain unpopulated, Chin (1985) calculated 8.8/km^2 for Kenyah on the upper Baram. Dove (1985) calculated 11.5/km^2 for the Kantu'. For the Bidayuh, Geddes (1954) wrote of unequal and constrained village territories, giving rise to frequent dispute. The regional density of the Bidayuh area was in the range 10–19/km^2 by the 1991 census, higher than in the other areas. Colfer, Lian, and Padoch refrained from calculating population densities. The meaning that can be given to population density is uncertain because although village territories may have extended notionally from the river to the

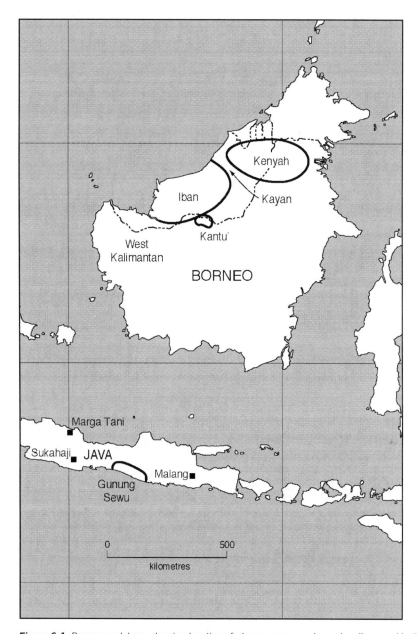

Figure 6.1 Borneo and Java, showing location of places, areas, and peoples discussed in the text.

watershed in the days before their claims were confined by timber concessions, areas more than about 1.0–1.5 km from streams navigable by canoe would seldom be entered by cultivators. In some areas there was no recognized inland boundary to a territory.

• All the people described depended first on rice and followed much the same agricultural calendar. Notwithstanding differences of more than a month in normal timing, site selection, slashing, felling, leveling, burning, planting, weeding, and harvesting occupied 9 or 10 months of the year within the same 11-month period, centrally dependent on burning during spells of dry weather that occur mainly between July and early September.

- Their dependence on rice, however central it may be to modern culture, may be historically recent. The upland rices they grow developed late in the evolution of the domesticated rices.[3] Visser (1989) believed rice to be only a sixteenth-century introduction in Halmahera to the east. Earlier crops probably included root crops and may have been dominated by Job's tears *(Coix lachrymae-jobi)* and millet, especially the latter (Dove 1985). Both are still widely grown in the Indonesian region. There have been further crop inventory changes in more modern times.

- All farmers used knives, axes (increasingly supplemented by chain saws), and digging and dibbling sticks as agricultural tools and woven basketware for containers. Dry land was nowhere tilled, and there were no working livestock. Use of herbicides to reduce labor time spent on weeding has become widespread since the 1970s, although there has been limited adoption of chemical fertilizers.

- Except for one of Padoch's groups who farmed a lowland area, all worked uplands with little flat land, sometimes none, within forests of very high biodiversity.

- All cleared fields in secondary forest of different ages as well as in old forest (also called primary forest or primeval forest in much of the literature, although it may not always be so); some groups worked only secondary forest and none worked only old forest.

- All these people lived, or had earlier lived, in longhouses in which the apartment of each family (nuclear or extended) opened onto a common veranda; all managed their farming as families, not as corporate groups; and although there was much joint work, no farms were owned or worked by groups.

- For convenience of joint work and protection of crops from wildlife, Borneo families in all areas commonly made fields closely adjacent to each other so that several hectares of land were cleared at once; decisions on what type of land to cultivate therefore were made by agreement among a group of households.

- All groups recounted a history that included voluntary and enforced migrations, wars, and epidemic disease, leading to major fluctuations in the population of any given area, especially since 1800. This continued until very recent times, including the 1980s and 1990s among the East Kalimantan Kenyah described by Colfer, Peluso, and Chin (1997).

- Spiritual and magical beliefs governing farming activity had much in common, and where they seem to be of little modern importance, they were much more central in a recent past; even in areas where people have fully converted to Christianity, spiritual beliefs have not died.[4]

- All Borneo inland peoples have participated in trade for at least a thousand years, selling a range of forest produce for valuables that are mainly of Chinese origin; they have also traded among themselves for tools, especially iron. In modern times all but the most remote have planted rubber as a cash crop, and some have diversified into other cash-earning activities.
- All undertook gathering, fishing and hunting, cash cropping, and off-farm work where available, and all these activities contributed to livelihood support; nonetheless, all activity remained centered on rice production. Although cash-earning activities were everywhere of increasing importance, even the most recent reports still described the basic systems in the same terms as did the earlier reports.

There were important differences in the structure of society. Kenyah and Kayan had a well-marked class distinction between aristocrats, lower aristocrats, commoners (the majority) and, formerly, slaves. Iban had no such class distinctions, and within each longhouse, core groups of closely related families enjoyed no special privileges (Freeman 1955). Kantu' had a headman, recognized by government, and other "persons of rank," but they enjoyed no privileges. Kenyah and Kayan aristocrats had the right to call on the labor of commoners and slaves and used to have important roles in ritual and decision making. Despite this differentiation, their farms seem always to have been managed in the same way as those of everyone else. Modern decline in the significance of ritual practices and prohibitions, and access by commoners to cash incomes that made some of them more affluent than aristocrats, greatly reduced the significance of social class.

There were some differences in farming practices. They mainly lay in the following:

- The cropping–fallow regime and the age of forest growth cleared for use
- The reported range of secondary crops and tree crops
- The significance of swamp rice
- The rules of access to land

Each of these is discussed briefly in turn because these are the areas in which diversity most clearly emerges.

DIFFERENCES IN THE CROPPING–FALLOW REGIME AND LAND SELECTION

Old forest, retaining no traces of former cultivation, offers the best soil conditions for new cropping and, if all the biomass is burned, the most abundant ash, but clearing it is hard work. The large trees, once felled, dry slowly and burn much less readily than younger growth. A main advantage is that many fewer herbaceous weeds spring up in the clearing. It is often possible to crop the land for a second year, clearing only a few months' regrowth, although a major increase in weed infestation sometimes results. This always occurs if a third crop is taken. Pioneering in areas not populated within living memory except by hunter–gatherers, all the Borneo farming people have used old forest, and some continued to do so even a generation after settlement.

Freeman (1955) regarded use of old forest as a cultural preference among inland Iban of the Baleh valley and castigated the destructiveness of farming such land for 2 years. In a sparsely peopled Kenyah area, Chin (1985) found 9 percent of fields in a second year of cultivation coming both from old forest and old secondary forest, perhaps 40 years old. Kantu' made 24–38 percent of their fields in old forest in 1975–76. Not only did they never cultivate rice a second year, but they also reported to Dove in the 1970s that their Iban neighbors never followed this practice (Dove 1985). Kenyah who had settled downriver from their original village in East Kalimantan sometimes cultivated a field twice in the 1980s. They also had a preference for cutting old forest because the less rainy climate made burning of old trees easier than in other areas (Colfer, Peluso, and Chin 1997).

Padoch (1982), working in areas of both ancient and recent Iban settlement, the former without significant remaining old forest, noted the frequency of fields made in forest of different ages to compensate the advantages and disadvantages of each. A field cleared from forest of different ages will have some land from older forest, richer in nutrients and less beset by weeds, and some land from younger regrowth that is easier to clear and on which a better burn usually can be obtained. Freeman made the same observation but did not stress it. Padoch found that 70–85 percent of households in an area of ancient Iban settlement farmed land drawn from more than one type of forest regrowth. The majority of fields in all areas except those of most recent settlement were made in mature secondary forest not less than 4 years old and often much older. Farmers did not select land on the basis of remembered age of previous use but on the type of vegetation and especially the girth of its trees, both indicators of soil condition. They also sought land as close as possible to the longhouse or to streams navigable by canoe. Where this was unavailable, some, the Iban and Kayan in particular but not the Kantu', made secondary settlements away from the main community (Freeman 1955; Rousseau 1977; Dove 1985). In modern times there has been substantial off-farm employment for men, especially in the timber industry, so there has been a tendency to use younger forest in which more of the clearing can be done by women and that is closer to home and more accessible. The consequence has been a reduction in the use of older secondary forest and a progressive shortening of the fallow period (Lian 1987).

Shortening the fallow and cultivating the same land too frequently create problems, as Freeman (1955) found among the Baleh Iban. The farming people also were aware that unwanted changes can arise quickly. Lian (1987) noted that valuable wild products occurring only in fallow of certain ages became scarce among Kenyah along the Tinjar River in the few years after a shorter cycle was adopted. More seriously, Chin (1985) focused attention on the weed content of secondary forest, not only of different ages but also with different histories of use. To the Kenyah of the upper Baram, *Scleria* species were the most vicious and ineradicable weeds. They persisted even 30 years into the period of secondary growth to infest new fields made after this seemingly conservative length of time. In addition, they appeared more persistent in sites that had been cleared three or more times in the past; forest on such land was given a specific name and it was believed that crops grown after clearing it would not be abundant. Similarly in northeastern India, Ramakrishnan (1992) has noted that arrested successions can emerge not only after shortened fallows but also if longer cycles of 10–30 years are constantly imposed on the same site.

Given the preceding discussion, it seems remarkable that land in the Batang Ai basin in what is now Sarawak, first settled around 1600 A.D. by Iban people and continuously used under shifting systems since that time, continued to be productive in the 1970s. There were few specific management adaptations that were not also found in areas of much more recent occupation (Padoch 1982). Field sites were very seldom used a second year, but the mean period of fallow was only 7–8 years. *Old forest,* using the same Iban term as is used by the Baleh Iban, meant forest aged 25 or more years, not forest that might be primary, as Freeman consistently called it. Patches and breaks of forest were left among the fields to provide seeding banks for regeneration, but fields were larger than in the Baleh. Padoch did not discuss weeds at any length, so a comparison with Chin's observation is not possible.

THE REPORTED RANGE OF SECONDARY CROPS AND TREE CROPS

The most important crop for all the Borneo farmers was rice, of which they grew several varieties within two main types: glutinous and nonglutinous. Lian (1987) listed 28 varieties grown by Kenyah of the Tinjar valley. Kantu' planted "a minimum of forty-four different, named varieties" (Dove 1985:159). In two East Kalimantan Kenyah villages, 22 and 29 varieties were planted in 1980 (Colfer, Peluso, and Chin 1997). The preoccupation with rice has been paralleled in the writings of the professional observers. Thus Freeman (1955) mentioned, in passing, only nine "catch crops" other than rice grown by the Baleh Iban. Padoch (1982) listed 24 found at her three Iban sites and believed Freeman's list to have been too small. This suggestion gained indirect support from the findings of Cramb (1985) in the Saribas Iban area; Cramb listed 26 subsidiary crops, among which maize and manioc both were important. Dove (1985) included 32 in his glossary of Kantu' Iban terms for flora and fauna and wrote in greater detail about 12 of them.

For the Kenyah of the upper Baram, botanist Chin (1985) listed and analyzed

89 cultivated plant types, with almost 200 varieties, belonging to at least 95 botanical species. Seventy-four of the 95 species provided food, and 3 were specifically cultivated for cash. Lian (1987) listed 33 by Kenyah names and stressed the importance of nonrice crops grown in the fields. He valued their 1985 production on the land of eight farmers, at local prices, at 47 percent of the combined value of rice and these subsidiary crops taken together. Cramb (1985) obtained a comparable figure of 42 percent among the Saribas Iban, noting that the contribution might exceed half in years of poor rice harvest. Among the Kayan, Rousseau (1977) named 20 subsidiary crops other than fruit trees; some were grown in minute quantity. The Kantu' planted some crops among the rice, some in distinctive areas of the field, and manioc and sweet potatoes sometimes were planted in small parts of an old field after the rice had been harvested, to continue into a second year (Dove 1985). The most detailed information in this group of sources is for the East Kalimantan Kenyah villages (Colfer, Peluso, and Chin 1997). Many nonrice crops were planted in the rice field, some at the same time as the rice, some later. These crops persisted as whole gardens into the first year of the fallow, and there was also new planting in these gardens. A partial taxonomy, using local nomenclature, listed 53 crops other than rice (Colfer, Peluso, and Chin 1997).

Dove described all Kantu' subsidiary crops as relishes, including even starchy manioc in this category. Freeman said very little about them. Rousseau's (1977) concise discussion of Kayan farming in an area between the Iban and the Kenyah, used in conjunction with the details in the other major sources, helps bridge a gap. Manioc was of major importance as a supplementary food, fodder for the pigs, and an insurance against failure of rice. It was planted in the same fields with the rice, was used to fill gaps, and could be harvested at any time during the year. Kenyah farmers who failed to get a good burn or whose crops seemed unlikely to prosper for other reasons would increase their planting of manioc while the rice remained in the ground (Lian 1987). This practice may well have been more widespread.

The building of farm huts was common to all systems, and in Kenyah and Kayan regions more specific mixed gardens were made around them. Here were grown taro, leaf mustard, tobacco, chilies, eggplant, sugar cane, betel leaf, cucurbits, millet, Job's tears, maize, sweet potatoes, and a range of beans. Among this splendid mixture of Asian domesticates and New World introductions, some, including manioc, probably were introduced only within the last century. Only if they could be protected from pigs were such mixed gardens made near the longhouses, but they were common around the old East Kalimantan Kenyah village of Long Ampung, where a 1991 survey enumerated 79 species (Colfer, Peluso, and Chin 1997). Chin (1985:134) described Kenyah farm gardens in the upper Baram as "intensively cultivated" and enduring for up to 2 years. Lian (1987) mapped farm gardens in the hills around a Kenyah longhouse in the Tinjar valley and took me to some of them, but I saw little to suggest intensive management. Rousseau (1977) stated that little effort went into the secondary food crops other than tobacco and manioc. Padoch (1982) wrote only of intercropping. Although no systematic conclusion can be drawn, there was certainly a large variety of crops. The variety was important and it represented an element in agriculture not so strictly bound by the calendar as is rice, but maintenance of crop biodiversity took up only limited space in the management of either land or time.

There was less reported variation in practice with regard to valued trees, in particular the illipe nut tree *(Shorea macrophylla),* durian, jackfruit, mango, and a few others. These occur in clusters in the wild, where they may represent the sites of former settlements (Peters 1996), but were planted around longhouses and in old fields at the beginning of the secondary succession. The illipe nut and durian, where the latter could be sold while still fresh, became valuable cash crops in recent years. The trees remained owned by the person who planted them, or who found and marked them in the wild, and rights were heritable (Chin 1985; Lian 1987; Sather 1990). In whatever environment they were found, their location was very precisely known.

Substantial new planting of fruit trees, mixed with nonrice crops, was reported only for the East Kalimantan Kenyah villages, where it was desirable to establish a claim to land by means of continuing improvements. Government intervention in the 1970s had the object of developing wet rice, but the means of water control were not available. Instead substantial home gardens on something like the Javanese model, discussed in chapter 8, were set up to satisfy the government's wish to see more permanent forms of agriculture (Colfer, Peluso, and Chin 1997).

Rubber, introduced in the 1920s and 1930s, became a major addition to the tree stock in all areas by the period 1960 to 1990, although tapping was variable through time. Rubber trees often were treated more as a bank on which to draw in times of need than as a regular cash-producing activity. Rubber trees rarely were planted in a regular manner, growing up among other secondary growth in the old field sites in which they were sown. Volunteer seedlings were insurance for the future. Lian (1987) recounted a revealing dispute in which he was involved after inadvertently damaging some small seedlings during a night expedition hunting mouse-deer. The seedlings in question were volunteers from much older rubber, cleared for cropping only 2 years previously.

Two other tree crops had a different significance. They grew wild and were also planted. These were the lowland sago palm *(Metroxylon sagu)* and the upland sago *(Eugeissona utilis).* The latter grows particularly around natural springs and seepage lines in the hills and was a main source of starch food for the remaining Punan hunter–gatherers. Both had value as emergency foods for the farming populations, but neither was regularly consumed by them. These several tree crops blurred the distinction between field, successional growth, and wild, and they gave ongoing value to the land many years after the last cultivated crop was taken.

THE SIGNIFICANCE OF SWAMP RICE

The Bidayuh ("land Dayak") of western Sarawak cultivated upland rice and subsidiary crops in much the same way as did the Iban, Kenyah, and Kayan. In the late 1940s they also cultivated wet fields, providing a higher yield and more reliable production of rice (Geddes 1954). Although Geddes's account lacks full quantification, it seems that almost all suitable swampy land was used and that in his day (1949–50) it provided from a quarter to more than half the total rice harvest. The cultivation methods were very much as in upland cultivation: the swampy areas were cleared and burned, and the only site-specific management consisted of digging simple irrigation channels and damming the main stream to divert water into the field. The period between successive uses of the same wet plot was about half that between uses of dry land, the latter estimated by Geddes to be 10 years or a bit less.

Padi paya, or swamp rice, was first fully described by Dove (1980) and has since been recognized much more extensively in Borneo. In Sarawak, Padoch (1982) provided the first account for a western Iban area of a system simpler than that of the Bidayuh; it included no irrigation ditches, although use of a seedbed and transplantation were described, as in Bidayuh. The system was a wetland adaptation of the dryland system, involving slashing, felling of any larger trees, and burning. On *paya* land, burning was immediately followed by a crude tillage with a knife, creating a nearly weed-free wet field. Fields might be used in successive years or undergo only about 3 years' fallow (Dove 1980; Padoch 1982, 1988a). The cultivars are different from those in the uplands and may be more closely related to historically early domesticated varieties. Dove (1985) detailed different methods of sowing and transplanting for the Kantu' Iban, who thus achieved a much higher density of rice plants than on dry fields. Between 30 and 40 percent of Kantu' fields, by number though not by area, were *paya* in the mid-1970s (Dove 1985).

There have been no reports of *padi paya* among the Kayan or the Kenyah or elsewhere in northeastern Sarawak and adjacent Kalimantan. Yet both the Tinjar and Baram Kenyah made many of their cleared dry fields on alluvial land and after 1960 used government aid to experiment with fully elaborated wet rice production *(sawah)* on such land (Chin 1985; Lian 1987). Carefully leveled, irrigated, and bunded rice fields have long been made by the Lun Dayeh–Kelabit, who are isolated from all other *sawah*-using people, deep inland on each side of the border between Sarawak and East Kalimantan (Padoch 1985, 1986). The simpler *padi paya* system remained widespread, and in the early 1990s I saw it used in western Sarawak and adjacent West Kalimantan almost wherever it could potentially be established in the area. There seemed to be a range of technological levels.

Not far from where Geddes worked at the end of the 1940s, I saw wet rice fields that had recently been cleared from swamp forest. The forest had been burned and stumps and logs remained in the field, but the plots were better leveled than in *padi paya,* and some had small bunds that could quickly lead to conversion into a full *sawah* system (Brookfield, Potter, and Byron 1995). Geddes's account of Bidayuh practices in the 1940s, a minor elaboration on what Padoch later described for a nearby Iban people, could have represented a stage along the road of transition. Padoch, Harwell, and Susanto (1998) noted that some of the progressively enlarging area of wet rice in the land of a village settled at much higher population density in West Kalimantan was made on sites formerly cultivated as "swamp swiddens," or *padi paya*.

THE RULES OF ACCESS TO LAND

For the Iban, Kantu', Kenyah, and Bidayuh, land once cleared became perceived as the heritable property of the household or lineage group whose members first cultivated it, however long a period might pass between successive uses. This was not so with the Kayan Rousseau (1977) studied. In all areas, a territory was recognized for each village community, whether the village was a single longhouse or more than one house. The boundaries were mutually agreed with neighboring villages. Movement of houses continued, and until at least very recently whole communities have come and gone over distances of several kilometers, with much larger migrations in only the recent past.

As important as the conditions of land claim were the conditions of loss of that claim: in all areas, when communities or holders moved away from the immediate area, the land became as if it were primary forest, available to all. When people moved away permanently, their houses as well as their land could be occupied by others. In areas of long settlement within a context of continued mobility, this happened many times. Not finding the inequalities in access to resources expected to result from generations of inheritance following Iban rules, Padoch (1982) stressed the importance of being able to borrow land with minimal formality. Lian (1987) similarly noted the growing importance of borrowing among the Kenyah of the Tinjar river. Dove (1985) referred not only to borrowing but also to sale of land among the Kantu', but he did not elaborate.

Chin (1985:134) did not mention borrowing but insisted that individual claims to land existed. What he wrote about site selection was quite possibly relevant to more than his own area:

> Invariably, informal discussions regarding the general location of the present year's farms would have taken place during the previous year. In this informal manner, and especially as most people tend to be followers rather than leaders, the majority would have come to a consensus as to the general location of the following year's farms. . . . As it is considered desirable to have farms adjoining one another, numerous discussions to sort out and to agree upon the final location of each other's new farm sites are held. . . . Usually, a final discussion (involving the whole house) is held sometime in March, to decide and settle all outstanding issues regarding farm sites for the year.

Thus the law of the land became accommodation between people who cooperated in many aspects of daily life, as we saw among the much more closely settled Chimbu of Papua New Guinea in chapter 1. Where resources were scarce, as in the case of *padi paya* land among the Bidayuh (Geddes 1954), and in areas that have been occupied a very long time, not all households were able to get the resources they sought (Padoch 1982). Nonetheless, the desired diversity in land holdings generally was achieved, and there is little reason to dwell too closely on the formal mechanisms by which the land of one household became transferred—in effect for a period of many years—to another. As is increasingly found as research advances, the rules of land tenure and transfer, obtained by ethnographers and others, are of less consequence than what actually happens in practice.

CONCLUSION ON BORNEO

The substantial differences, within a broadly comparable set of basic systems, reflect cultural differences between peoples, but they also reflect a history that includes significant change in the past few hundred years. Changes in crop inventory are of particular importance from the standpoint of agrodiversity, but warfare, migration, and population loss through disease have also impinged differently on the people of different areas and societies. A lot of the diversity revealed in this section is detailed, but detail is what agrodiversity is all about. Larger differences and bigger changes arise when the Borneo peoples are compared with those of another region.

■ Borneo in Perspective

SEEKING AN APPROPRIATE COMPARISON

My earlier intention in planning this chapter was to compare the Borneo literature with that from South America, where most shifting cultivation populations are at densities lower than in Borneo. The South American literature is hard to use, especially for an author without field experience among the Amerindians. It was therefore with some comfort that I read the analysis of Moran (1996), who warned that comparative human ecology of shifting cultivation in Amazonia is in a poor state.[5] More important in the Latin American literature as a whole is the question of successional fallow management, and I use it from this point of view in chapter 8. I conclude this chapter with a more regional comparison to show, within the context of the same general region as Borneo, how environmental and cultural differences are reflected in agricultural differences. This is found among the Hanunóo and Buid of Mindoro, Philippines.

THE HANUNÓO AS CLASSIC CASE

It probably would be safe to assert that Conklin's (1957) report on the agriculture of the Hanunóo is the most oft-cited item in the whole literature on shifting cultivation, although it is probably read much less often than it is cited. The quality of Conklin's reporting is appropriately recognized by the fame his monograph, and the Hanunóo, have achieved. Availability of more recent accounts makes possible some discussion of change, a theme otherwise stronger in the next two chapters than in this. Two accounts with a central focus on agriculture and land have been written about the adjacent but further inland Buid people (Lopez-Gonzaga 1983; Gibson 1986); both made frequent reference to the Hanunóo. Postma (1974) and Miyamoto (1988) gave direct information on the Hanunóo but were less concerned with agriculture.[6]

Conklin's work was done in the late 1940s and early 1950s (1949, 1954, 1957). Missionaries, teachers, and government were beginning to enter the area, although Conklin (1957) wrote, enigmatically, only that they were not present when his work began. He did not mention the lowland settlers already beginning to invade the land of the upland people on a large scale (Postma 1974; Lopez 1976; Lopez-Gonzaga 1983; Gibson 1986). There had been centuries of contact between inland and coastal people, involving trade that left the Hanunóo with the Chinese gongs and jars that were valuable to them; such traded objects also were valuable to the people of inland Borneo (Conklin 1957). In the 1950s the Hanunóo produced a surplus of rice and some other commodities that they used in trade, both with the coastal peoples and with their Buid neighbors. Visits to the coast, mainly by young men in search of temporary work, were not uncommon. However, trading was on a very small scale and there was little money in the economy.[7] Only a few Hanunóo were fluent in the national language (Conklin 1949). Hanunóo were not Christian, the first enduring mission not being established until 1958 (Postma 1974).

THE HANUNÓO SYSTEM AND THE BORNEO SYSTEMS: COMMON ELEMENTS AND DIFFERENCES

The Hanunóo occupy an upland area in the southeast of the island of Mindoro, about halfway between the equator and the tropic of Cancer. Although no month is dry, there is a much clearer distinction between wet and less wet seasons than in Borneo. The forest is less diverse and lower than that of Borneo, and it burns more readily but produces less ash. The timing of agricultural practices therefore was more relaxed; clearing might begin at any time during a period of more than a month, and burning could be spread over as much as 3 months. In Conklin's time, there was no continuing individual claim to land once farmed, and, largely because of the long productive life of fields, often it was not even possible for farmers to select land in close proximity to that of desired neighbors, as Borneo farmers did. Clearings were scattered and were cut and burned individually. Houses were built in small clusters, and location of settlements changed from time to time. Augury and ritual were at least as important in site selection as in Borneo but not so significant in the matter of timing.

There was the same preference for older secondary forest, and not many fields were made in old forest, of which there was little except in areas protected because they were the abode of spirits. An escaping fire was a rare event in Borneo, where few precautions were taken to avoid it. In Mindoro it was a serious risk, and wide firebreaks were made around every field. Sometimes breaks were also made between fields and in areas that were to be protected, such as those with houses. For the Buid, the same risk was described but there was no mention of firebreaks around the burned clearings still made in the late 1970s (Lopez-Gonzaga 1983). Hanunóo still made wide firebreaks in the late 1970s and early 1980s (Miyamoto 1988). Fields were fired with care, but the burn often was incomplete, so secondary burning, avoided if possible in Borneo, was normal practice by the Hanunóo. Because fires are less intense in Mindoro, it was easier to preserve useful trees, even bananas, in the burned area. Stumps were used to provide supports for the vines of yams and some other climbers.

Rice, though not the largest element in diet, was of central importance to the Hanunóo, less so to the Buid. Although rice was the major first crop, it was interplanted with maize. Much more interplanting than this was normal, and in this respect the Hanunóo system was sharply contrasted with those of Borneo. Conklin (1957) listed 87 crops planted in the fields, and on the basis of his observations wrote that "by the end of April, five or six basic crop types are already represented in an average swidden. From May through August, 20 to 25 more are added, and from September through January, an additional 15 to 20 to bring the average total to between 40 and 50 for the first year of the swidden cycle" (1957:85). In addition, some plants were always put in fields separate from the grains.

Weeding was far more a laborious activity with the Hanunóo than in any of the Borneo areas, especially to get rid of numerous herbaceous plants, *Compositae,* twiners, and creepers. Conklin showed that the landscape was a patchwork of forest of different types and grassland; the large amount of bamboo present suggests deflected successions. Hanunóo were familiar with the risks of soil erosion and exhaustion. Three concentrated spells of weeding, each taking 2–4 weeks, were normal, the last also involving thinning of the rice. Maize was harvested first, then came the rice harvest, carried out by hand, with little use of knives. It involved elaborate arrangements for group labor, feasts, religious rites, and careful observance of practices designed to propitiate the rice people, or spirits. The harvest period extended over 3 months in the early part of the drier season toward the end of the calendar year. The shifting cultivation practices of the southern Buid, as detailed by Lopez-Gonzaga (1983), were very similar except that although rice was the main first crop, it had much less cultural significance than to the Hanunóo, and propitiation of the rice people was far less important.

Up to the rice harvest, some of the differences between Hanunóo and Borneo practices clearly were responses to different environmental conditions, not least the lower and more open nature of the forest and the admixture of grasses. Other variations suggest a marked difference in the basis of livelihood. What took place in southern Mindoro after the rice was harvested represents a much sharper contrast with Borneo practice. The rice straw was cleared, piled, and burned, with a further weeding, to favor a new life for the field dominated by yam vines, young bananas,

papaya, manioc, and pigeon peas, all "towering above the dense dark green carpet of entwined sweet potato stems now covering the last traces of grain stubble" (Conklin 1957:121).

The fields continued in production into the following year. They might be planted again with crops other than rice, progressively converted into tree crop gardens (for fruit trees, palms, and bamboo) still under management for some years, or permitted to return to a fallow cover. The root crops, together with bananas, provided more of the diet than did the rice. Root and tree crop land occupied two or three times the area of first-season grain on most farms (Conklin 1957). As Lopez-Gonzaga (1983) suggested for the Buid, it may be that the strong Hanunóo emphasis on root crops owed much to the vulnerability of rice and maize to damage by hurricanes. Conklin witnessed none during his period of fieldwork, but there were six in 1978 and 1979 alone. In this region they are an annual risk.

Over time, Hanunóo fields overlapped one another, so that through a 20-year cycle any given piece of land might have been through four or five different sequences of use and fallow (Conklin 1957). Hanunóo managed plants, not land, and this was Conklin's central message. Management also included the fallow stage. Some areas might be encouraged, by renewed firing, to become grassland to provide fodder for small herds of cattle and thatch for houses. A few areas might be left a long time to recover into old forest, which, together with the reserved sacred areas, was important to Hanunóo for durable wood, rattan, medicinal plants, and wild plant food. Most, whether or not they went through the tree crop garden stage, would be left to develop secondary forest, 20 years or more in age, from which new fields could be made most readily. This took place under a general population density of around $10/km^2$, rising to $25-35/km^2$ in the more closely occupied parts of Hanunóo territory.

■ The Forces of Change

A NEW WORLD IN MINDORO

Disturbance during World War II and a smallpox epidemic in 1948–49 had lesser impact in Hanunóo than among the Buid (Conklin 1957; Gibson 1986), but population probably was reduced by the time of Conklin's fieldwork. Since the 1950s the experience of Hanunóo and the adjacent Buid people has been dramatic. Its dominant element has been massive invasion of their country by settlers from the Visayas and Luzon in search of timber and land for farms and pasture. Most Hanunóo and some Buid tried to retain the agricultural and social system they had evolved, but with growing difficulty as pressures on them increased, and there were important changes.

When Buid refugees returned to their homes in the 1950s and 1960s they found that Visayan and other Christian immigrants to Mindoro, who had been unable to find land in the coastlands, had already occupied the lower parts of the valleys, leaving them only the hills. Hanunóo had little valley land attractive to farmers who wanted to invest in irrigation and other improvements, but large areas were taken over by squatters who used shifting methods without rotation. The areas of grassland enlarged quickly, which in turn attracted cattle owners (Postma 1974; Miyamoto 1988). Lacking either a concept of individual land holding or any sound knowledge of the political system, both groups were easily deprived of their land by squatters and also were exploited in trading or employment relations with the newcomers (Postma 1974; Lopez-Gonzaga 1983; Gibson 1986). Many became indebted tenants on what had been their own land. Starting as early as 1958, when Conklin's report was still very new,

> [Hanunóo] free choice of plots to cultivate, according to their well-developed and systematic methods of swidden-farming, has slowly been curtailed. Not only are they kept out of all land in the coastal plains and lower foothills, now often fenced in by the landgrabbing lowlanders, but the cleared forest lands have, in turn, become pasture lands leased to big cattle ranchers. (Postma 1974:25)

With the reduced ability to select land, cultivation cycles were shortened (Miyamoto 1988; Lopez-Gonzaga 1983). Rice yields declined so steeply in Buid that rice was given up and maize became the dominant crop. Hanunóo persisted with rice, central to their whole system of beliefs, but suffered increasing hardship (Gibson 1986). Maize growing was increasingly commercialized, especially in Buid, where on many holdings two crops a year came to be taken, and a number of farmers bought wooden plows and carabao, aiming to clear the land permanently (Lopez-Gonzaga 1983). The same probably unsustainable innovation came also to Hanunóo (Miyamoto 1988).

By the 1980s, the Buid and Hanunóo were becoming peasant farmers, either as tenants or on small holdings between which there was sometimes land dispute; they engaged increasingly in commerce. The integral system of shifting cultivation was still practiced by some in the early 1980s but with increasing erosion of important elements. At least up to that time, both Buid and Hanunóo had been able to preserve cultural identity, but an acceleration of change can perhaps be read into the detailed case material on conflicts and their resolution provided by Miyamoto (1988). Conklin's classic account is now a historical document. The system he described no longer enjoys the conditions it needed to remain sustainable or even survive. Given that the Hanunóo system retains such a place in the literature, it is important to stress this regrettable fact.

THE NEW WORLD IN BORNEO

In both Indonesian and Malaysian Borneo, the timber industry intruded into the regions used by shifting cultivators in the 1970s and 1980s. The cultivators gained employment and some income, but they lost access to land. In Indonesian

Borneo but not so much in Malaysian Borneo, the logging industry has been followed first by plantations of fast-growing timber, then more and more widely by oil palm plantations, some very large (Potter and Lee 1998). Although they have attempted to strengthen their claims to land by increasing rubber planting and adopting wet rice, many shifting cultivators have lost access to large tracts and have seen their valued trees cut down or the fruit poached. In Indonesian Borneo, settlers from other parts of Indonesia became the main workforce on the plantations, and the shifting cultivators have been under increasing pressure. There have been outbreaks of murderous violence between indigenous people and migrant settlers.

Sources based on work done in the 1970s and 1980s only touch on these external impacts, although one of the latest of them (Lian 1987) already indicated their early effect. Since that time, the people described by Lian—his own natal community—have given up their former riverside longhouse in search of a more secure cash income. At the time Lian wrote, shifting cultivation was an efficient provider of a varied diet and a cash income. The resettlement alternative was unattractive by comparison. Yet Lian (1987) already insisted that the shifting system per se was not an integral component of the culture of the inland people.

In East Kalimantan external interference has proceeded from logging, through the entry of transmigrants, to the modern establishment of industrial wood plantations comparable in their demands on indigenous land to the massive oil palm plantations of other parts of Kalimantan. Such conversions permanently prevent access by the local people. The large-scale forest fires of 1983, repeated in the 1990s, also affected Kenyah communities. When Colfer returned to East Kalimantan in 1995, she found that some of the people known in the 1980s had given up the unequal struggle to sustain their former agricultural system, shortening fallow cycles disastrously. They were coming to rely on incomes from wage labor for the industrial timber estate and transmigrants. Even these were not necessarily secure, for the planted *Acacia mangium* was not doing well, and the transmigrants had no experience in managing Borneo soils (Colfer, Peluso, and Chin 1997). Elsewhere in Kalimantan, farmers were struggling to hold onto their land against the oil palm plantations by converting land to permanent uses (L. Potter, personal communication, 1999). But an account written in the early 1990s of relations between the forest-dwelling people of Borneo and the two states is now overtaken by events (Brookfield, Potter, and Byron 1995); its guarded optimism now has to give way to the pessimism voiced by Potter and Lee (1998). It was not mere pedantry to write this chapter, even the parts using information from the 1980s, in the past tense.

Notes

1. The manner is which perception can be distorted emerges clearly from my own experience in 1994. An article introducing the value of agrodiversity, published in a leading but semipopular environmental journal, was given cover story status but with the editorial add-on title "Sustainable slash and burn," printed on the background of a lurid photograph of a burning forest (Brookfield and Padoch 1994).

2. Dove (1985) distinguished the Kantu' from Iban living around them. The latter moved in from Sarawak in historical times, whereas the Kantu' have remained throughout orally remembered history in present-day Kalimantan. They speak an Ibanic language but have a long history of hostile relations with the Sarawak Iban.

3. Chang, Loresto, and Tagumpay (1972) and Chang (1976) detailed the differences between the upland rices and the historical varieties grown in wet sites. Chang (1989:132) summarized the differences and their evolutionary significance: "When rice was introduced into areas where the soil had poor moisture retention and the water table was rather low, the upland type that evolved had early maturity, lower tillering capacity, and deep and thick roots. In mainland Southeast Asia, these types are known as the hill rices. These upland rices have become markedly differentiated from the ancestral lowland type in that they grow better in an aerobic soil. The root system of drought-resistant upland rice is comparable with that of wheat." The conclusions are based mainly on morphological characteristics (phenotypes). It is to be expected that they will be modified by the results of work on the rice genome (chapter 14).

4. The common belief was that a spirit abides in all physical things, alive or otherwise. Appropriate rituals had to be performed at all principal stages in the agricultural cycle, and great stress was also placed on omens, the means by which spirits make their will known. Complex sanctions and prohibitions involved frequent and expensive offerings to propitiate the spirits. Not only crop failure, but also human sickness and death could be brought about by failing to propitiate the spirits or going against their expressed will. Serious delays in agricultural activities could come about through adverse omens. Acceptance of Christianity has not eliminated spiritual beliefs, but it has greatly reduced the crippling of productive activity caused by the need for complex ritual observance (Jensen 1974; Rousseau 1977; Sather 1990; Lian 1987; Dove 1996).

5. Descriptions of shifting cultivators in Amazonia recount basic cultivation methods in which there is usually less emphasis on securing a good burn than in Borneo; there is greater variation between the behavior of individual farmers. In addition, there seem to be greater differences in cropping practices between groups, especially in regard to intercropping; there is much variation in weeding practice. Although manioc is the principal crop in most societies, bananas, maize, and sweet potatoes sometimes are more important. There is now a substantial literature including studies by Carneiro (1960), Harris (1971), Flowers et al. (1982), Hames (1983), Johnson (1983), Beckerman (1983, 1987), and Balée (1994), among many others. Some of these writers do generalize, but it would be unproductive to follow them in doing so.

6. All the post-Conklin writers described the Hanunóo as the Mangyan, the Hanunóo–Mangyan, or the Mangyan–patag (lowland Mangyan), which is what the Buid call them. In the Hanunóo language, the word *hanunóo* means "true" or "genuine." They identify themselves as Mangyan, a term also used in the Philippines to describe all the people of the Mindoro interior. The Hanunóo and southern Buid are distinctive in several ways, most particularly in use of a writing script of possibly Sanskrit origin, used by etching on bamboo. The Buid are also described as the Buhid, the term Conklin used, based on the southern dialect of their language (Lopez-Gonzaga 1983; Gibson 1986).

7. Although there can be no doubt about this observation, which is well quantified by Conklin (1957), Lopez-Gonzaga (1983) suggests an important role for the Hanunóo as intermediaries between the coastal and urban markets and at least the southern Buid. Until the 1950s, all Buid trading contact was through the Hanunóo, involving barter or the use of beads rather than money. Cocoa already was exported from Buid territory by this means in the 1920s and 1930s (Lopez-Gonzaga 1983).

Chapter 7

Alternative Ways to Farm Parsimonious Soils

■ *Citemene* and *Fundikila:* Northeastern Zambia

COPING WITH A REGION WITH POOR RESOURCES

This chapter continues with shifting cultivators but in a different context and with a different theme. I focus on one specific region in south central Africa in which alternative agricultural strategies have been available, and both have been applied, to manage some very poor soils. Very high-temperature fire is generated in one, and in the other only ash from grass burning is incorporated into tilled soil. The region is largely mantled in *miombo* savanna forest, one of the largest continuous vegetation systems in the tropical world, occupying most of the plateau from Tanzania to Angola, almost from coast to coast. Within this, I concentrate on the Bemba people of northeast Zambia and their immediate neighbors, whose agriculture and land have been studied for 60 years. Agricultural systems related or similar to those of the Bemba region occupy a much wider area, but the Bemba area is one on which both human and natural scientists have written at length. Information with historical depth is therefore available.

I begin by describing the region and its scant resources. *Miombo* is a deciduous savanna woodland, with important areas of grassland. Over great areas it is dominated by species of just three tree genera, all members of the Caesalpinioid subfamily of Leguminosae, *Brachystegia, Julbernardia,* and *Isoberlinia.* Green, with many flowers among the underlying grassland during the rains, the *miombo* becomes leafless in the long dry season. The interlocking canopy reaches a height of more than 15 m when mature, with shrubs and extensive grassland beneath. On lower slopes of soil catenas is an open woodland dominated by *Combretum* spp., with a tall grass understory. Together with *chipya* savanna, these vegetation types are repeated over a very large area. Their nature and distribution respond to nutrient and water limitations and probably have been greatly affected by millennia of dry season fires (Lawton 1978; Frost 1996).

The plateau on which the *miombo* is developed is an upland with low relief, with shallow valleys among wide interfluve areas. In terms of modern soil taxonomy, there is an unsurprisingly close relationship to climatic zones modified by parent material, with Oxisols in the higher rainfall zones and light-textured and deeply leached Alfisols in the lower rainfall zones; both are developed on the ancient basement complex and granites. Ultisols occur over the small areas of intrusive basic rocks, and the few clay accumulations have Mollisols (Frost 1996). Low levels of organic matter and low-activity clays give most of the soils a low cation exchange capacity and low pH. Nutrients, especially phosphorus, tend to be fixed in the clay fraction, and there are potentially toxic levels of aluminum. The soils are therefore lean in available nutrients and, except in more favored patches, do not even support continuous forest. In detail, modern scientific study finds the distribution of soils to be closely related to geomorphic processes (Dalal-Clayton and Robinson 1993), as long ago proposed by Milne (1936), Trapnell and Clothier (1937), and Trapnell (1943). Variations are related to parent material, catenas, and the density of soil biota, especially termites (Webster 1965; Watson 1964; Jones 1990). Characteristic catenas run from deeply weathered plateau soils down to the seasonally saturated shallow valleys *(dambos),* which are peaty in places.

The whole *miombo* region has a high density of termites, which accumulate nutrients in their mounds while leaving a more impoverished soil between the mounds. Swift et al. (1989) estimate that aboveground production of litter from *miombo* is in the range 3–5 tons/ha per annum, which is not great, and much is captured by the termites or consumed by frequent fire (Trapnell et al. 1976).

Only a minority of *miombo* tree species associate with nitrogen-fixing bacteria. Scarcely any Caesalpinioids do, even though they are leguminous. The presence of ectomycorrhiza therefore may be critical in the development of a woodland type that is so widespread over such an extensive area. *Miombo* contains many trees that are obligately mycorrhizal and others that can make good use of the fungus infection (Högberg 1989). Notwithstanding the low quality of the soils and the lack of forest-type recycling, the leaves, branches, and stems of the *miombo* trees hold a good but variable store of the scarce nutrients (Lawton 1978, 1982; Högberg and Piearce 1986; Strømgaard 1988a, 1988b; Högberg 1989). The *miombo* also provides resources of honey and beeswax, caterpillars, and, not least, edible mushrooms. These are produced seasonally by its abundant ectomycorrhiza in sufficient quantity to form an important element in the diet of the people.

TWO WAYS OF MAKING A LIVING FROM THE LAND

Two principal farming systems, generalized by the northeast Zambian terms *citemene* and *fundikila,* both old, have been developed by the plateau people for using this land, necessarily creating improvements to obtain nutrients from a parsimonious environment. The improvements that these systems offer can be long lasting and permit good yields for some years on soils of very low intrinsic quality. Principal sources on these farming systems include Richards (1939, 1951, 1958), Peters (1950), Trapnell (1943), Allan et al. (1948), Trapnell and Clothier (1937), Allan (1949, 1965), Strømgaard (1984, 1985a, 1985b, 1988a, 1988b, 1989, 1990, 1992), Moore and Vaughan (1994), and Sugiyama (1987, 1992).

The early descriptions by anthropologist Audrey Richards and ecologist Colin Trapnell are still the most comprehensive. Because the initial work was so good, it inspired a later generation to follow it up and explore history. By the mid-1990s, it was possible to view change in northeastern Zambia over a whole century. Comparative discussion through time by Moore and Vaughan (1994) also enabled the significance of the contrasted agricultural strategies to emerge. Setting out to review change since the time of Audrey Richards's work in the early 1930s, they used a large quantity of "gray literature" available only in Zambia and in unpublished theses. Except where specific mention is needed, the sources are used without individual reference in the following discussion.

The most obvious contrast between the two systems is that *citemene* makes major use of fire, whereas *fundikila* relies mainly on tillage. Early external observers praised the latter while scorning the former, regarding it as a particularly wasteful form of shifting cultivation. The anthropologist who made *citemene* famous did not dissent from a prevailing view that it was one of the most primitive forms of bush cultivation and in one place characterized its practitioners as "a mass of inefficient and indifferent cultivators" (Richards 1939:327). *Citemene* makes use of fire almost more intensively than does any other agricultural system, and its study is therefore the right place in which to examine the relationship of the soil to fire. In a period of human history in which the use of fire is a matter of controversy and of major concern among conservationists, a closer examination is of value. Box 7.1 summarizes some of the scientific evidence.

Box 7.1
What Does Fire Do to the Soil?

The burn quickly creates a farming space and provides a sudden flush of nutrient ions to the soil. Against this, almost all nitrogen, carbon, and sulfur in the biomass and litter—but not in the soil humus—are lost into the atmosphere. Nitrogen loss may soon be offset. Most burning takes place late in the dry season, where there is one. When the dry season ends, dramatic increases of inorganic nitrogen occur, the amount in direct proportion to the duration and intensity of the preceding dry period. This nitrogen flush was first fully described by Birch (1960). There are several causes; one of them is that an active microbial population builds up very swiftly when rain resumes, quickly using up the available carbon. Nitrogen is then added to as the microbial population dies (Sanchez 1976).

Potassium, phosphorus, magnesium, and other nutrients that were in the biomass remain in the ash, but some of this is quickly dispersed by the wind and another part is washed or leached away before it can be taken up by plants. What is not lost is added to the soil, although some of it, the soluble potassium in particular, soon is leached away. The burn also sterilizes the surface soil of plant pathogens and of shallow-rooted weeds. Soil bacteria are killed, but their population is replaced quickly. The most positive effect in almost all of the few sites where the effect of the burn has been scientifically measured is that it gives a sharp lift to the pH. This change more readily permits the taking up of a number of nutrient ions by plants and lowers the ability of toxic elements to damage plant growth.

The effect of the burn differs strongly from place to place, and varies greatly according to the intensity of the fire. Nye and Greenland (1960) effectively summarized the state of knowledge in the 1950s, but the first serious task of monitoring and measurement was undertaken in the early 1970s. On an Oxisol in Amazonia, Carl Jordan undertook a 4-year experiment (Jordan 1987). Following local Amerindian practice, his assistants cut the site selectively, leading to an incomplete burn. The pH quickly increased from 3.9 to 5.4, then declined through the remainder of the 4 years. A substantial amount of organic matter remained on and within the soil. At the end of the 4 years the soil retained sufficient nitrogen, calcium, magnesium, and potassium, but not phosphorus, to have supported further crop growth and did support an immediate burst of secondary vegetation. Yet crop yields had declined rapidly after the second year. Because of the early date of the work, available phosphorus was not measured. As Jordan suggested, its initial abundance then scarcity may have been critical.

In the late 1970s and early 1980s the Dutch Royal Tropical Institute supported a major project of monitoring and measurement at sites in Sarawak, Sri Lanka, and Thailand. These represented equatorial, subhumid, and monsoonal conditions, respectively (Andriesse and Koopmans 1984/85; Andriesse and Schelhaas 1987a, 1987b). Beginning with laboratory work on soil samples, it was found that temperatures had to exceed 200° C before pH increased sharply (Andriesse and Koopmans 1984/85). Actual field measurement of the effects of fire at the three sites did not give such unambiguous results (Andriesse and Schelhaas 1987b). The uniform positive finding was of a 50–300 percent increase in available phosphorus and large increases in exchangeable calcium, manganese, and potassium in the topsoils, accompanied by rapid increase in pH. Very different burning temperatures were obtained, and the role of heat emerged as ambivalent. In northeastern India, Ramakrishnan (1992) analyzed the nutrient budget of soils exposed to low-intensity and high-intensity burn. He found that heat has an important effect in releasing nutrients from nonexchangeable forms to become available to plants. He encountered the same massive increase in available phosphorus as did Andriesse and Schelhaas (1987b) but in this case throughout a soil column of 40 cm depth. It built up during the 30 days after the burn (Ramakrishnan 1992).

In savanna, the principal value of fire is in clearing the grass. There is much less nutrient addition, and the effect on pH is different. Tillage, to uproot the grass and aerate the soil, becomes essential and has particular value where organic matter is dug into the soil (Nye and Greenland 1960). Where fire is important for nutrient enrichment in areas without forest, sometimes it is used in burning the soil itself, together with its organic matter. This generates a high temperature, improves pH and available potassium, releases phosphorus, and increases nitrogen availability by reducing the carbon/nitrogen ratio (Roder, Calvert, and Dorji 1992). The practice was used in premodern Europe and in many other areas (Kerridge 1967; Steensberg 1993). It was used extensively to convert poor pasture into arable land when prices were high in the seventeenth and eighteenth centuries, but its longer-term effects were not satisfactory (Kerridge 1967).

CITEMENE: THE 1930S AND AFTERWARDS

Citemene uses branches from trees cut over an area five to ten times larger than the plot to be cultivated. It uses both the nutrients thus collected and concentrated and the heat generated by burning them to create a transient fertility sufficient to support several years of cropping. The term *citemene* derives from the Bemba term for a cut area, meaning the whole area within which the burned garden forms a small part (Richards 1939). Although other terms are used, including variants also based on the word *citeme* (to cut), the widely understood term *citemene* is used here singly and without qualification. In the literature it is sometimes *citimene* or *chitimene.* The system has a large literature, most of it specific to the Bemba, although other people practice similar systems across northeastern and northwestern Zambia and in southern Congo. As first described in the 1930s, it had several forms, among which the principal was and still is the large-circle *citemene.* The briefer modern accounts, especially those by Strømgaard and by Moore and Vaughan, indicate that the basic system had not significantly changed half a century later.

Tree cutting was carried out during the dry season. In a collective activity, young men climbed the trees to the highest branches and cut most of the branches with axes. They left the pollarded trunk and principal stems to regenerate. Men and women, mainly the latter, then collected and carried the branches to a central circle or oval about one-sixth to one-tenth the size of the whole cut area. Here women closely stacked the branches and their leaves to a depth of more than half a meter. In the 1930s,

Piling the branches is the hardest work a Bemba woman does. . . . The women lift boughs, often fifteen or twenty feet long, stagger as they balance the length on their heads, adjust themselves to the weight, and start off with the load. All morning they come and go in this way. There is a good deal of skill re-

> quired in the stacking of branches. They must be laid with their stems to the
> centre, so that they radiate out in a roughly circular or oval pile. The branches
> must be planted evenly one on top of the other until they reach a height of
> about two feet. If any portion of the garden is only lightly covered with boughs,
> the soil beneath it will be only partly burnt. (Richards 1939:293)

The operation continued over several weeks during the dry season, and the
stacked material then dried during the hottest period of the year. Just before the
rains, and usually at dusk to minimize spreading of wildfire, these stacked circles
were burned, generating high temperatures that reduced all the stacked material to
ash and heated the surface soil. Working in three Asian areas, Andriesse and Schelhaas (1987a) found that the concentration of nutrients is much higher in the
branches and leaves of trees than in the trunk. *Citemene* use of branches seems to
have made a lot of sense (see also box 7.1). Moore and Vaughan (1994) indicated
that the practice was the same at the end of the 1980s.

For a few years, the concentrated burning effectively transformed the soil
through the combined effect of both ash and heat, sharply raising the pH. In two
examples measured in the 1980s, an estimated 5000 kg of dry material was stacked
for burning on 0.1 ha and 29,400 kg on 0.47 ha (50 and 63 tons/ha, respectively).
The intensity of burning such a biomass raised soil temperature at a depth of 5 cm
to 150–250° C. This led to an increase rather than a decrease in organic matter as
well as a sharp rise in pH (Strømgaard 1992). The initial per-hectare contribution
to the upper 50 cm of soil was equivalent to inputs of 300–450 kg each of phosphorus, potassium, and calcium. The potassium was leached rapidly, but the available phosphorus, which reached a maximum only some weeks after burning, persisted at greater-than-ambient levels for several years. These beneficial results of a
high-temperature burn are in contrast with a finding by Andriesse and Schelhaas in
Sri Lanka (1987b). Piling up of the vegetation for a second, very hot burn led to
gross heterogeneity in soil fertility on the plot and, on the patches, to such high
pH—over 10 in three samples—as to create strongly alkaline conditions that were
inimical to crop growth.

Strømgaard (1992) tracked soil nutrient status on four plots at intervals from
before burning in 1981 until 1990. Two of the plots were normal *citemene* plots,
cultivated for 4 years; the other two were experimental, not cultivated. Nitrogen
was depleted rapidly. pH declined during the cultivation cycle but less rapidly in
the uncultivated experimental plots. Organic carbon increased after burning.
Available phosphorus increased dramatically at all depths to levels 10 to more than
150 times preburn levels, reaching a maximum 40 days after burning. Where the
soil was mounded, available phosphorus remained high and even increased again 4,
5, and 9 years after burning. One plot, burned 16 years before 1981, still had high-
er-than-ambient concentration of phosphorus in depth. The continuation of beneficial effects was the most remarkable consequence of this version of the fire-using
system.

The main first-year crop, and the crop most specific to *citemene* among the Bemba, has throughout the record been eleusine or finger millet, broadcast and covered with lightly hoed ash and soil. Before the millet was sown, at least in the 1930s, gourds, pumpkins, cucumbers, and manioc were planted around the edge of the burned field. Sorghum and sometimes cowpeas and maize also were interplanted so that mixed cropping created a field harvested at different times, quickly becoming covered and thus limiting the impact of rain-splash erosion. Close to streams, a bushy fish poison also was planted. Because the heat of the burn destroyed the seed bank in the soil, no weeding was needed in the first year. Gardens were formerly fenced against wildlife, especially wild pig, but this labor-intensive activity was given up after hunting reduced the problem of field invasion by wild animals. Only a minority of fields was still fenced in the 1980s.

In the 1930s the field commonly could be used for 4 years; 5 years was not unusual, and sequences of up to 6 years, and even 10 years on patches of better soil, were recorded (Richards 1939). A common sequence after millet was groundnuts in the second year, millet in the third, and beans in the fourth. This was not a universal pattern, and the system has adapted to increasing use of manioc and maize through time. A range of subsidiary crops was interplanted and, although there were no rotations as such, the successions were designed to optimize use of the declining bank of fertility and help sustain it longer by the use of legumes such as groundnuts and beans. The garden continued to produce through a large part of each year.

Generalizing largely from the writings of Richards and of Trapnell (1943), Moore and Vaughan (1994:39) wrote,

> There were (and are) a number of other possible sequences, including allowing the perennial sorghum planted with the millet in the first year to spring up again with the groundnuts in the second year and then to take possession of the garden in the third year. In some instances, beans intercropped with other crops were grown in the garden on mounds (*mputa*) in subsequent years. It was unusual to find two years of successive millet crops, but this situation did occur if the soil was good enough or if the first crop had failed. In such cases, the garden was then planted to beans or sorghum on mounds in the third year. There were also cases in which millet was followed by the annual sorghum or kaffir corn (*amasaka*) or by further cassava [manioc] plantings.

In his monumental effort to give order to complexity, Trapnell (1943) distinguished three large-circle systems, as well as a small-circle system and contact systems in which mixed practices were encountered. Distinctive field types also were classified.

The modern work led to an evaluation of *citemene* different from that of the severe critics who first wrote of the system in the 1930s and 1940s. A particular and continuing value of *citemene* as a management strategy was now seen as its flexibility and ability to incorporate different cropping systems and sequences, providing each household with a range of gardens at different stages (Moore and Vaughan 1994). Rather than declining soil fertility alone, the ultimate limiting factor on the length of the cultivation cycle has more often been lack of labor, especially female labor, to work four, five, or more fields simultaneously. The productivity of the system also was evaluated in new ways. Even though it was grown always in mixed stands, measured first-year *citemene* millet yields were comparable with those obtained using industrial fertilizer on experimental farms in the region or elsewhere in Africa. Yields of crops grown in second and subsequent years were below this standard, and variable, but respectable by comparison with data from elsewhere on the continent (Strømgaard 1985b).

The pendulum of opinion may have swung too far. These yield comparisons are impressive, and many commentators have been impressed. A more appropriate basis of comparison would reduce the *citemene* yields to about one-sixth to one-tenth, on average perhaps one-seventh, to account for the contribution of the biomass of the much larger area cut to supply the burn. McGrath (1987) has pointed out the error in energy-based comparisons between shifting cultivation and permanent systems that fail to take account of the energy store contained in the biomass, incorporated into the shifting cultivation field. Taking account of this stored energy, shifting cultivation is energy-intensive, with an energy efficiency far lower than that of industrial agriculture. This would be especially so in the case of *citemene*. The early observers regarded *citemene* as a system of low efficiency, and on this proper basis of comparison they were entirely right, its remarkable short-term productivity notwithstanding.

FUNDIKILA AND IBALA

Mounding with hoes was an element of *citemene* in all years after the first. It was the basic method of land preparation in other systems practiced by the plateau people. Most ubiquitous were the village gardens, found as part of all plateau farming systems (*ibala* in Bemba). Richards (1939) did not use the term *ibala,* common in later literature, and wrote only of the making of mounds *(mputa)* on the land around villages. In her time, most Bemba relied on millet and grew little manioc, whereas their western and northern neighbors already relied principally on hoe-cultivated manioc, not using *citemene.*

As described in the 1930s, the village gardens were worked outward from the circle of huts, with additions made each year. Many were simply mounded *(mputa),* although some mounds were composted by burying the cut grass under their own turves. Mainly the domain of the women, village gardens have tended to be undervalued in a literature that, with two notable exceptions, has given limited visibility to the role of women in the farming system. The principal exceptions are the works of Richards and Moore and Vaughan, all women. The latter discuss in detail changing genderization of farming practices through time.

Even in the 1930s, some village gardens extended outward, without much visible distinction, into fields that have become increasingly important and extensive since that time, especially for maize cultivation. Such fields have always been mounded or ridged by hoe. In terms of management, they linked the mounded village gardens to a grassland management system described in recent years as *fundikila,* which seems to have been elaborated particularly by the Aisa Mambwe in a now little-treed area north of the Bemba region. Trapnell (1943) described this system well, but it was analyzed and related to *citemene* only in the 1980s by Strømgaard (1988a).

After the taller grass was burned in heaps at the end of the dry season, mounds were hoed up over the remaining grass and ash, slice by slice, into mounds so that the grass mat was first buried under its own turves, then by a deeper-cut second turf layer. The mounds therefore incorporated a compost of the grass. As described in the 1980s, the first planting was beans with manioc, with the legumes leading to chemical improvement in the soil. In a second year the mounds were opened and spread, and the flattened land was planted to millet, maize, and manioc. In a third year it might again be planted flat with groundnuts or be mounded anew for maize and manioc, then again beans even into a fifth year, when flat land could still produce millet. At the end of the sequence, calcium and phosphorus in the top 10 cm still were well above ambient levels, and cultivation could continue until weeds, particularly a wild relative of finger millet, *Eleusine indica,* became too invasive. This weed is indistinguishable from millet in the early stages of growth, so it cannot be cleared adequately. Sugiyama (1992) described a somewhat different use of the system in southern Bemba. *Fundikila* is not a permanent land use system, but it enables land to be used for longer than does *citemene.*

Already in the 1930s, *fundikila* supported many more people on a comparable area of land than did *citemene.* Allan (1965) found a regional density of almost 15/km^2 in 1938 and more than 20/km^2 in 1960, more than four times the densities of *citemene* areas. *Fundikila* was not new in the 1930s and probably has been used for at least two centuries. It seems to belong to a group of grassland management systems also common in southern Tanzania, of which the best known is the pit system of the more densely peopled Matengo uplands. Allan (1965) described this system in the 1960s, and M. Stocking (personal communication, 1996) confirmed that it retained the same characteristics in the 1990s. Grass was cut and laid in a grid across the hillside, then women hoed the squares thus formed down to the subsoil and threw the topsoil over the grass, all of which was incorporated into the soil, without burning. Half the land was sown with maize and half with pulses, and in a second planting the cropping was reversed. In the next year new pits were formed where the old banks intersected, and new banks thus were made over the old pits. Although also an effective method of water and erosion control, the Matengo system had the same element of grass-composting as *fundikila,* and it has seemed to be very durable.

In northeastern Zambia, semipermanent cultivation already took place along floodplains in the 1930s. The most common crop was maize, at some risk from both flood and drought. Seasonally wet colluvial soils along the shallow valleys (*dambos*) were cultivated only where not peaty, but mounded maize gardens already were made on some *dambos*. The most labor-intensive use of soils in the 1930s was the tilling and pulverization of the soil in small plots close to villages for cultivating the indigenous Coleus or Livingstone potato *(Coleus esculentus)*. Trapnell and Clothier (1937:23) interestingly speculated that domestication of this crop from regional wild relatives possibly generated "the original form of ridge cultivation extensively employed for root crops in sandy dambos and plains" of western Zambia. A wide range of practices were adapted to site conditions, most of them detailed in the initial ecological surveys of northwestern and northeastern Zambia (Trapnell and Clothier 1937; Trapnell 1943).

■ Farming Systems Across Space and Through Time

EYES FRONT, NOT BACK

Trapnell, Richards, and their contemporaries in the 1930s and 1940s were worried about the future, but they gave only limited thought to change through earlier time. Like other scholars throughout Africa at that time, they identified farming systems with particular tribal groups. Although they recognized that farmers could change their practices under pressure and that one group might then copy another, they wrote of the systems mainly as though they were timeless and traditional. Only one study during this early period provided some specific detail on change. This was Peters's (1950) account of the small-circle *citemene* of the Lala people south of the Bemba, on the Serenje plateau. This system used an impoverished forest, and trees were not pollarded but cut down to waist height for all their timber. Although very large areas were cut, the burned patches were small. They were cropped no more than 2 years and sometimes only 1 year. Peters described most *citemene* here as being cut for the second or third time but at an average interval only half the full regeneration time of the trees. He also provided some quantitative data on change, summarized in box 7.2. Small-circle *citemene* was seen to have only a limited future by its observers (Peters 1950; Allan 1965). The whole area was subsequently transformed, but in ways that made it more productive (Moore and Vaughan 1994).

Box 7.2

Quantitative Evidence on Change Through Time in Bemba Agriculture

Quantitative data with which to underpin a review of change in use of different management practices are scarce. Necessarily setting aside the site-specialized practices, three main elements are widespread in the plateau systems: the *citemene*, the village gardens, and the larger *ibala* and related gardens that extend the village garden ridging, mounding, and green manuring—more correctly, composting—of *fundikila* over wider areas.

For the small-circle *citemene* system on the Serenje plateau, in 1946, sample data collected by Peters (1950) yielded average burned garden areas in four small groups, distinguished by approximate age of the trees cut, as 0.06, 0.07, 0.11, and 0.10 ha per head of garden family. From 63 to 71 percent of this area was under eleusine millet. Separate samples gave 0.17 and 0.11 ha per head under subsidiary gardens, suggesting that the combined area of the latter was already at least as large as that under the declining small-circle *citemene*.

Peters also recounted that air photographs of 1930 show much larger clearings than existed in 1946, with all burned-field patches within them fenced. The rapid decrease in size was attributed to village splitting, related to reduction in the extent of large blocks of well-regenerated *miombo*, which may or may not be the correct explanation (Peters 1950). This is a vast change: in 1930 more than half of 82 cleared areas exceeded 42.5 ha, and only 16 years later, the average size of cleared area was less than 7.0 ha. Unfortunately, the implications are not fully discussed. Because Peters's monograph is still one of only six internationally published studies of *citemene* based on field survey, there is nothing else in the available literature of the period that would indicate whether this sort of change was more widespread (Peters 1950).

In 1981, in a community on the northern plateau that was selected as a base for the study of *citemene*, Peter Strømgaard surveyed the land of a single garden family. The family used large-circle *citemene*, *ibala*, and *fundikila*. It had 1.43 ha under *citemene*, of which 0.53 ha was new that year (involving cutting of an additional 2.80 ha), and 0.34 ha under the village garden systems (Strømgaard 1984). This village was 30 km from the nearest town.

Measurements made in 1986 at three villages close to the northern plateau regional capital yielded only 10 percent of the farmed area under *citemene*. *Ibala* under manioc and *fundikila* each occupied 35 percent, 16 percent was cultivated by other methods, and 4 percent was cultivated by tractor (Moore and Vaughan 1994, citing a locally published study by Holden).

CITEMENE IN CRISIS?

Almost all writers from the 1930s to the 1980s saw *citemene* as being in crisis. To Richards, the crisis was caused by the withdrawal of male labor to work in the mines; remote Zambia supplied labor to the mines of Katanga (Congo), Zimbabwe, and, already, South Africa. Zambia's own mining industry began seriously only in the late 1930s. Richards hypothesized that the strictly male job of tree climbing to make *citemene* would perish because of labor shortage, believing the shortage of agricultural labor already to be causing incipient malnutrition. In fact, as Moore and

Vaughan (1994) showed, women alone could manage the second and subsequent-year *citemene* gardens *(icifwani)* without male labor. Not until the 1980s was it discovered through quantitative measurement that new *citemene* were not made every year but commonly only in every second year. Because this finding came so late, it cannot be said that this was already the practice in the 1930s (Strømgaard 1984).

To ecologists, the problem was the rapid diminution of good-quality *miombo* forest. When use of aircraft and air photographs began seriously at the end of the 1920s, it was soon seen that most *miombo* forest bore the marks of former use, looking pockmarked by old *citemene* circles. In some areas the forest had gone beyond this stage and was being replaced by grassland. Strømgaard (1984:83) used air photography of 1966 and 1974 to compare areas under *citemene* at these two dates, for a sample area, and noted that "it appears that some areas, originally cultivated, and in 1966 in various stages of regrowth, had returned to forest in 1974." This suggests that the individual pockmarks seen in the 1930s and 1940s probably did not endure long.

Ecologists and government officers supposed that growing pressure on the landscape was arising under an unchanging traditional system being used by more and more people as population grew. Developing a simple formula that later enjoyed widespread use, Allan (1949, 1965) calculated population carrying capacities under the different forms of *citemene*. He concluded that the population densities critical for sustainable use of the system had already been exceeded in the 1940s. The notion of carrying capacity was borrowed from animal ecology and rangeland management. It was assumed that when the capacity was exceeded, degradation would be the result. The concept has been overturned even for rangeland management (Scoones 1992; Turner 1998). It can have no validity in the context of modern nonequilibrium ecological interpretation. For humans, it always disregarded the potential for changes in land management and other strategies.

Yet numbers and population densities have continued to increase. Though still low by African standards, in northeast Zambia they are at least three times what they were in the 1940s, and there has been no ecological or social disaster (Moore and Vaughan 1994).

The *miombo* forest needs a long undisturbed period for full regeneration. Peters's (1950) estimate that full recovery would take at least 35 years has gained support from modern research (Lawton 1978; Strømgaard 1988b). Some areas have indeed become short of good *miombo* for making *citemene* clearings over the past 50 years. The big error of the colonial period, and beyond it, was to believe that people would not change their cultivation practices or their cropping systems without compulsion or at least very heavy advice, which they received constantly.

This is strange, and it shows how little it was then realized that African farmers can quickly adapt and innovate. Trapnell (1943:39) remarked that the system now called *fundikila* made "maximum use of available organic matter [and] may, in effect, be regarded as belonging to a different realm of agriculture" from *citemene*. He also commented on the major changes that must have taken place to incorporate

American crops during the preceding three centuries (Trapnell and Clothier 1937). Roberts (1973:142), a historian, later concluded that spread of mainly American crops, slow before the late eighteenth century, was greatly accelerated by growth of long-distance trade for ivory and, later, slaves. By stimulating the adoption of new crops, especially maize, manioc, sweet potatoes, groundnuts, sugar cane, and, later, rice, trade "made possible a kind of agricultural revolution."

Although Audrey Richards never gave up her view that Bemba agriculture was extremely backward, she introduced evidence that plateau farmers could cheerfully and very rapidly make changes by themselves. After describing adoption of near-permanent cultivation with heavy use of tillage, groundnuts, and a short fallow by the Bisa neighbors of Bemba on a confined island within Lake Bangweulu, she went on to write,

> I do not know whether such an adaptation would occur among the Bemba with their far greater reliance on millet for their diet and their traditional interest in the crop, but I saw a case of conscious adoption of the Bisa practice by a Bemba village near their border on a treeless plain. . . . Here, the headman told me, the villagers had agreed by common consent to give up their ifitemene for the first time in their lives "because we have used up all the trees and we said 'Well! then let us turn into Bisa now,'" i.e. live on cassava [manioc]. (Richards 1939:328)

MODERN CHANGES

In recent years, both the small-circle *citemene* and the compost mound and crop rotation system of the Mambwe have been interpreted as two among several adaptations, under different conditions of population density and *miombo* shortage (Strømgaard 1989). During the 60 years between the 1930s and the 1990s many changes occurred in farming systems, of which two became particularly widespread. Manioc, still uncommon among Bemba in the 1930s, became common in the years after World War II. During those years, the last generations of colonial administrators were very concerned to avoid what they saw as impending famine. They put heavy pressure, including compulsion, onto plateau farmers to grow manioc. But it would have spread in any case. The people of one almost deforested area were seen as facing nutritional and ecological crisis in the 1940s. Only a few years later they had given up small-circle *citemene* in favor of building large mounds and ridges for manioc; there had also been extensive new planting of fruit trees. The horrified alarm of one generation of colonial observers became high praise of an advanced farming system by visitors of the next generation, without the change being noted for the spontaneous response that it was. This is one of several examples of successful change, reconstructed from 40 years of good records by Moore and Vaughan (1994:88), to supplement observation and information; as they remarked, "There was, it appeared, life after citemene."

Other widespread changes were more directly the result of government policy. Bans on *citemene* were imposed after independence in 1966, with some effect in areas close to the roads. In the 1980s, hybrid maize was introduced as a major cash crop to provide employment for men who, especially during economic downturn in the 1970s, could no longer be absorbed in the mines and towns. The main maize variety initially used was developed from American hybrids in 1950–54 for use mainly by settler farmers in Zimbabwe and those holding land along the railway line and supplying the mining towns of the Copperbelt in Zambia. After Zambian independence in 1960, most of the settlers departed, and the double-hybrid variety was distributed to African farmers (Pearse 1980).

It also became important in the diet. Except in the limited areas where plowing was adopted, hybrid maize was grown on land mounded with hoes. The effect of use of this system for cash cropping was that male labor and crop ownership became more important, giving men a stake in farming beyond initial tree cutting and mound preparation. Maize fields (*faamu*) were prepared in several ways, which included simplifications of both *citemene* and composted *fundikila*. Often, the sites of old *citemene* fields were used. Maize was adopted as an intendedly permanent crop, without rotations but with inputs of commercial fertilizer by those who could afford it. A decline in yields soon was reported (Moore and Vaughan 1994), and Sugiyama (1992) found that *faamu* seldom yielded for longer than 10 years. The enduring change was that maize became a male crop, at all stages, whereas there was no such distinction before. A consequence was greater involvement of women in the remaining *citemene* cultivation, seeking to protect the range and variety of foods and other plants they needed for household use.

MULTIPLE AGRICULTURAL STRATEGIES[1]

The modern dynamism is impressive but no more so than what must have taken place spontaneously in the remoter past. The early ecological and anthropological work took note only of change within the colonial period. In Bemba, a typical village in the 1930s consisted of 30–50 houses; some of its *citemene* often were at a distance of several kilometers, and many people also maintained temporary houses, or *mitanda*. This latter practice was of much concern to the administration, the missionaries and the chiefs alike, who wanted people to live more accessibly. Periodically it was prohibited. Every 5–10 years, and from 1920 to 1940 only rarely at longer intervals, the whole village was relocated. There were many reasons, often unrelated to exhaustion of trees or the soil. At the time of which Audrey Richards wrote, they divided mainly for social reasons. Bemba villages were essentially locality clusterings of interconnected matrilocal family groups, under chieftainship, with a lot of mobility. The death of a chief required regrouping around a successor, and this might take several years. In recent years, villages have relocated less often than before. Most villages moved only within a limited distance, but except in the much more static modern roadside communities there was emphasis on proximity to good resources of mature *miombo* in selecting a site.

Yet only some 40 years before the time of Richards's fieldwork, in a period of warfare before colonial control was established at the very end of the 1890s, many Bemba lived in large, stockaded and fortified villages, with up to several hundred houses. *Citemene,* as a system for widespread use, must have been hard to maintain during a period when so many people were gathered into large settlements. Food shortages, triggered by natural disasters in the 1890s, were among the reasons why the Bemba swiftly dispersed into smaller villages to escape from the control of their dominant chiefs once warfare ceased (Roberts 1973, 1976). It also seems likely they wanted to escape constant tillage, and make full use of the *miombo.* Richards (1939:304) made much of the drudgery of tillage, considered by Bemba of the 1930s to be "hard and unromantic work," unlike *citemene.* Bemba continued to resist new efforts to control and concentrate their places of residence until these unavailing attempts were given up quietly in the 1940s. The agricultural patterns of the plateau, carefully described in the 1930s, therefore belonged essentially to the colonial period and had already undergone significant change within that period. Before the colonial period ended, they were changing further.

CHANGE, HISTORY, AND SUSTAINABILITY

The limited quantitative data on the changing farming systems of northeastern Zambia are brought together in box 7.2. Beyond these crumbs we have to use impressionistic accounts, made easier by the fact that the principal writers, from Richards and Trapnell onward, traveled widely in the course of their research. It seems abundantly clear that by the 1980s, *citemene* had declined and even vanished from some areas, whereas *ibala* and *fundikila* had expanded greatly. Unfortunately, the early descriptions gave little detail on *fundikila* except where it was dominant, and modern descriptions have not provided adequate detail to draw clear distinctions between *fundikila* and *ibala,* leaving only an implication that the latter excluded systematic composting of the grass. The probability is that both these systems were at all times much more important than use of the term *subsidiary gardens* suggests. Their expansion was associated with the spread of manioc and maize and with a greater fixity in village location. Moore and Vaughan (1994:43–44) wrote,

> The spread of green-manuring (fundikila) is further evidence of a move toward permanent or semipermanent cultivation. Described by Trapnell and others as a feature of the Northern Grassland (or Mambwe) system, this has been a major cultivation practice in some areas for well over 100 years, and probably for much longer.

The relegation of non-*citemene* systems to an apparently subsidiary status in the third decade of the twentieth century can be explained readily. There was a lot more *citemene* in the 1930s than in more recent years, and probably more than in the late nineteenth century as well. The distribution of the two main systems probably was the product of the dispersal and mobility of settlement that had been made possible by colonial peace, coupled with availability of a fairly rich *miombo* resource that had grown up in the precolonial decades. On the origin of the systems there is only speculation, but their flexibility, interchangeability, and capacity to yield a mixed range of crops for several years on poor soils are the marks of systems that must have arisen through experimentation over a fairly long time. It should be recalled that agriculture entered southern Africa only with Iron Age farmers, not earlier than 2000 BP (Marshall 1998). Everything has happened within that time span, or only a part of it.

Citemene and *fundikila,* the latter more lastingly than the former but with much greater labor, both created improved soil conditions for periods of a few to many years. They were ways of modifying a very poor soil to ensure crops, not of creating an enduring agricultural environment. It is too much to claim, as Schultz (1976:13) did, that *citemene* and *fundikila/ibala* together "are adaptations which make these systems relatively independent of soil fertility." Variations in soil fertility at microscale have been vitally important in determining the productivity of village gardens, where farmers were bound by customary allocation practices to the area backing onto their own houses, irrespective of quality. On *citemene,* differences in soil fertility showed up in the later years of a sequence. Variations in soil fertility, moisture, and structure determine where farming can endure and where it cannot. From the 1930s it has been clear that these variations were well understood by the users of the land (Richards 1939). Basically, the limitations of generally poor soils, and a slowly regenerating *miombo* woodland, controlled what the plateau farmers could do within their own resources. The options they developed were very ingenious, very resilient, and remarkably flexible but had only finite possibilities.

■ Some Concluding Remarks About Work on Shifting Cultivation

It is not often in the shifting cultivation literature that we can find comparative information across space and through time of the amount available for *citemene* and *fundikila* in northeastern Zambia. It is even less often that good scientific information comes along with good ethnography and good history. This chapter makes rather clear how much can be added to understanding by such a conjuncture. The literature on shifting cultivators as a whole suffers from lack of long-term information. Although some writers continue their observations over several years, and even a decade or more, most students of shifting cultivation are forced by resources to rely on only 1 to 2 years' observations, hardly more than a snapshot. History is sketchy and, although many writers without specialized training in the relevant disciplines do make a serious effort to understand environmental dynamics and present them, the results are mixed.

Work of this type has built up a great store of information in the past, but it seldom says much about the dynamics of change and rarely adds findings of more than local interest. The rich and dynamic history of agricultural systems is not often captured, and an impression may be gained that systems either are in equilibrium or have departed from some previous equilibrium. Yet agricultural systems are never in equilibrium; they are constantly in dynamic adaptation to harsh and unpredictable natural and social changes, to variations and new knowledge and opportunities (Niemeijer 1996). Especially in view of the major changes that have taken place in and around most shifting cultivation areas and are accelerating in many of them, the time for snapshot ethnographic studies of shifting cultivators probably has run its course. A new and more diachronic approach, using all available means to capture the patterns of change, is needed.

Note

1. The title of this section is drawn from Moore and Vaughan (1994:89).

Chapter 8

Managing Plants in the Fallow and the Forest

■ Introducing the Management of Plants

Up to this point, most of the discussion has been about managing the soil, and plant management has taken second place. How farmers manage the system after field crops have been harvested has been briefly introduced in chapter 6, principally in relation to some Kenyah farmers in Borneo and to the Hanunóo of the Philippines. Management of plants beyond the field, in the forest and the fallow, is an important part of agrodiversity, one of the four themes introduced in chapter 3 as resource management as a whole.

THREE SUBTHEMES

This chapter draws on a wide range of material. Unlike chapters 6 and 7, it focuses on no one region, although material from Latin America receives most emphasis. Three subthemes must be explored. First, successional vegetation is an important source of food, medicines, wood, and a range of products used both in subsistence and for commerce. It shelters wildlife that is important to the people for meat and sometimes skins and plumes. Often there is active management of the successional growth to ensure that desired species can flourish in the competitive environment of plant recolonization. Second, cultivated forests are created in some parts of the world. They range from forests in which only a few key species are actually planted, leaving everything else to natural process, to highly complex agroforests in which everything is created by the farmer. Often, it is not easy for an observer to know what is natural and what is planted, and errors are common. Third, plant management can be used in conjunction with soil management to modify environments, creating the conditions for variation or improvement in agricultural systems. This differs somewhat from the soil manufacture discussed in chapter 5 in that there is a much more specific role for plants. I begin with the first of these subthemes. Management of the successional fallow is a topic that, until lately, has been pursued mainly by researchers in Latin America rather than in Asia and Africa. Discussion therefore begins with work done in Amazonia and Central America.[1]

■ Managing the Successional Forest in Latin America

HISTORICAL ECOLOGY: AMAZONIA

In Latin America there is a distinctive historical context, arising from the huge mortality among indigenous populations that followed European conquest and settlement. In almost all parts of Latin America, large populations were wiped out by disease, enslavement, and warfare or were greatly reduced in numbers. In some areas of Amazonia population reduction continued through the twentieth century. The consequence is that modern forest occupies areas formerly cultivated or managed more intensively than they are today. This greatly affects forest composition. In Amazonia and its margins there are very wide areas populated at lower density than inland Borneo or even the most sparsely peopled regions of intertropical Africa.

Balée (1989, 1992, 1994) emphasized what he and others call historical ecology. Using data from various estimates and excluding some land types from consideration, he estimated that at least 11.8 percent of Brazilian Amazonian *terra firma* forest is of cultural origin. Large areas are monospecific or dominated by a small number of species that have value to people and are there because people in the past have managed the forest in their favor (Balée 1989). Balée's was almost certainly an underestimate. Denevan (1992) denied that there were any truly virgin forests at all in the American tropics. Drawing on modern work in paleobotany and environmental archeology, Roosevelt (1999:372) argued, for the whole of the Amazon basin, that "salient ecological and biodiversity patterns traditionally interpreted as natural are strongly associated with evidence of prehistoric human interventions."

Although this is part of a new general and worldwide view of nature as being in very large measure the creation of disturbance, both by naturally occurring and human events (Cronon 1996; Zimmerer and Young 1998), it is a particularly significant view in the Amazonian context. The catastrophic decline in numbers emptied land formerly occupied by large farming and foraging populations. The abundance of black earth made by generations of village cooking fires *(terra preta do Indios)* in the Amazonian forest, mentioned as a manufactured soil in chapter 5, is strong supporting evidence (Smith 1980; Moran 1996). In the areas most severely affected by population decline, an agricultural regression has been proposed as group numbers became too small to sustain farming, leading to narrowing of the cropping range (Balée 1992). Some peoples have largely or totally given up cropping, relying on hunted and gathered foods obtained on migratory treks, but the resources on which they mainly rely are those created by former agricultural activities.[2] Through successive and varied forms of occupation over 12,000 years, plant species still present have been redistributed and modified but, in the important absence of domesticated livestock until modern times, have survived even intensive forms of use. What has been extinguished is the mosaic of populations and cultures that modified the landscape, so that "cultural diversity has been considerably more fragile than biological diversity" (Roosevelt 1999:386).

CULTIVATION IN THE CONTEXT OF NATURAL FOREST PROCESSES

This aspect of Latin American research interpretations is not paralleled in accounts from Asia, the Pacific, or Africa, even in areas where there have also been large population reductions in historical time. Another Latin American departure is stress on viewing the cultivation stage not as the center of all activity but as being "in the midst of forest processes that are slowed for a few years so that a crop can be taken" (Alcorn 1989:70). This different view of farming is strong in Janis Alcorn's work in the Huastec area of northeastern Mexico (1981, 1984a, 1984b, 1989, 1990). Here, in the 1970s, patches of managed forest formed elements in a complex that also included shifting fields, commercial crop fields, and house gardens. Plants were managed as individuals when the vegetation was not being cleared en masse for crops. In a multiple-use system in which production spaces shifted through time, species remaining in the same physical location might be tended at one time, neglected at another, and at still another either cleared away or, if of sufficient worth, transplanted. Species of value at one period, such as coffee, may become of less value at another.

Certain of the species described were in the full sense domesticates, but many were spontaneous. Food, timber, fuel, medicine, and cash crops were among a total of more than 300 managed species. The favored sites for managed forest plots were on slopes, where they had the additional functions of preventing erosion and holding water. Although many of them were on land that had at one time been cleared for crops, this was not true of all. This work has had a strong historical dimension, for it formed part of a rising interest in Middle American systems possibly related to those practiced by the prehistoric Maya (e.g., Harrison and Turner 1978; Remmers and de Koeijer 1992).

A less-intensive "tolerant" management of forests for a range of species, but especially the *açaí* palm *(Euterpe oleracea),* of high market value in the nearby city of Belém, was described for the Amazon estuary of Brazil by Anderson (1990) and Hiraoka (1995). The latter discussed the establishment of these forests on land formerly cultivated for field crops. *Açaí* is a spontaneous member of the secondary succession and has been favored by selective weeding around it, by clearing away or ring-barking tall trees (the boughs of which would interfere with the crown of the palm), and by cutting out vines as well as cutting trees for fuel. The resulting forest was not wholly dominated by *açaí*. Many other spontaneous plants were tolerated and a number of other cash, food, and fruit crops were planted, but it had both lower biomass and species diversity than unmanaged forest. Peters (1996) pointed to it as a striking example of sustainable, low-input forest management.

THE KAYAPÓ SYSTEM

The work of Darrell Posey (1983, 1984, 1985a, 1985b, 1991, 1992, 1993; Anderson and Posey 1989; Hecht and Posey 1989) on diversity management by the Kayapó Amerindians, in a forest–savanna transition area in the middle Xingu basin of south central Amazonia, made an important contribution to the understanding of managed forests. It provided almost the strongest statement about skilled and purposive plant management in the whole shifting cultivation literature. Posey's emphasis throughout was on indigenous knowledge itself, but he reported how Kayapó used their knowledge to manipulate ecosystems in remarkable ways. They both exploited and created patchiness to maximize biological diversity. They "not only recognize the richness of 'ecotones,' they create them" (Posey 1985a:156). This intensive management is the more surprising given that Kayapó are among Amerindian people who have been greatly reduced in numbers. In a recent period they have undertaken extensive trekking for months at a time away from their villages, moving between different resource islands. The population density within their reserved area is less than 1/km^2 (Parker 1992). Unfortunately without being specific about the aspects of Kayapó management concerned, Anderson and Posey (1989:172) wrote that "some of the management practices described above are rarely followed nowadays . . . due to profound changes caused by the recent arrival of goldminers and timber companies." The ethnographic present therefore was being used, with the usual obfuscatory result. In reporting this work, it is safest to leave everything in the past tense but in this case without any specific time period to which to refer.

Fields were managed in concentric circles of different crop types and reportedly were both mulched and fertilized with ash obtained from different tree types, each for its own virtues (Hecht and Posey 1989). The pH of some ash is given as over 10, suggesting that it could be harmfully alkaline unless well mixed with more acid soil (chapter 7, box 7.1). Kayapó carefully planted their fields so that crops complemented one another rather than competed. Old fields continued in use for several years, first under root crops and bananas, then under a variety of plants useful to the Kayapó; they then became secondary forest rich in planted and protected trees of value to the people and endured in production for several years. Kayapó also made small fields in gaps in the forest and along trailsides to provide the necessities of life for wandering groups. They repositioned plants and created pockets of improved soil. They also managed forest islands in the savanna, in ways discussed later in this chapter.

■ Managed Successional Fallows in Amazonia and Southeast Asia

EVIDENCE FROM THE PERUVIAN AMAZON

From work done in the 1980s, considerable detail is available on the managed successional fallows of the Peruvian Amazon (Denevan et al. 1984; Padoch et al. 1985; Denevan and Padoch 1987; Denevan and Treacy 1987; Padoch and de Jong 1987; Padoch and de Jong 1995; Alcorn 1989; Unruh 1990; de Jong 1996). More firmly than in previous work, it was established that on the land of tribal Amerindians and mestizos of the riverside, fields were only gradually given up to forest regrowth and rarely were abandoned suddenly. They were specifically managed far into the fallow period by selective weeding and replanting.[3]

Denevan and Treacy (1987) studied the land of one Amerindian shifting cultivation population, the Bora, settled where they now are since about 1930. As described in the mid-1980s, they would cut and burn, sparing valuable tall species. They depended most heavily on manioc but cultivated at least 16 other annual crops. They interspersed pineapples and seeds and seedlings of some 37 trees, mainly fruiting trees, among the crops. Some fields were monocropped with staple foods, others interplanted with a variety of subsidiary crops. Separation of near-monoculture of main crops from polyculture or intercropping of subsidiary crops is described for several Amazonian tribes (Harris 1971; Beckerman 1987; Hecht and Posey 1989).

Somewhat different subsequent histories followed, with successional diversity evolving more rapidly from the interplanted fields. A Bora third-year fallow field contained at least 20 cultigens and had developed multiple canopies, mainly of fruit trees. Management still included periodic weeding, replanting of manioc, and coppicing to sustain unshaded space. By 5 and 6 years, field sites became successional orchard fallow while forest regrowth crept in from the margins. By 8 years, at least 60 spontaneous plants had become established, shading out fruit trees except where the land around them was still weeded. Some of the spontaneous plants had value for their wood. By 9 or 10 years only a small amount of orchard remained among the secondary growth. After 12 years, no management continued and the land became fallow under secondary growth. It still contained fruit, which could be gathered fallen and also attracted game animals that were hunted. Denevan and Treacy (1987:40) concluded,

> The life of a field is one of sequential utilization rather than simply planting–harvest–abandonment–fallow. Harvesting proceeds from grain-producing annuals (rice and maize) to root crops and pineapples to fruit trees and unplanted

useful trees and vines. The concept of "sequential variation" is probably more useful here than "phased abandonment."[4]

There was little marketing from the remote Bora site. Two other communities demonstrate the differences introduced by access to markets. Tamshiyacu, on the Amazon above Iquitos, is a small town, but most of its people were farmers in the 1980s. Almost all the same crops were grown as in the Bora region, but market price determined what was selected for emphasis (Padoch et al. 1985; Padoch and de Jong 1995). After clearance, planting went on for more than a year so that the field became first a mixed fruit orchard, then finally a largely monocrop stand of a tall fruit tree, the heavily yielding *umarí (Poraqueiba sericea),* the fruits of which were in sustained high demand in Iquitos. Fewer trees were encouraged or protected than among the Bora, but clearing was more purposive and was continued for longer to keep the ground clear for collecting falling fruit. The managed stand might last 30 years. Further from Iquitos, at Santa Rosa, management of successional fallows varied much more between individuals, according to their personal goals. Intensification resulted mainly from decline in high-forest resources within easy reach (de Jong 1996). Yet where high-forest resources were still perceived as adequate in 1989, even close to Iquitos, successional management was of a far lower intensity, and as many fruit species were collected from the high forest as from the secondary successional woods (Pinedo-Vasquez, Zarin, and Jipp 1995).

THE ECOLOGY OF MANAGED SUCCESSIONAL GROWTH

Commenting on the ecological aspects of this work, in some of which he participated, Unruh (1990) summarized an extensive set of findings in a way that has wider implications. He paid particular attention to the importance of protecting valuable plants when forest is cleared for agriculture. "Seeds of parent trees which are left uncut in the swidden may have a greater chance of surviving a burn, and would, below the parent tree, have greater access to nutrients for longer periods of time than regrowth plants elsewhere in the swidden" (Unruh 1990:191). Elsewhere in the paper, he noted that a managed canopy continues to be more open than the canopies of unmanaged sites, making possible a much more pronounced understory in which most of the valuable plants live. Constant slashing of vegetation inhibits transfer of phosphorus and nitrogen from the leaf to the woody tissue and instead enriches the ground with these limited nutrients. This is especially significant because the slashing takes place mainly around the useful successional fallow plants. Concentration of phosphorus is particularly important (Unruh 1990).

When the forest is again cleared, valued species are again preserved and the process continues. Thus an altered forest evolves progressively through time. There is good reason to suppose that these practices have great antiquity. Although there is a risk of arguing in a circular manner, it could be suggested plausibly that successional fallow management may be one way in which the monospecific stands, or rich concentrations of a single species within the forest, so widespread in Amazonia, might have come about.

SUCCESSIONAL FOREST MANAGEMENT IN SOUTHEAST ASIA

In southeast Asia the study of fallow and forest management has been less central than in Latin America. The century-old but increasingly crowded out *talun-kebun* system of Java, in which short periods of cropping are separated by long periods under a productive tree crop cover, rests on total planting of the "fallow" (Igarashi 1985; Palte 1989) but has been little studied from this point of view. In one *talun-kebun* area that I saw in 1997, close to Bandung, the "fallow" was almost wholly planted bamboo, itself harvested for sale. The shifting cultivators of Borneo, discussed in chapter 6, certainly obtained an important part of their subsistence and other needs from the successional growth. Drawing on evidence obtained over much of two decades, Colfer, Peluso, and Chin (1997) analyzed forest use in relation to successional stages, from the rice field itself to old secondary forest. Food came mainly from the youngest fallow, and medicinal plants came from established but still young secondary growth. Older secondary forest provided fruit and a range of nontimber forest products. Old forest was the main source of both timber and wildlife. Management was not discussed in detail.

Over most of the region, it is still believed that fallow management is fairly new, a response to very modern pressure on swidden farmers to eliminate their "wasteful" fallow-using systems and adopt "permanent" cultivation in their place. The impression from most of the literature remains that, although products are obtained from the fallow and certain shade-tolerant tree crops and vines are planted in the fallow, the soil restorative agronomic function of the fallow period is its overwhelmingly dominant role. Yet it was of the worked-over forests of southeast Asia that Spencer (1966:39) wrote that "it is possible that old forests are not secondary forests or even tertiary forests, but forests of some number well above three." There is a good deal of evidence to support a view that foraging and farming spaces have migrated across the forest as people have moved from place to place and populations have risen and fallen. Selecting, cutting, planting, and weeding, gatherers and farmers have wrought many changes in species composition. Peters (1997) demonstrated how any sustained use of forest can alter the conditions of plant and animal reproduction so that the harvest of any type of plant tissue necessarily will affect the species involved.

There may have been enrichment in species of value to people. From Kalimantan, Peters (1996) reported that old and mature *Shorea* species, long ago planted for illipe nuts around villages and in successional fallow, form dense groves within what later generations often perceive as undisturbed forest. Durian is found almost throughout Borneo, and everywhere it occurs in concentration, it may represent the site of a former settlement (Peluso 1996). The ecology of fallow management is insufficiently studied in southeast Asia. Although there are fragments of information in much of the literature, only one or two cases have been examined in any detail. Chin (1985) provided an important discussion not confined to the Kenyah of the upper Baram, and more recent studies are those of Colfer and Soedjito (1996). Several of these findings are brought together by Colfer, Peluso, and Chin (1997).

The number of wild and fallow species described as being used by villagers across southeast Asia ranges from a few score to several hundred (Conklin 1957; Kunstadter 1978). Padoch (1982) was early in drawing more specific attention to use of successional fallow by the Iban and other Bornean groups and to the diversity of their agriculture. Later, writing in a semipopular book, Padoch (1988a:23) remarked,

> Sarawak's Iban, like other Borneans and shifting cultivators throughout the tropics, regularly go to their old swiddens to find greens for their soup pots, fruit for a snack, wood for their cookstoves, and materials for constructing their houses. Some of these economically important products are remnants of the crops that were planted, some are spontaneous invaders that came in after active cropping ceased, and still others may be plants that were in the fields before they were cleared and that escaped the burn. Many of those not deliberately planted have probably been helped along by some occasional management: a little weeding is often done around a medicinal plant, or a potentially harmful liana may be cleared from a useful timber tree.

DIVERSIFICATION AND FALLOW IMPROVEMENT

Since the 1970s, agroforestry has become a major topic of interest among agricultural scientists. Although the knowledge has been slow to be translated into their practice as experimenters and advocates, agroforestry soon came to mean more than alley cropping with a small range of introduced nitrogen-fixing species (Nair 1989). In the 1990s there has been a surge of interest in the agroforestry practices of indigenous farmers in the southeast Asian region, especially in Indonesia and among the multilingual "minority" peoples who inhabit large upland regions between India, China, Myanmar, Thailand, Laos, Cambodia, and Vietnam. Peluso and Padoch (1996:133) were writing of more than just Borneo when they remarked that "swidden-fallows and other woodlands are far more explicitly managed than is commonly understood."

Under the new spirit of learning from the farmers, the International Centre for Research in Agroforestry (ICRAF), with other partners, sponsored a pioneer regional workshop on indigenous fallow improvement at Bogor, Indonesia, in 1997. The papers presented described practices ranging from simple forms of fallow management to very complex agroforestry. They were grouped according to whether the improvements had either or both of two main purposes. The first was described by Malcolm Cairns, the organizer, as generating more effective fallows, in which the biological efficiency of the fallow process is improved, and the same or greater benefits can be achieved in a shorter time. Innovations lead to improvement of the soil by nutritional enrichment.

The second was described as creation of more productive fallows in which fallow lengths stay the same, or are made longer, as the farmer adds value by introducing perennial economic species. This means introducing or encouraging what are, in effect, additional crops. In some cases these are already present in the cultivation period and extend into the fallow; in others they partly replace the fallow with a long-term crop (Cairns 1997). Systems developed by farmers, using both indigenous and introduced plants, are quite often being observed by outsiders for the first time, having been totally ignored in the past. Malcolm Cairns (personal communication, 1999) called attention to the use of the wild sunflower *Tithonia diversifolia,* an introduction from tropical America, as green manure in parts of the Philippines and possibly also in Thailand. Fast-growing *T. diversifolia* shades out grasses, scavenges a range of nutrients, and produces copious leaf litter. Information on its use was readily obtained from farmers but not from agricultural experts.

Not surprisingly, in view of the biases of scientific observers, there is more historical information on improvements that have an economic rather than conservationist purpose; money is more visible than soil improvement. Although most of the information still has only limited historical depth, it is certain that the planting of tea both under forest and in secondary successional forest is several hundred years old in upland southern China (Guo Huijun and Padoch 1995). In the same region, a long history has also been established for interplanting a range of cereal, tuberous, cash, medicinal, and oil-bearing crops among Chinese fir *(Cunninghamia lanceolata),* itself of high value for its timber (Menzies 1988). This system was imported by the British downriver from southwest China into Myanmar (Burma) and India in the mid-nineteenth century. It is the origin of the well-known *taungya* system of authoritarian management, permitting cropping among young teak and other valued timbers planted as a monoculture in place of clear-felled forest (Tapp and Menzies 1997).

The use of planted trees for firewood and construction wood is widespread. The fast-growing *Cassia siamea* is commonly planted in fallows and as a hedgerow and coppiced for its prolific supply of wood. Much greater benefit is derived from the Nepalese alder *(Alnus nepalensis),* closely planted among crops in northeastern India and also managed by coppicing. The alder fixes nitrogen, grows rapidly to shade out weeds, has a high rate of litter fall that recycles both nitrogen and phosphorus, and is long lived. Large areas of seeming forest in the Naga region are in fact alder woodland, much of it with crops among the trees (M. Cairns, personal communication, 1999). Use of alder permits management with only 2 years of fallow after 2 years of crops (Ramakrishnan 1992; Guo Huijun and Padoch 1995; Guo Huijun, Xia Yongmei, and Padoch 1997; Cairns, Keitzar, and Yaden 1997). Cash crops can be grown among alder with significantly higher yield than in control plots (Sharma 1997).

Although this is unlike the successional fallow management of Amazonia, it is management of a comparable order. Farmers who use alder, leucaena, legumes, and other fallow improvers are "hitching a ride on a multiplicity of processes observed in nature," to use a happy expression attributed to Paul Richards (unpublished, cited in Fairhead and Leach 1996:207). This means that farmers are harnessing and directing ecological processes that occur naturally rather than attempting to innovate over these or override them. They are using these processes and managing them. Not one of these many innovations was introduced by outsiders; all were developed by the farmers themselves, some of them only recently but others over a long period of time (Brookfield 1997).

They are also using their knowledge and skills to maintain a wider range of income options. In southeast Asia, just as on the floodplain of the lower Amazon in South America, farmers combine agriculture, agroforestry, and natural forest management. Managing a succession of forests, fallows, house gardens, and fields, they not only improve their security but also reduce ecological risk (Pinedo-Vasquez and Rabelo 1999).

■ Complex Multistory Agroforests in Southeast Asia

FOREST GARDENS

Diversion from both arable and forest land use to long-enduring or permanent agroforestry is a particular feature of the southeast Asian region. This second subtheme takes us an important step away from simple successional fallow management. The diverse home gardens and forest gardens of Java are the best known examples (Soemarwoto and Soemarwoto 1984; Wiersum 1982; Karyono 1990), but their intensive use puts them in a class apart. More extensive are the agroforests in areas of lower population density, but up to $200/km^2$, located closer to remaining natural old forest (Michon and Mary 1990; de Foresta and Michon 1993). Though well scattered through the Indonesian archipelago, these are found mainly in Sumatra and Indonesian Borneo (Kalimantan). Mostly these forest gardens succeed dryland cultivation, not only for rice but sometimes for bananas, as in a case in Kalimantan (Smith 1997). Although the main purpose may once have been to ensure a range of foods for subsistence, it is now generally commercial, and success depends on integrating commercial tree crops within a multipurpose forest structure.

Where rubber is planted, it is often with the intention of clearing it again at the end of its productive life. Elsewhere, the outcome of creating a forest garden is a long-lived managed forest composed principally of useful species of which not more than three to five are the commercial crops around which the managed forest is designed. In Sumatra, a common core tree is *Shorea javanica*, planted from nurseries, which is tapped for its resin, *damar. Damar* is extensively tapped from wild trees in the region, but the species is scattered. One common method of extraction involves making a box-shaped hole in the trunk, within which a small fire is made to stimulate the flow of resin (Peters 1997). Although this does not necessarily kill the tree, it does it no good, and shortage of wild sources long ago stimulated domestication and planting.

The managed forest system is old: cinnamon has been grown in such a system for several centuries, coffee since the mid-nineteenth century and nutmeg and cloves within the twentieth century (Michon and Mary 1990). There is a mixture of deliberate planting and toleration of spontaneous individuals, often with limited on-site management, although it has been suggested in Kalimantan that a repeated cluster of obligatory mycotrophs are deliberately planted close to the highly valued durians to create an environment that will ensure their strong growth (Smith et al. 1997).

TEMBAWANG IN BORNEO

Complex agroforests of less commercial intent are found in Borneo, especially in western Kalimantan, where I have seen them in all their diversity in company with Christine Padoch and Charles Peters, who described them (Padoch and Peters 1993). They are found quite widely in a region with no substantial surviving high forest. At least 300 years old and continuing to be made, they are named from the term for an old settlement site, *tembawang*. They originated as fruit gardens surrounding villages, relocated when the trees grew big. They now contain up to 74 different species of fruit trees. One sample transect yielded 44 tree species, of which 30 produce edible fruits or shoots (Padoch and Peters 1993).

Tembawang have been developed and are sustained by a combination of deliberate planting, casual planting, and volunteer growth spared in successive weeding around the durian, langsat, rambutan, mangosteen, rubber, illipe nut, sugar palm, and constructional wood trees, to mention only the principal plants. Several species of each genus, and their wild relatives, are consciously conserved and planted. The length of the investment horizon is indicated by the continued planting of Borneo ironwood *(Eusideroxylon zwageri),* valuable as construction timber, which takes a century to become fully grown. *Tembawang* grow in a landscape that also includes protected but used modified forest. There are also short-term forests of marketable species, principally rubber, which are cyclic in that they both follow and precede clearance for cultivation. Not many fire-cleared fields are now made in this region of Borneo, and their role is increasingly taken over by progressively extended wet rice fields, another form of transformation (Padoch, Harwell, and Susanto 1998). The *tembawang* maintain genetic diversity, conserve soil and water, and meet both domestic and commercial needs; they appear highly sustainable. With the extending wet rice, they support population densities from 50 to 90/km^2 (Peluso and Padoch 1996).

Although the biodiversity of *tembawang* is somewhat less than that of the natural secondary forest, the effect on the soil is comparable. In a well-designed sample study elsewhere in West Kalimantan, Lawrence et al. (1997) found that soil organic matter, cation exchange capacity, total nitrogen and available phosphorus, silt and clay content, and pH were comparable in secondary growth and *tembawang* of equivalent age (at least 15 years old), with calcium, magnesium, and sodium actually higher in the *tembawang*. Under a rubber fallow, by contrast, all elements except potassium were considerably lower. Yet it is the latter type of managed fallow that is expected to be cleared again, after some years, for cultivation.

VILLAGE FORESTS AND HOME GARDENS IN JAVA: A SYSTEM UNDER PRESSURE

The village forests of Java have many similarities with the agroforests of Kalimantan and Sumatra but tend to be more exclusively composed of valuable species. All were derived from original species in the forest they have replaced, both by natural selection and planting (Christanty 1990; Michon and Mary 1990, 1994). In examples studied in the 1980s, some 250 species were represented, growing in a forest-type multilayered structure at mean densities of around 800 individuals per hectare. With overlap at different heights, the tree crown cover equaled up to 200 percent of the plot surface. Management was minimal in these gardens, which were designed "to provide all the products necessary for daily life, except for the staple food" (Michon and Mary 1994:46). A wide variety of medicinal plants also was cultivated.

The village forests were adjacent to the home gardens around houses, which were more open, with more use made of the ground. This was kept clean by slashing, and after the trash was burned it was then dug into the soil, in which a range of shade-tolerant crops was grown. In modern times, as villages have grown, more and more of the village forests have become home gardens. In addition, many have concentrated increasingly on commercial production of fruit, nutmeg, cloves, and coffee. Whereas the village gardens used to have a social function in that they were open to all and the poor were allowed considerable rights of collection, this is ceasing to be the case, and private interest now extends down to the ownership even of individual trees. Recent changes have included a shift to monocropping and threaten diversity and the maintenance of a germplasm bank (Michon and Mary 1994).

■ What Is Natural and What Is Human-Made?

CREATING FOREST IN SAVANNA

The multistory southeast Asian agroforests are remarkable, but their uniqueness probably lies principally in the very high proportion of useful plants contained within them. In Europe, woodlands have been managed, often intensively, for centuries, and "result from long-running interaction between human activities and natural processes" (Rackham 1986:64), and this is also true of most other areas of old and complex agriculture. In the tropics, on the other hand, it has until very lately been the habit among scientists and officials to regard the interaction between people and the vegetation as a one-way process of degradation from a supposedly climax state. Officials, and even forest scientists, have sometimes mistaken created forests for remnants of formerly more extensive natural stands. Yet many examples reviewed earlier in this chapter provide strong evidence of deliberate management. Two examples in the modern literature describe situations in which woodland found to be managed is so similar in composition to wild forest as to permit observers to hold different views about its real origins. One is in Amazonia, one in west Africa. A closer look at both will illustrate this second subtheme.

THE FOREST ISLANDS *(APÊTÊ)* OF THE AMAZONIAN KAYAPÓ

The Kayapó system of plant management in the forest–savanna transition zone of the Brazilian Amazon, discussed earlier, contained one aspect that has achieved very wide notice. As described by Posey (1985a, 1985b; Anderson and Posey 1989), it involved the deliberate creation of forest islands *(apêtê)* in the grassland close to the village where his work was concentrated. Beginning with selection and fertilization of a planting spot, often a former ant mound, these islands were described as growing naturally but were added to and managed to create a diverse artificial ecosystem abundant in plants of value to people and wildlife. Plants were brought into the *apêtê,* coming from a very wide area; they were collected by Kayapó in the course of their treks, some said to be from the land of distant people several hundred kilometers away. Enlarged over time and deliberately modified once mature by clearing of a central patch within them, *apêtê* formed a complex of diverse spaces rich in useful plants and planting spaces. They were private property, one *apêtê* per extended family. This constituted "ecological engineering [which] requires detailed knowledge of soil fertility, micro-climatic variations and species niches, as well as the interrelationships among species that are introduced into these human-made communities" (Posey 1991:4).

The *apêtê* have become famous as models of agroforestry management, even as models for reforestation of land degraded to grassland. Although their deliberate creation may have belonged mainly to a recent period now ended, when warfare called for concealment strategies (Posey 1993), they were consistently described as current in Posey's writing. Yet their very existence as human creations has been questioned. Parker (1992) comparatively inventoried one of Posey's near-village *apêtê* and a forest island well removed from the village and without trace of human intervention. He found the species composition very closely parallel. Checking specimens against herbarium material in Belém he found that "the species found by Posey and Anderson in their study area were precisely *those commonly found in such areas,* the disturbed margin of a savanna and forest transition" (Parker 1992:416, emphasis in original). That is, all are present locally and not necessarily brought from any distance at all. They all occur in naturally regenerating forest islands, which Parker believed the *apêtê* truly to be, without any necessary human aid.

It would be neither fruitful nor appropriate to recapitulate Parker's arguments or imply any judgment on them in this book. Most comments in the subsequent literature are cautious. Balée (1994) merely noted the inapplicability of Posey's model to people he studied further east, although the same plants are present. Moran (1996) called on the need to demonstrate statistical differences between plants growing in "managed" areas and those in the naturally occurring adjacent secondary successional forests. In a region of the world in which manipulation of forest content by farmers is well established, many of Posey's regional colleagues find his construction credible, even if not necessarily in all details. In modern thinking about vegetation change in the forest–savanna boundary zones, transitions from one enduring state to another are coming to be seen as pathways in which a wide range of causes is operative, giving more space for deliberate as well as inadvertent human intervention. In the case of forest extensions into savanna, the most critical conditions seem to lie in the presence of initiating trees, the elimination of fire-prone grasses, and the exclusion of fire (Belsky and Amundson 1992). All three conditions can be produced by human intervention.

FOREST ISLANDS IN WEST AFRICA

In the Kissidougou prefecture of Guinea Republic, west Africa, both the fact of enlarging forest islands and the deliberate and partly inadvertent means by which this is brought about have been established painstakingly by anthropologists James Fairhead and Melissa Leach (1996). The forest islands of Kissidougou are broad belts around the villages that lie at their heart and are many times larger than the villages. Sometimes, they extend to merge with the forest islands of nearby villages. They are large, and I have seen them clearly from a regular commercial flight between Abidjan, Côte d'Ivoire, and Conakry, the capital of Guinea Republic. Fairhead and Leach incontrovertibly established the enlargement of the forest islands in the period since 1950 by comparing air photographs and satellite imagery of different dates. Their existence and extension over a longer period are supported from oral history and travelers' accounts going back into the nineteenth century.

Yet until very lately, the forest islands have been thought of as relics of an older forest continually being reduced by an expanding savanna, mainly through clearance and fire. Their species composition is closely similar to that of remaining gallery forest and forest in the more humid regions to the south. The presence within them of species also found in the savanna has been interpreted as evidence of invasion. In their establishment, initial planting of specific tall-growing shade trees around the village was a necessary step. Exclusion of fire, to protect the village, was achieved by low-heat burning early in the dry season and by tethering cattle in the adjacent savanna, selectively grazing to promote less combustible grasses and distribute forest germplasm. Once a woodland became established, other trees could be planted in its shade, including interlocking thorny species grown for defense in earlier times and fruit trees in modern times. Soil conditions were enhanced, not least by the deposition of refuse and human wastes within the forest island.

After about 50 years, the species composition had become very diverse, characteristically in the three-story pattern of the natural forests that they came to resemble closely. Fairhead and Leach (1996) cited locally published surveys of forest islands; in one of these the species count in all three stories together totaled 188 in an old part of the island and 67 in a newer extension protected from fire since 1958. Species typical of savanna are progressively displaced by forest species over time. Forest islands are closely linked to settlements and to social life. Whereas "outsiders have treated forest islands as an archive of past vegetation, inhabitants, by contrast, find in these forests an archive of past habitation and sociality, as well as a landscape feature necessary for and shaped to meet present-day needs" (Fairhead and Leach 1996:113). As population density has increased, so has the area of forest.

■ Using Plants and Soil in Conjunction

DISPLACING SAVANNA BY CULTIVATION IN WEST AFRICA

I turn finally to the last subtheme, the use of plants and soil management together for environmental modification. Two examples from the recent literature are detailed, one of them from the same area as that of the forest islands just discussed. Almost throughout the west African literature, as well as that of other climatically marginal regions, the creation of savanna at the expense of forest, or savannization, is viewed as consequence of the extension of cultivation. Transformation of the forest–savanna transition zone in southern Ghana after the extensive planting of cocoa from the 1890s onward, and then of food crops since the 1940s, has been widely described in these terms (Hill 1963; Amanor 1994; Gyasi et al. 1995). Even in the view of farmers themselves, the process has been accelerated by the more extensive modern use of hoe cultivation, in place of slashing vegetation to ground level by bush knife (cutlass in west Africa). The hoe breaks up the seed and root bank of forest species in the soil. All the available evidence from southern Ghana suggests that the farmers there are right, and we will follow them into the assessment and management of their degraded landscape in chapter 9. From the experience of Kissidougou farmers, Fairhead and Leach (1996) offered a variant view that, probably without sufficient warrant, they proposed for wider adoption.

As described, the process was as follows. Early in the twentieth century, savanna land grew mainly upland rice, with 2–4 years of cultivation separated by long periods of grass fallow. In modern times, most rice is grown in swamp sites, which have been enlarged and improved. The uplands have become the site mainly of groundnut and manioc cultivation. What is most important is that, unlike the grains, these crops are mounded, with the grass mulch composted in the mounds. Preferred garden sites are old village locations, where deeper and richer soils are found. Mounding practices are extended beyond these limits into savanna, which in time becomes like the old village sites in terms of soil conditions. Infiltration and water retention are improved, and horizons are mixed partly by tillage and partly by enhancing termite activity. Soils become more workable and oily, with an increase in clay content. The savanna grasses are replaced by shorter grasses and, because the land between the mounds is not tilled, tree germplasm and seedlings are not destroyed and benefit from improved soil conditions. In time, a mixture of savanna and forest species is displaced by the latter so that a forest thicket becomes the fallow in place of grass. Aware of these dynamics, Kissidougou farmers consciously manipulate them to enrich the productivity of their land.

These findings are not yet tested by controlled experiment. However, the consequence can be established firmly. The air photographs and images demonstrate an extension of general forest cover, beyond the forest islands, in what was formerly bare or lightly wooded savanna. Insofar as it may be proper to describe savanna as a degraded landscape, the integrated vegetation management practiced by these African farmers is a highly comprehensive reclamation (Fairhead and Leach 1996). Importantly, it operates not through direct intervention in the vegetation, as by making edaphic changes in the soil through cultivation, which facilitate a natural process of reforestation. Plant cover and soil management are thus linked, and each reinforces the other.

USING PLANTS TO HELP RECLAIM LAND IN TIDAL AMAZONIA

Most species of riverine and coastal vegetation trap sediments and build up the height of soils. Wilken (1987) showed how Mexican farmers have made use of this property in reclaiming land, and it has been used widely in reclaiming tidal swamps. Padoch and Pinedo-Vasquez (1999) provided detail on how farmers in the tidal lower Amazon have used two widely occurring palms to raise land above the height of the twice-daily tidal floods. The mouth of the Amazon admits no salt water, but the freshwater flow is tidally raised by about 2.4 m at springs and 1.7 m at neaps on the north shore. Annual riverine floods rise higher than this and bring down large quantities of sediment, but they occupy only a limited part of the year (Anderson, Marques, and Noguiera 1999).

Farmers on the north shore use two trees to raise the height of field and garden plots above the level of high tide in the space of 3–11 years. The burutí palm *(Mauritia flexuosa)* develops hydromorphic roots and grows new ones whenever old ones are covered by sediment. Farmers cover roots with leaves and repeat this when new roots emerge. The *açaí* palm forms clumps and sprouts. Farmers thin and prune the palms about twice each year, then cover the ground around and between palm clumps with the slash to encourage new sprouts. Sediment from the annual high flood therefore is trapped and accumulation of organic matter further raises the soil. Once the ground surface is no longer covered daily by floods, flood-intolerant species colonize the site, and the land can be used for fields, or for extension of house gardens in the case of *açaí*.

This process, described from observation in the 1990s, is possibly old. Digging into the soil revealed alternating organic and sediment layers, suggesting progressive buildup by the methods described. Estuarine farmers also plant on raised platforms and on rafts. Although nothing is known for sure, these may have been among the methods used by the densely settled farmers in the delta region in preconquest times (chapter 4).

■ Conclusion

In modern science, fields and crops are the domain of agricultural science, and trees are the domain of forestry. Only the very modern growth of interest in agroforestry breaks down this dichotomy. Farmers do not draw a distinction between the trees and the crops, and they treat the whole farming space as one. The rather late date of scientific discovery of integrated management practices, as discussed in this chapter, owes much to the slow breakdown of the categories of modern science when applied to farming systems. The integration of field and wood is central to agrodiversity. It is also central to proper soil fertility management and erosion prevention. There is much more to the land rotation farm than the simple crop–fallow dichotomy that shifting cultivation is supposed to be. This chapter's entry into the management of a more complex set of dynamics therefore takes us forward into the final chapter of part II.

Notes

1. There are two important early studies of successional fallow management in the Asia–Pacific region. These are the early work of Conklin (1957) on the Hanunóo and a distinctive study by Clarke (1971) on the fringes of the New Guinea highlands. Before this, well ahead of its time, was an important general survey of wild plants in relation to agriculture in Papua New Guinea by Treide (1967).

2. The importance of inherited resources emerges very strikingly in the case of the babassu palm *(Orbignya oleifera)*, the seeds of which can survive fire to form pure stands over large areas, creating a major economic resource (Hecht, Anderson, and May 1988; Anderson, May, and Balick 1991).

3. Denevan and his colleagues wrote of "swidden–fallow agroforestry" or "successional agroforestry," and the first of these terms in particular has come into common use. Because I have decided not to use the word *swidden* (introduction), I use their second term.

4. The project of which this study formed a central part was described in detail by Denevan and Padoch (1987). Working in a field with no previous examples on which to draw and having only a limited time for fieldwork, the multidisciplinary project unfortunately was not able to provide data from control plots, they noted with regret.

Chapter 9

Coping with Problems:
Degraded Land, Slope Dynamics, and Flood

■ Degraded Land

SOIL EROSION: HOW BAD A PROBLEM IS IT?

Chapter 8 ended on a very upbeat note, and it would be good to close the discussion of land rotation cultivation systems with the ingenuity and high quality of farmers' management improving the state of their resources. Unfortunately, to do so would be to disregard an enormous modern literature describing the severity of land degradation, especially soil erosion. Much of that literature identifies shifting cultivation practices and related traditional methods as major underlying causes. Population and livestock pressures are seen as the principal modern triggers. How can this literature be reconciled with the evidence presented in this book, and by writers cited in it, on the extensive development of effective practices for physical soil management and biological management of the land cover? If the situation is as bad as described in west Africa, for example, by scientists such as Lal (1993, 1995), how has an even moderately productive agriculture endured this long? These issues must be addressed before we can move on, and this final chapter of part II concerns how farmers cope with a range of regularly occurring problems.[1] I begin with land degradation.

Writing on this topic ought to be easy for me because 14 years ago I wrote my share of a joint book on land degradation and society (Blaikie and Brookfield 1987). Unfortunately, very little of it anticipated a discussion of agrodiversity. Land degradation has since been variously defined by different authors, succinctly by Stocking (1995:225) as the "aggregate diminution of the productive potential of the land." In a wider context, it has recently been described as "any form of deterioration of the natural potential of land that affects ecosystem integrity either in terms of reducing its sustainable ecological productivity or in terms of its native biological richness and maintenance of resilience" (Scientific and Technical Advisory Panel Expert Group 1999).

Within this broad ambit, there is strong emphasis in most of the literature on soil erosion. Oldeman, Hakkeling, and Sombroek (1990) brought together the estimates of a large number of national and regional specialists through an expert consultation. Their figures indicated that 15 percent of the world's land surface was suffering from human-induced land degradation, within which 56 percent was caused by water erosion and 38 percent by wind erosion. Pimentel (1993) wrote more alarmingly that 30–50 percent of the earth's land surface was suffering from degradation.

There is real evidence that matters have become much worse in recent decades as new land has been brought into cultivation, forests cut, roads built, and new cropping practices introduced all over the developing world. Mechanized plowing, accompanying monoculture, has been particularly instrumental in damaging soil structure. Chemicals used as pesticides can affect soil fauna, reducing infiltration capacity. With deep plow-pans created by tillage, the conditions permitting erosion are increased.[2] There is now clear evidence that crop yields are gravely reduced by losses of surface soil, within which is much of the soil organic matter. Without appropriate management, such losses can reduce productivity dramatically within a few years to decades, the time depending greatly on soil and slope conditions (Stocking 1987, 1996; Lal 1995; Tengberg, Stocking, and Da Veiga 1997; Lu and Stocking 1998). To most modern writers, soil loss by erosion is most of the problem. A high proportion of the literature, even in specialized journals such as *Land Degradation and Development,* concerns soil erosion, and many writers continue to use the terms *degradation* and *erosion* almost interchangeably. Regrettably, Blaikie and Brookfield (1987) were no exceptions.

Scientific and official concern with soil erosion is a product of history as well as observation (Anderson 1984; Leach and Mearns 1996; Grove 1997). It evolved in the context of economic depression in the 1930s and was influenced greatly by the widely publicized dust bowl problems in the United States and an alarmist international literature that followed these events. In Africa, alarm was also generated by a series of droughts, leading to the first fears of desertification, and by growth of colonial populations. By the 1940s the real or perceived threat of soil erosion had pervaded thinking in much of the world. The wider problems of soil management, of central importance to an earlier generation of specialists, were subordinated. Nye and Greenland (1960) reviewed the problems as a whole, but during a critical period for policy in rural development and conservation, erosion itself has been the dominant consideration for a great number of specialists. Largely through the work of environmental and agricultural consultants, hired to propose solutions to stated problems rather than analyze what is actually on the ground, it remains a dominant consideration to this day.

Yet there are questions, increasingly raised by the contradiction between the doom-laden scenarios and the reality of operative farming systems. A number of recent reviews query the severity of the problem. They do not suggest that accelerated erosion does not exist, but they do suggest that it has been greatly exaggerated both in the literature and in common perception (Pagiola 1999). Soil eroded from slopes does not necessarily travel very far; this is one reason why the widely used soil loss measurements can so massively overstate the problem. The sediment delivery ratio, which measures the sediment exiting the catchment in proportion to the measured field soil loss rate, typically is 0.1 for small catchments, 0.05 for major catchments, and 0.01 for large river systems. These and other problems with use of soil loss measurements are reviewed by Stocking (1996).

Very widely, eroded soil may lodge at the foot of slopes and in basins, creating advantageous new environments for farming. This consideration, slightly dramatized, underlay the refusal of a director of agriculture (talking to Brookfield in Fiji in the 1970s) to take seriously dire statements about soil erosion in his country. He said that it had the beneficial effect of moving soil from the hillsides, where it was worthless, into the valleys, where it could be used. Erosion of deep old soils may rejuvenate them, and in dry regions, where moisture availability usually is the limiting factor, shallower soils overlying water-retentive rocks may provide better conditions for agriculture than deep soils. In the eastern Mediterranean, where there is a longer history of intensive farming than anywhere else in the world, extensive erosion "may cause soils to become unsuitable for cereals (annual grasses) but favor the cultivation of woody plants such as the vine and olive" (Blumler 1998:230).

SOIL LOSS AND FERTILITY MANAGEMENT

For many farmers, soil fertility depletion is much more immediately obvious a problem than soil loss, and they are often reluctant to make the major investments in erosion control measures advocated by specialists. Erosion, as one set of hill farmers in Thailand saw it, is not the major perceived problem, less important to them than weed infestation, insect pests, and water shortages. Fertility depletion was much more widely recognized (Pahlman 1992). This is far from unusual. Reporting on alley cropping experiments to manage soil conservation on sloping land in the Pacific islands for the International Board for Soil Research and Management (IBSRAM) in 1992, Geoff Humphreys and I repeatedly encountered the view that erosion was a minor problem by comparison with soil nutrient depletion, on sloping and flat lands alike (Brookfield and Humphreys 1992). At that time IBSRAM was concerned mainly with the management of problem soils: acid soils, Vertisols, and soils on sloping lands, prone to erosion. In the Pacific island region, the last of these was their main concern. Our finding therefore was received without enthusiasm.

An emphasis on soil fertility as the central issue was restored to some degree during the 1990s through an emphasis on nutrient balances, measuring or estimating the balance between nutrient inputs and outputs in a given agricultural system. The approach is not new but has been put into new contexts by work originating mainly at Wageningen Agricultural University in the Netherlands and then applied by others (Stoorvogel and Smaling 1990; Smaling, Fresco, and de Jager 1996; Scoones and Toulmin 1998, 1999). Nutrient budgets can be analyzed at scales from the farm to the nation, although with large error factors. Importantly, flows can be fairly stable through time, making it possible to evaluate them in relation to stocks of nutrients, high or low, in determining the conditions of any area. For infield/outfield systems, flows are almost always negative in the outfield but often positive in the infield, as in the Fouta Djallon case discussed in chapter 5. A recent policy-oriented discussion based primarily on nutrient balances brings the issue of fertility management back to center stage, at least in the African continent (Scoones and Toulmin 1999). It is a refreshing change from the overemphasis on land degradation and soil erosion in most of the literature.

The basic considerations that underlie fertility depletion are set out in chapter 5. There is a big difference between what farmers see and what scientists see. In an important paper, Kiome and Stocking (1995) found that although Kenyan farmers with whom they worked were well aware of erosion, they meant principally gullies and were responding to an orchestrated supply of external information about erosion problems. They saw no direct connection between gullies and their own farming difficulties. They remained unaware of sheet erosion, responsible for enormously greater productivity loss than the more physically visible and more localized forms of damage. Yet the management practice most of them were using, the one Kiome and Stocking found to provide the best returns for investment, trash-lines, was fairly effective in coping with sheet erosion. To the farmers, this was a means of production, not of soil and water conservation.

Farmers' awareness of physical and biological changes, for easily understandable reasons, is biased toward those that are more obvious. Soil loss must become severe to be readily visible. Degrading processes come to them through decline in the rewards for their work, in the quality and quantity of the plants they use, and growing insecurity of production and livelihood. Until something serious takes place, the most sensitive indicator to farmers is a decline in their crop yields. The problem with trying to use this fact analytically is that data on farmers' crop yields through time are mostly aggregate data at district level, collected through sample surveys or estimates; although they are usable where there are clear and obvious trends, they contain a large error component. A particularly effective use of aggregated production data at district level was devised by Tiffen, Mortimore, and Gichuki (1994). They used and interpreted this record to show how the Akamba of Machakos, in Kenya, have successfully adapted management of a degraded landscape as population has grown. They demonstrated that since 1930, the volume of output has increased markedly, especially on a per-hectare basis, despite extension of farming into areas of lower potential. The Machakos case is examined further in chapter 11.

At a more local level, farmers' recollection has been used by a number of modern researchers, sometimes to good effect. Yet few small farmers keep records, and recollection as a source of data is subject to exaggeration of more bountiful yields obtained in the remembered past. A few writers have reacted against reliance on soil loss measurements and estimates by analyzing their inaccuracies and, quite often, irrelevance (e.g., Stocking 1987). Severe criticism has attached to the still widespread use of the Universal Soil Loss Equation (USLE) to estimate soil loss widely across the world under conditions of cropping, soils, farming practice, and climate far removed from those of the midwestern United States, where it was generated. The equation's founder warned against such extrapolation more than two decades ago (Wischmeier 1976). But it continues to be used in many countries, with or without modification.

In the new intellectual climate of the postmodern era, since the mid-1980s there has been a strong tendency to contrast the often poor and even damaging results of top-down interventions with the sustainability of indigenous management methods. Well known studies by Chambers, Pacey, and Thrupp (1989) and Pretty and Shah (1997) are good examples. The term *land husbandry* has to some degree taken the place of *soil conservation* to describe an approach that promotes stewardship by the people who will benefit from the actions being taken (Hudson 1992). It goes along with a new approach to analyzing environmental crises, reexamining them in a historical context, well exemplified in collections edited by Roe (1995) and by Leach and Mearns (1996). This wave is an important corrective, but there are problems in using farmers' perception where land degradation is concerned.

It is far beyond the scope of this chapter to examine the vast literature on land degradation, even on soil erosion alone. To do that would take another and different book. Instead, I present one almost unique study that specifically analyzed farmers' own unaided response to declining soil productivity. Because it describes both how farmers see the problem and how they try to cope with it without benefit of advice from external experts, I review this unusual example in some depth. The material concerns some very enterprising west African people who, a century ago, established a major agricultural industry that, for them, has since collapsed.

■ Coping with Degradation in Southeastern Ghana

THE PEOPLE AND THEIR BACKGROUND

The case concerns the Krobo and Akuapem people of southeastern Ghana. Principal sources are Amanor (1994) and Gyasi et al. (1995). The former wrote only about the Krobo, the latter mainly about the Akuapem region but also about the Krobo. The stories they tell are not the same, but in effect they offer two perspectives on a set of events. These people have a history of land use totally different from that of the Kissidougou people of Guinea discussed in chapter 8. Mainly between about 1890 and 1920, they spread from their homeland on a prominent ridge north of the capital, Accra, to colonize a wide valley developed on granite and sandstone to the west. This part of Ghana occupies a break in the continuous wet forest belt of west Africa, with lower rainfall and naturally semideciduous forest. The country had been sparsely occupied by survivors of a more numerous people reduced by warfare, flight of the defeated, enslavement, and sale to the plantations of the New World. A large region, much of it cultivated before the seventeenth century, became reforested, but with oil palms *(Elaeis guineensis)* among well-established forest trees.[3]

Krobo began the colonization in the mid-nineteenth century, with the economic purpose of supplying a rising international market for palm oil.[4] When the market for palm oil became unstable, they looked for other tree crops. On the ridge south of the Krobo, a few enterprising Akuapem were already growing introduced coffee and cocoa by 1880. Organized migration off the ridge began in the early 1890s, with chiefs and some individuals buying land to set up cocoa farms with income obtained largely from sale of wild rubber (Hill 1963, especially at pp. 161–77).[5] Rapid colonization led to extensive plantations of cocoa that, mainly through the production of these enterprising farmers, made Ghana (then the Gold Coast) the world's leading producer of this crop by 1911. What these farmers achieved in response to an economic opportunity was remarkable and is one of the clearest examples of small-farmers' responsiveness and ingenuity. In regard to their descendants' ability to cope with degradation, however, the story is somewhat different over fully half a century. Only in recent years have indications of successful adaptation begun to emerge.

COPING WITH THE FIRST DISASTER

The deciduous forest had to be thinned for cocoa and for the crops that were grown between the young cocoa plants. Over time there was steady loss of forest species as cropping of the land beneath the canopy, including long-lived tree crops, ensured that the forest species were not reestablished. In this seasonally dry area conditions were marginal for cocoa and became more so as the forest was thinned. The swollen-shoot disease that began to afflict Ghanaian cocoa in the 1930s took hold first and most strongly in the pioneer region. By 1940 it was already a disaster and by 1950 little productive cocoa remained in the less humid regions. The farmers had to change to a food-crop orientation, adopting rotational fallow in place of the agroforestry system of the preceding half-century. They turned to maize for the national market as their main cash crop. Shade crops, in particular manioc, were the initial basis of weed management, so actual weeding was not often necessary. Yams were grown wherever there were trees to support their vines, and taro germinates spontaneously from residual corms when the land is cleared for cultivation. This pattern still exists, but yams and taro are fewer. The last 20 years have witnessed declining yields and replacement of nutrient-demanding crops by more tolerant crops. Maize production has declined and weed infestation has become serious.

Migrants from the east and north have added to population numbers, becoming first laborers then sharecropping tenants on Akuapem and Krobo land. Government agencies continued to favor cocoa and palm oil farmers, and no assistance was given for other crops. More recently, national economic restructuring has ensured that prices of agricultural produce on the national market have remained low. In the 1990s, few farmers still produced cocoa, and most made palm oil only for domestic use. In addition, they tapped palm wine, the distillation of which became a major cash-earning activity with the emergence of some specialized distillers.

Bush fallow, formerly of several years' duration, already was being shortened in the 1960s. The reduced semideciduous forest had been opened quickly to clear land for maize, and by 1990 only patches and individual trees remained. In some areas firewood grew short. Wide tongues of savanna intruded into the farmland, especially after extensive fires that burned large areas during a major drought in 1984. Manioc became the principal food crop, with hardier, bitter varieties replacing the more succulent varieties reportedly grown in the past. Reduction of the area and duration of the successional fallow led to loss of an increasing number of tree species and of many supplementary food and medicinal plants. To some degree, imports from the savanna regions took their place.

A DETAILED REPORT FROM THE EARLY 1990S

Based on fieldwork in 1990–91, Amanor (1994) examined the problems of degradation on the land of five Krobo communities. Degradation was perceived by the farmers in terms of reduced yields, especially of maize, and of unwanted changes in vegetation. Managing escaping fire remained a serious problem. In addition to the tongues of savanna, deforested areas were being invaded by the introduced weed *Chromolaena odorata*. This grows very rapidly and chokes out even the manioc.[6] Some farmers concentrated efforts in remaining forest patches or on the more water-retentive but heavy soil in the valleys; some left grassland unused, hoping forest would regenerate, which it failed to do. Only more recently had they turned to the hard work of clearing grassland, especially *Panicum maximum* grassland in which rigorous cultivation, with protection from fire, after about 3 years can permit pioneer forest species to become established. Weeding necessarily became a major activity.

Vegetables, especially leguminous cowpeas *(Vigna unguiculata)* in a quasirotation with manioc and maize, together with tomatoes, beans, and eggplant, were adopted as new cash crops. After introduction to this area in about 1980, cowpea became the second cash crop to maize by 1990 (Amanor 1994). On the shores of the seasonally rising and falling Lake Volta, sweet potatoes and some vegetables were grown by flood-recessional methods, using the same site each year. Total crop biodiversity even increased as the stock, which still included most crops grown in the past, increased. A survey in 1993 listed 22 tree and field crops in an Akuapem area, and a 1991 list for the more diverse Krobo area included more than 30 (Gyasi et al. 1995; Amanor 1994). The lists still included cocoa and oil palm, together with manioc, maize, yam, plantain and banana, taro, rice, sweet potato, beans, cowpea, eggplant, tomato, pepper, okra, sugar cane, groundnut, pineapple, and a range of fruits including citrus, papaya, avocado, mango, coconut, and pear. In addition, the fallow or successional vegetation provided a large number of timber and firewood species, household provisions of various kinds, medicines, condiments, and spices as well as small quantities of wild food and diminishing numbers of game (Amanor 1994).

Individual farmers also developed new clearing techniques. They slashed the grass to form a mulch and encouraged small pioneer forest trees under it. There were several individual variants of these practices, leading into farmer-managed fallows. A small minority of farmers, among whom Amanor concentrated his inquiries, undertook more substantial agroforestry innovations using indigenous species; some involved selective forest regeneration. A number of trees were conserved in clearing, or seedlings were protected and encouraged. Widely, and also in Guinea (chapter 8) an important tree for agriculture was *Newbouldia laevis,* which farmers found beneficial to yam cultivation. Several reportedly beneficial species are now being studied by the People, Land Management and Environmental Change (PLEC) group in southern Ghana. The innovators used mainly pioneer forest species found to be beneficial to soil and crops, often dispensed with the burn, and incorporated a range of crops. On one farm "all the elements characteristic of [standard] alley cropping, with the exception of row planting, are present" (Amanor 1994:211). Amanor particularly noted that innovations were to be found especially in the most degraded areas, on which the rotational fallow system was most evidently failing. Manuring and hand-watering were introduced patchily, and a minority of more affluent farmers used chemical fertilizers.

DISCUSSION OF A SET OF PARTIAL CHANGES

Degradation is here described as the farmers saw it in the early 1990s, as a serious livelihood problem, the cause of additional work, and a challenge to their ingenuity. Soil loss was given emphasis by Gyasi et al. (1995) and by Owusu-Bennoah (1997), but Amanor (1994) made no specific mention of it. Owusu-Bennoah (1997) reported heavy loss of sand content from cultivated soils and probable clogging of the surface soil by washed-in clays, leading to increased runoff. In 1996, I walked over substantial quantities of sand washed out of fields and down a Krobo track after a heavy overnight rain but did not see what had happened to the clay and silt. Farmers were much more aware of breakdown of soil structure and loss of seed stock in the soil through heavy and repeated use of hoe cultivation on the land of tenants (Amanor 1994; Gyasi and Uitto 1997).

Analysis of soil samples from sites with different recent histories of land use and cover did indicate changes in pH, texture, and organic matter content, the presence of other nutrients (Amanor 1994; Gyasi et al. 1995). Organic carbon content of cultivated soils averaged 37 percent below that of forest soils, and nitrogen content was 46 percent lower (Owusu-Bennoah 1997). But organic carbon even in Amanor's cultivated samples still averaged 1.7 percent, low but not desperate. Loss of physical structure probably has been as serious as loss of chemical fertility.

With the single exception of the introduction in rotation of leguminous cowpeas, most of the spontaneous innovations in southeastern Ghana up to the early 1990s applied more to biological management of the plant cover than to crop and soil fertility management. Livestock were seen more in terms of income generation than in relation to the whole farming system; manuring was casual, there were few livestock pens, and nothing resembling the intensive use of manure in the *tapades* of the Fouta Djallon (chapter 5) had emerged. Composting was not developed even in backyard farming, where soil regeneration practices were still uncommon. Farmers' response to land degradation had not yet led to any comprehensive change in farming technology.

■ Managing the Dynamics of Steep Slopes

SOME CONSIDERATIONS IN STEEP SLOPE MANAGEMENT

There are situations in which management of dynamic processes in the environment is so critical that it warrants special discussion. One is the management of dynamic processes on steep slopes. The topic continues the discussion of degradation because erosion is a serious problem on steep slopes. Although some writers claim to have seen slopes as steep as 70° under cultivation without terracing, a comparison of actual measurements probably would leave the world record somewhere close to 55°, at least for slopes more than a few meters long.

Topsoil moves down steep slopes very readily by simple soil creep, but where the subsoil has good mineral fertility, this latter is then exposed to weathering. Although it lacks an organically enriched topsoil, the new soil may produce good crops. There are parts of the world in which rilling and gullying are major problems on steep farmed slopes, but these forms of erosion are more common on slopes gentler than about 25°. The principal form of actual slope failure in mountains is mass movement in the form of slumps, landslips, and landslides. Contrary to a widely held belief, this is not necessarily any more common on cleared land than on slopes that are still forested. Although there is no doubt that runoff is much greater on cleared slopes, leading to surface erosion, major rains affect all land, whether forested or not.[7] Mass wasting is of several types, differing widely in their effects on agriculture. There are the hugely destructive mass debris flows, which can run for kilometers; there are big, deep-seated landslides, some of which run into debris flows; and there are minor landslips, both planar, mainly on valley sides, and rotational, which occur very widely in unstable ground. Whereas the larger events can wipe out big areas for years, many of the smaller events are manageable and may even have beneficial consequences.

A brief return to Papua New Guinea will help explain. Slopes of 30° to 40°, some several hundred meters long, are widely cultivated in the highlands of Papua New Guinea with what seem to be very simple management methods, almost all used in a land rotation context (Humphreys and Brookfield 1991). Some slopes in this range are managed with only downhill ditches, dug down into the subsoil to prevent soil saturation, which might lead to major slope breakdown; they rarely lead to gullying. Quite common, especially in the Chimbu area discussed in chapter 1, are low wicker fences, which check soil creep and also mark boundaries between plots planted by different women. Also having the purpose of marking boundaries, but in this case between the enduring land claims of different families and groups, are more durable live fences of *Cordyline fruticosa*. Where these lie across the slope, they build up an accumulation terrace 1–2 m high of the type discussed from Vanuatu in chapter 4. In limestone country in the Baliem valley of Indonesian Papua, dry stone walls have been built, but these probably were built as much to stack the stones as to protect the soil because they were also built where they were not needed for conservation purposes (Mitton 1983). Nor have they been always effective for conservation. In a limestone area near the modern town of Wamena in the Baliem valley in 1959 I walked over the remains of limestone walls in an area now almost bare of soil, with patches of limestone pavement.

There are many small rotational landslips, and these can be advantageous where they are not so frequent as to destroy fences and damage crops. There is an extensive belt of country on late-Mesozoic marine shales beneath a major limestone outcrop, an association repeated by faulting over a span of more than 200 km. The area I studied in Chimbu included a large tract of such land. This belt of 15°–30° slopes is in nearly continuous motion, and minor rotational slumps occur frequently. The pedoturbation of these small rotational slumps creates numerous pockets of deep, friable, and moister soil, which Chimbu farmers use for mixed gardens growing crops that need more moisture than the ubiquitous sweet potato (Humphreys and Brookfield 1991; Brookfield 1996b). In such cases, minor slope failure creates patches of deeper soils with gentler slope, forming the basis for diversity in farming spaces. It is only where much larger failures take place that lasting damage is done.

MANAGING A HIGH-ENERGY ENVIRONMENT IN NEPAL

The mountains of Papua New Guinea are a high-energy environment, producing a lot of natural mass wasting, but they pale into insignificance alongside the mountains of Nepal. Monsoonal rains are highly erosive. A high proportion of mass wasting takes place in gorges with sides too steep to be cultivated. Carson

(1985) found that of a calculated 1320 tons lost over a year from one 63 ha catchment, 1100 tons were derived from only 6 percent of the area. Loss from the 57 ha of cultivated and pasture area (especially the latter) totaled only 220 tons, 3.86 tons per hectare per year. Although not insignificant, this is well below many soil loss rates measured elsewhere.[8]

Wherever possible, by diverting water from small rivers or from springs, even on 30–45° slopes, farmers of the middle hills of Nepal make flights of irrigable terraces growing rice in the summer and wheat or other dry crops in winter. The lower slopes of whole valleys are filled with flights of such terraces. On sloping land that cannot be irrigated they make dry terraces that range from almost level to as much as 25°, cut back each year to gain mineral fertility from the weathered subsoil exposed in the terrace riser. Farmers grow maize, legumes, and vegetables in summer and, where there is sufficient soil moisture, millet and buckwheat in the dry winter. There are few unterraced arable patches, and little land lies fallow. The basis of the system is a large livestock population, mainly cattle and buffalo, that pasture on uncultivated land but are stall-fed with grass and with leaves gathered from heavily used forest land. The bedding and manure are applied to the irrigated terraces and all unirrigated terraces that lie close to the homestead. An immense amount of work goes into constructing, extending, and maintaining these systems.

A paper by Johnson, Olson, and Manandhar (1982) is one of the few works in a large literature in which the actual slope management experiences of individual Nepalese farmers are discussed.[9] They described how farmers perceived erosion damage and went about repairing it. The greatest disaster was the unheralded collapse of a section of irrigated terraces, usually involving a large landslide. If there was warning, farmers might leave a terrace unirrigated, even uncultivated, for a season or two until they could accumulate resources needed for repair. If there was no warning or if the preventive measures failed, the terraces collapsed and had to be rebuilt. Lower terraces along the rivers quite often were flooded, and bank erosion could undercut their walls, causing collapse. Vulnerable riverside terraces were protected by bamboo fences, built out into the river.

Comprehensive and labor-intensive management was needed to defend and maintain the system. When major repair became necessary, it was done within a few years. Within the same catchment, landslides that had been photographed in 1978 were entirely reterraced when the site was rephotographed in 1987, and the shape of terraces all around was changed (Ives and Messerli 1989). In early 1984, I saw one area that had been mapped as a long and wide landslide mudflow, with developing gullies, in 1978–79 (Kienholz et al. 1983). Already, it was fully reterraced.

Almost every farmer in the area studied by Johnson and her colleagues had experience of serious damage, and exceptional rainfalls could overwhelm their ability to maintain the systems, leading to a heavy task of repair and reconstruction over the following 1–3 years. Farmers "draw upon a fund of widely known control measures practised with varying degrees of efficacy throughout the region. . . . Many farmers distinguish between ideal measures which might be adopted given the availability of resources, and feasible measures which are the ones they must select" (Johnson, Olson, and Manandhar 1982:184–85). Because resources varied between people, damage was less readily prevented and less effectively repaired by those with fewer resources. "This suggests that the overall effect of 'random' landslides and floods may result in increased disparities between rich and poor" (Johnson, Olson, and Manandhar 1982:188).

SLOPE MANAGEMENT IN PERSPECTIVE

It is as easy to overemphasize terracing in reviewing slope management as it is to overemphasize the larger and more obvious irrigation works in writing about water management. Terracing is both soil conservation and water conservation, and aspects of the role of terraces in these two contexts are set out in box 9.1 and in the following section. There is a common tendency to lay stress on the soil conservation aspect. Terraces may be built up from below or may be cut out of the slope, but unless the topsoil is deep, to cut them level will expose subsoil low in organic content and of lower productivity. This has been the undoing of many programs for the mechanical construction of bench terraces on hillsides in developing countries. As a device to improve production, they often do not work, and if they are less than perfectly constructed, spillage over low points can actually increase the erosion hazard. Wilken (1987:117) remarked of flat terraces that if they were the ultimate in slope management, they were also "the ultimate in slope artificiality and instability." Yet enduring terraces can have long-lasting value. Investigating soil in an anciently constructed irrigated terrace in Peru, Sandor and Eash (1995) found the soils to have better structure than adjacent uncultivated soils and improved nutrient content, including phosphorus, even after four centuries of abandonment.

Box 9.1
Terracing and Soil Conservation

Terracing always and everywhere has the effect of decreasing the slope angle and length of a hillside to create more stable arable land. But terracing is not always well described. The surveys of Spencer and Hale (1961) and Donkin (1979) still dominate the literature, most of which overemphasizes stone-walled terraces. Insufficient attention is paid to less durable earth or wood construction. Nor is enough attention paid to accumulation terraces *(taluds)* developed behind live barriers. These do persist, as we saw in the case described in chapter 4. Not all terrace systems have been described in all parts of the world, and some areas described as terraced have only rudimentary slope management. In Africa, excellent surveys by Sutton (1984), Grove and Sutton (1989), Adams (1989),

Sutton (1989), and Adams, Potkanski, and Sutton (1994), among others, reveal the range of complexity.

Where stone walls are made, the risers of terraces usually are much steeper than elsewhere, approaching vertical in the carefully constructed and extremely durable terraces made under Incan rule, late in the precolonial period, in the mountains of Peru. These latter terraces were built by conscripted labor, often on very steep slopes, and specifically for producing maize for the granaries of the Incan state. They are the most elaborate terraces anywhere, sometimes with stone staircases and spillways, and were so well built that a large proportion still endure after almost 500 years, with or without subsequent use (Donkin 1979; Zimmerer 1996). Simpler and ancient peasant terracing survived alongside the masonry works of the Incas, on the lands left to the peasants to cultivate for their own subsistence.

Jan Nibbering (1991a, 1991b, 1997) described transformation in the limestone uplands of the Gunung Sewu, southern Java, from an initially sparse occupation in the early nineteenth century. He traced the effect of landscape modification on land use. After a period of severe land degradation, a stable land use employing terraces was adopted spontaneously. Rounded hills about 50 m high are separated by valleys, in which soil accumulates. Landscape-wide shifting cultivation, together with cattle raising, had already begun to give way to more intensive forms of land use by 1900. Terracing began about 1900 in the valleys, then on the saddles between the valleys, and after 1930 on the hillsides.

By the 1940s, hillside terracing had put an end to the recurrent flow of nutrients into the valleys. What had been a mixed cropping of upland rice, with some wet rice in the valleys, together with maize, manioc, and tobacco, increasingly gave way to a mixture of manioc with upland rice in the valleys and of manioc with maize on the hills. Pasture for cattle vanished when the hillsides were terraced, and livestock were thereafter stall fed. Their most important product became their manure, applied almost wholly on the valley fields. A famine in 1963 drew a measure of external intervention, but it also led to perfection of the terraces, even though a fallow period was still used on the hillsides. Cultivation in the valley bottoms became permanent, growing upland rice, manioc, maize, and groundnuts. Chillies, beans, and cowpeas were cultivated, and trees were grown for saleable wood and charcoal for a market made accessible by improved roads. None of the terracing work, the key to the greater sustainability of the system, had been instigated externally, here or in other upland areas of southern Java. Inorganic fertilizer has been used since the 1970s. Yields have improved, but it remains an open question whether this is more a result of fertilizer use than of better soil conservation. The best conserved lands also received most inputs of both manure and fertilizers by the late 1980s (Nibbering 1997). Population growth, a major factor up to the 1950s, has since slackened because of large emigration.

Slope management always presents both problems and opportunities. Farmers who take advantage of the latter, as in all cases, are very likely to seek ways to cope with the former, once the problems become evident and some successful innovator can demonstrate how to manage the slope dynamics of a particular site. There is no need to follow Spencer and Hale (1961) in postulating either an evolutionary development of terracing or a global diffusion from a supposed area of origin between the Mediterranean and Caspian seas. It is unwise to assume that the idea of terracing has to be imported to an area where people were previously ignorant of this technology. Terracing, like diverting water to bring it to places where it is needed, is a commonsense adaptation to local needs and conditions. These practices have arisen spontaneously in many areas. Their diversity reflects site and social conditions, and the scale of the constructions, whether minor or major, is related to the larger agricultural and social systems.

Temporary within-field small fences (Chimbu), hand-built *fanya juu* (thrown up) terraces (Kenya), or grass-mulched squares (southern Tanzania) on a hillside are as much slope management as are Ifugao rice terraces or ancient works in Peru. Although there is great contrast in the durability of the structures, the agricultural systems they support or have supported may all have long duration. Downhill ditches may not look like good slope management to an agricultural engineer, but they may be effective in achieving their purpose: getting rid of water that might otherwise cause slope failure. The conditions of the site, and the materials and means available, need to be taken into account in any evaluation.

■ Managing Water

TERRACES AND WATER

Whether or not it includes irrigation, terracing has a soil moisture conservation benefit, and for quite a lot of terracing this, rather than irrigation or soil conservation, may be the primary objective. Surveying the available evidence, Sutton (1989:103) remarked for Africa that "terrace cultivators are and have been more concerned with adequate watering and drainage of their fields in the general sense—that is through natural and assisted agencies—than with artificial irrigation devices of a formal sort." This seems equally true in western parts of the Yunnan plateau in China, where many terraces are neither leveled nor bunded, and few have diversionary systems to bring them water. Only if a structured irrigation component is included is it normal to ensure that surfaces are flat and, although some terraces slope backward to retain soil and moisture, many are sloped forward quite deliberately to permit runoff.

In areas where dry terraces are walled with stone, the space behind the walls may be filled by hand, but many—perhaps most—such terrace-building farmers are content to allow natural slope processes to fill the spaces. In a region of southern Java, stone lines, initially to stack the stones and then to check erosion, became terraces over time. Farmers used natural soil wash processes to fill the space behind walls that they progressively raised; "essentially, farmers have picked up, cut and moved the stones; Nature has moved the soil" (Nibbering 1997:175–76).

As we saw in the Ifugao case in chapter 1, builders of irrigated rice terraces have used water to transport soil into the spaces created by their terrace walls. In Java, it remains common practice to cut into the soft material of volcanic hillsides to establish or extend irrigated rice fields. The marginal channels are used to redistribute or dispose of the excavated material. Other material later slumps into these channels and is carried away to the rivers (Diemont, Smiet, and Nurdin 1991). This practice probably is an important contributor to the very high sediment loads carried by Javanese rivers, often cited as evidence of land degradation in this densely peopled island.

Water transport of soil certainly has a specific role in the case of cross-valley terraces, where the transport of soil into the protected space can take place with no specific diversionary measures. The range extends from simple silt traps, through check dams across gullies, to the larger works of some Mexican and Peruvian valleys (Wilken 1987; Donkin 1979). Historically, farmers of the Roman period in Tripolitania sought to manage the yield of ephemeral streams *(wadis)* by constructing low dams that trapped sediment and held runoff. They used natural process not only to conserve soil, but also to increase the water available for farming by about eight times (Vita-Finzi 1969). This system endured for several centuries.

FLOOD FARMING

In many semiarid areas are found practices that include little or no terracing yet still depend on the entrapment or diversion of water and the sediments and nutrients it carries. Rain falling on few days in any year, patchy where it occurs but sometimes very heavy, creates short-lived floods. Trapping of water and sediment is the only way in which to sustain any benefit. Except on principal rivers, major works are inappropriate. William Doolittle (1984) showed how individual farmers in the *arroyos* (flat-floored gullies) of northwestern Mexico first checked flows and trapped sediment with brush fences, second used stones, third went on to create low terraces protected by bunds, and finally moved into cooperative diversion of the ephemeral *arroyo* streams. Maintenance and construction took place simultaneously with cultivation, and there was a progressive increase in the value of the investment in land, or landesque capital, and of the soil.

Such efforts are not always successful. Sandor, Gersper, and Hawley (1990) found that simple cross-valley terraces used in prehistoric and historic times by Indian farmers in the American Southwest were not beneficial to the structure and content of the soil. Nor did the terraces cope adequately with the erosive power of concentrated flow in the *arroyos*. Flood farming and its improvements, with their great vulnerability to violent weather, must rank as among the most chancy of all agricultural enterprises.

THE AMAZONIAN FLOODPLAIN

A very different situation concludes this wide-ranging discussion and brings the review of land rotation cultivation and its relatives to an end. The people of the upper Amazon floodplain use shifting cultivation methods, but in an environment that is itself constantly shifting. The upper Amazon has an annual rise and fall in river level of more than 10 m, with occasional greater floods. The system drains the high-energy Andean mountain range and carries very large loads of sediment. Each flood erodes, transports, and deposits enormous quantities of material and can substantially change the topography of the valley floor. This results in building well-marked and high natural levees along the sides of the main river and its branches, with backswamps or lakes between the levees and the margin of higher ground. Within this pattern there is considerable variation in detail, and the elements in this variation may change from year to year and change much more in years of very high flood, such as the La Niña year of 1999 (M. Pinedo-Vasquez, personal communication, 1999). The whole river changes position in the widest, most unstable and anastomosed sections, where the Amazon floodplain reaches widths of 20–40 km.

There are three main landforms in the floodplain: the natural levees, their backslopes, and the silt bars or mud flats. There are also patches of sand in place of the silt (Pinedo-Vasquez 1999). The levees are flooded infrequently, and their higher parts develop forest communities if undisturbed. The silt bars are flooded annually, and the best areas have fertile, water-retaining soils and stay above water 5 or 6 months in each year. They are preferred for rice cultivation, whereas in the vicinity of Iquitos the levee backslopes are used for jute. Both are cultivated for 4–6 years, then fallowed. The backslopes have a recovery pattern similar to that of adjacent dry ground, but on the silt bars weeding practices and rice cultivation itself destroy the habitat for the dense floating grasses, which are the pioneer species and trap silt. Without this grass, the silt bars therefore accumulate sand and lose a lot of their agricultural value, not replaced until river migration across the floodplain has recreated the original environmental complex.

A somewhat different situation is reported on the meandering Ucayali, where there is greater stability. Here, the village of Santa Rosa has become famous as an example of agrodiversity (Padoch 1988b; Padoch and de Jong 1992). The people of Santa Rosa cultivated the high ground above the floodplain by a system of shifting cultivation with increasingly managed fallow and maintained substantial agroforest areas in and around the village (Padoch and de Jong 1987, 1989, 1990, 1991). They also derived a lot of their protein from fish in the river and lakes (Hiraoka 1995). Diversity on the floodplain was remarkable. The silt bars were used for growing broadcast rice, without land preparation. The harvest often was completed in standing water, and some of the crop might be lost at either end of the season. The sandy areas were used only for cowpeas or groundnuts.

The levees were used for five different types of production: monocultures of rice, maize, and plantain bananas, mixed shifting cultivation fields, and agroforests. Jute was not important here. The land had to be cleared by cutting and burning, and the heavy soils used for rice would yield only one rice crop. Maize was grown on sandier soils and followed rice. These crops depended on additions to soil fertility from annual or frequent floods. Plantains were grown on the highest land, not often flooded, and might remain in production as long as 15 years. In the diverse shifting cultivation fields on the levees, manioc was the dominant crop, with the same mixtures as were grown on the dry land outside the floodplain: plantains, squashes, papaya, taro, and vegetables. On the highest levee areas, where flood risk is least, agroforestry fields dominated by fruit and timber species were created.

Padoch and de Jong (1992) organized this complexity to distinguish 12 distinct agricultural systems (or farming spaces) practiced on both the floodplain and the dry land around this community. Among 46 households, 39 ways of combining the 12 production types were found in 1985. In 1986, a sample of 12 farmers included only 3 who followed the same strategy as in 1985; the other 9 had adopted new combinations. Changes in the floodplain environment, recent losses due to floods, changes in available household labor, and variations in both capital resources and preferences were identified as reasons. The power of natural forces in this case seems capable of overwhelming any deliberate management, and each major flood creates new conditions. Even so, the people who use the valuable resource of the floodplain make a risky investment in long-term crops and trees on the highest levees.

■ Discussion

A BETTER WAY TO LOOK AT LAND DEGRADATION ISSUES

Even on the silt bars in the Amazon floodplain, the cultivation practices associated with a modern cash crop, rice, have created the conditions for unexpected resource degradation. As in Ghana, it can take farmers some time to learn how to cope with problems that result from changes in the content or method of production. In chapter 12 we shall see more of these in northern India, where irrigation has led to waterlogging and salinization, and where introduction of a summer crop has created serious difficulties in soil management.[10]

A distinction must be drawn between short-term and long-term adaptations to degradation and deterioration of the resource base. The short term, as in the case of Nepalese terrace repair, may be as short as 1–3 years, but in this case people know from long experience what to do. Where this is not so, a learning period may endure for one, two or even more generations. Where farmers have entered a new area and are unfamiliar with how to manage it, unless they are able to learn quickly from others with more experience, they may largely or completely destroy the resource before they have been able to develop successful strategies through their own experiments. It is farmers like these who do the worst reported damage. Javanese transmigrants in eastern Borneo soon found that clearance with fire was creating large areas of *Imperata cylindrica* grassland. After learning from nearby Dayak farmers they reduced the amount of damage being done, but for up to a generation after settlement they had still only partly learned the Dayak methods of managing a shifting cultivation system that was, to them, entirely new (Hidayati 1994).

A prerequisite of successful management is an understanding of the dynamics of the natural system being managed and of the management process itself. After almost two generations, only a minority of the Ghanaian Krobo farmers described by Amanor (1994), discussed earlier, were on their way to achieving this understanding. The natural resilience and sensitivity of the ecosystem being managed is an important variable, for the conjunction of high sensitivity with low resilience leaves very much less time for adaptive learning (Blaikie and Brookfield 1987). Sensitivity may vary through time, especially under climatic variability, and an altered system may develop resilience once initial changes have taken place. Both natural and human-induced processes are at work, and they do not operate in a linear fashion. One simple way of viewing this is through an equation, relettered by the Scientific and Technical Advisory Panel Working Group (1999) from an original in Blaikie and Brookfield (1987):

$$NetD = (NatD + HD) - (NR + HI)$$

where

NetD = Net degradation

NatD = Natural degradation

HD = Human induced degradation

NR = Natural recovery

HI = Human improvement

THE ROLE OF INVESTMENT IN MANAGING DEGRADATION

Effective long-term management involves investing in the land, using more labor and capital than is needed just for the production of a crop. It also involves investment in organizational skills. Investment on and in the land is an issue carried forward into part III of this book. The physical capital purchased or created by farmers can be divided into three groups. There is the working capital of farm tools, livestock, and household storage and equipment. There are trees and deliberate modifications of the vegetation that endure for a long time, even centuries. There are physical works, which include manufactured soils as well as terraces, drainage and irrigation systems, and wells. Even among shifting cultivators living at low population density, farmers' management rarely uses only the avoidance strategy of leaving everything to natural repair. Everything farmers do modifies the environment, and it does so principally by creating the landesque capital introduced in chapter 3.

Chapter 6 mentioned the creation of stands of useful trees after cropping. In chapter 7, seemingly simple practices were seen to lead to long-continued modification of the soil. In chapter 8, the creation of managed forests was related to the manufacture of changed soil conditions. In this chapter it has been shown that few environments that present difficulties for agricultural use or are sensitive to degradation can be worked for long without investment to manage the land and its plant cover. Capital investment on the land is not a topic with a large literature. But we may carry forward the notion that almost all agricultural improvement involves the creation of landesque capital.

Land degradation is serious and may well have become more widespread in recent decades. Crop losses are caused by land degradation. A good deal of this has been going on for a very long time, and some whole regions have been made uninhabitable. Yet despite millennia of degradation, farming in Israel and Syria in the late twentieth century had again reached limits it had occupied in Roman times (Blumler 1998). The exaggerated alarm in the soil conservation literature seems likely to contain a measure of overstatement. It disregards ecosystem resilience, a topic on which not nearly enough is known. As Stocking (1996:154) remarked, "Environmental policies are much more convincing if one can argue that an environmental catastrophe is imminent." This is not to suggest that scientists and consultants behave irresponsibly, but it does accord with the broad conclusion of Leach and Mearns (1996) that the received wisdom is often less than wise and that alternative perspectives must be given full weight. They are given such weight in this book.

Notes

1. The chapter does not examine irregular problems or disturbances, often called hazards. In terms of interference ecology these disturbances are essential for the formation and adaptation of ecosystems, including agroecosystems (chapter 3).

2. Plows used in Europe in Roman times reached a depth of only 5–10 cm. Medieval and early modern plows reached about 18 cm. Mechanical plowing can now reach a depth of 40 cm (van Vliet-Lanoë et al. 1992). Hoe cultivation can dig down to about 20 cm and raise mounds and ridges as high as 40 cm above their base (Wilken 1987).

3. Principal sources on the history of southeastern Ghana and its people include Hill (1963), Kea (1982), Amanor (1994), and numerous articles cited by these authors.

4. Krobo developed a highly distinctive system for land purchase and allocation, called *huza*. It involved the buying of blocks of land by "companies" and its division into parallel strips starting from a baseline along or close to which individual houses were built; there were no villages as such.

The *huza* system is classically described by Field (1943–44) and Benneh (1970) and is most strikingly displayed in the numerous maps printed in Hill (1963). See also Hunter and Hayward (1965).

5. Patrilineal Akuapem copied the Krobo *huza* model of settlement, but the matrilineal people allocated blocks of purchased land among family members in a patchwork. This created the sharp contrast between mosaic and strip land use patterns that persists today. This contrast is discussed in detail by Hill (1956).

6. The bushy weed *Chromolaena odorata* is an introduction from the Americas to wide areas of southeast Asia and Africa. In Asian regions it competes successfully with *Imperata cylindrica,* a worse weed, though one with uses to the farmer. The African variety of *I. cylindrica* is less of a problem. *C. odorata* is very prone to fire, very aggressive, and hard to clear (though less so than *Imperata* grass), but it accumulates a range of nutrients and litters profusely, so there is sometimes benefit in the soil, notably in the pH and hence the cation exchange capacity. This property is very variably evaluated by farmers of different regions, but some southeast Asian farmers specifically manage *C. odorata* as a fallow cover.

7. Major rainfall always produces not only flooding but also erosion, even under thick forest. A problem in the latter is that when deeply weathered ground is saturated in depth, the weight of the trees on slopes can overcome the binding action of their root systems, leading to very large landslides that can bring great quantities of debris into the valleys. Major slope failures create large scars in many tropical forests as well as in cultivated land. I recall in 1980 the forested central mountain range of St. Vincent in the West Indies looking, from a few kilometers away, like the black and white keys of a piano after such a rare event. I have seen similar but less dramatic scars on forested mountain sides in southeast Asia and the Pacific, while nearby farmed land exhibited little visible evidence of damage. Even in England, rotational landslips are much more common in woodland than on open ground (Rackham 1986).

8. The plateau of Tibet has risen by as much as 4–5 mm per year through the past 100,000 years, continuing today (Liu and Sun 1981; Ives and Messerli 1989). This is steepening the Himalayan slope below it at a rate sufficient to account for a large part of several estimates of overall lowering by mass wasting, even one of 12 mm per year calculated for a catchment near Kathmandu by Caine and Mool (1982).

9. A survey up to the mid-1980s was presented by Blaikie and Brookfield (1987). Since that book was written there have been many additions to the Himalayan literature, including the proceedings of an important conference (Ives and Ives 1987) and a major book by Ives and Messerli (1989) that comprehensively reviews large subject areas.

10. Elsewhere in the world, both wetland and dryland salinization have sharply increased with modern irrigation and land clearance. The solutions to the problems arising from irrigation, which are far from modern, require centralized management and heavy investment. When resources are not available, irrigated farmland land is ultimately lost, an event first recorded in Mesopotamia between 2000 and 1000 B.C. Dryland salinization is more amenable to management at a local level (Ghassemi, Jakeman, and Nix 1995).

Part III

Paths of Transformation

Chapter 10

Who Has Driven Agricultural Change?

■ Introducing Part III

EMPHASIS ON AGRICULTURAL CHANGE

At this point in the book, there is an abrupt shift of emphasis. Three of the themes set out in chapter 3—adaptability, innovation, and management beyond the farm—have been developed in part II, and sustainability, in the context of adaptability, will be a main topic of part IV. The three chapters in part III are devoted to the underlying theme of long-term change. They are concerned with paths of transformation in the ever-changing patterns of agriculture and agrarian society. The evidence already presented confirms that the notion of long-term equilibrium in small-farming systems is invalid. Wherever we explore the past, we find that change has been a constant condition. This is especially so in the period since the sixteenth century, but chapter 4 suggested that although change before that time may have been slower, it has been present from the beginning. Not all change has been positive, and chapters 2, 6, and 9 all introduced cases in which farming systems were put under such severe stress as to need radical change. There are other cases, such as some mentioned in chapter 8, in which total collapse was the consequence.

WHAT CONSTITUTES AN AGRICULTURAL REVOLUTION?

These three chapters are concerned with positive changes on a large scale. To begin, we need to take critical note of a term that has become very common in the literature of both modern and historical change: *agricultural revolution*. It is applied to changes that took place over a span of one to three centuries in the past as well as to very rapid changes within the twentieth century. Some researchers, such as Grigg (1984:14), regard only the latter as meriting the term *revolution*, with everything before being evolutionary. A closer view of modern change often breaks it down into parts, always with a major surge after 1950 but with precursors of critical importance. Grigg probably was right, and anything much less than these modern changes probably is better described as progressive transformation. However, the term *agricultural revolution* is so deeply ingrained in the literature that it has to be used, though often with qualifiers. More important for this book than the pace of events are two other questions: what has been revolutionized, and what have been the main driving forces? My underlying purpose in this part of the book is to isolate the farmers' own role in guiding and implementing these changes. This will help us respond to a persistent view that hitherto static and unenterprising rural economies can be reconstructed only through external intervention.

Not all major changes have been regarded as revolutions. Sweeping changes in the wider system took place after maize was introduced into southern Europe in the sixteenth and seventeenth centuries and in Ireland after the potato became the major subsistence crop in the eighteenth century. Similarly, manioc has been a major crop in Indonesia, and even parts of Africa, only since the end of the nineteenth century. These were big changes, but few have seen them as anything more. It seems that to be described as an agricultural revolution, changes in field crops and practices must be associated with a complex of mutually supporting changes in the management and organization of production.[1] To qualify as revolutionary, such changes must have been of a major order, and it is entirely possible for change originating outside agriculture itself to have been the main driver.

The period since 1950 is unique. In the developed as well as the developing countries, this half-century has seen production increases at a historically unprecedented rate. There has been growing emphasis on marketable production, and control strategies, seeking to override natural biophysical diversity, have been applied where possible. At the same time there have been large shifts of population from rural to urban environments and major changes in the structure of most national economies. Many national populations have grown at unprecedented rates. Governments, which have for very long intervened to tax, command, and prohibit, have not ceased these activities but have increasingly provided services designed to enhance change. For farmers in Asia and Africa, there was major political change as colonialism gave way to national independence between the end of the 1940s and the end of the 1970s. Subsequently, there has been a new round of enforced integration into the international market, called globalization, with a loss of decision-making independence.

■ Bursts of Innovation and Incremental Change

HOW FARMER-DRIVEN CHANGE MIGHT COME ABOUT

The last half-century is therefore a difficult context in which to isolate the extent to which farmers have been able to drive, or at least guide, the patterns of change. There have been obvious differences between regions. The contrasting conditions in areas not closely affected by the Green Revolution and those that were closely affected are described in chapters 11 and 12. Some of the cases discussed in parts I and II also reflect these contrasts. In some of these cases there is evidence of bursts of innovation, most occurring in response to new conditions. The highlands of New Guinea as a whole after the seventeenth century and the more detailed response of one mountain-dwelling group to a new innovation in the 1950s are one such case (chapter 1). The switch between very contrasted farming systems in south central Africa and the earlier evolution of both these systems seems to represent another (chapter 7). Bursts of change such as those that must have taken place in these instances are basic to the understanding of how farmers can develop their own systems and determine, at least in part, their own destinies.

NEW COMBINATIONS OF PRODUCTION

Innovation, rather than just change, lies at the core of any comprehensive agrarian transformation. Innovations create qualitatively new elements in the farming system, and they should be the main focus of attention. Their effects can be thought of in the context of Schumpeter's (1934:66) often disregarded definition of development. He saw the key to development, as distinct from change without development, as "carrying out new combinations of productive means." In other words, economic growth, so often regarded as the indicator of progress, need not by itself be development at all. I applauded this view in an earlier book on a very different set of topics (Brookfield 1975).

Drawing what is relevant from Schumpeter's (1934:66) words, such new combinations would include

> (1) The introduction of a new good . . . or new quality of a good, (2) the introduction of a new method of production, which need by no means be founded on a discovery scientifically new, and can also exist as a new way of handling a commodity. . . . (3) the opening of a new market . . . [and] (5) the carrying out of the new organization of any industry.

Innovations occur in waves, in a sporadic manner. Where they are comprehensively taken up in the space of a few short years, it is clear that a revolution has taken place. More often, innovations occur, and are taken up, over a long period of time. Innovation often involves the creation of capital, whether landesque or of other forms, with which chapter 9 concluded. How can small farmers, almost by definition resource poor, achieve this? An idea first introduced in chapter 3 helps answer this question.

INCREMENTAL CHANGE AND CAPITAL FORMATION

In reporting his work on flood flow management in north Mexican *arroyos,* also discussed in chapter 9, William Doolittle (1984) developed an important model showing how a lot of new work can happen without any specific plan being imposed on people and their land. The work is done incrementally. "New fields and associated features are not swiftly changed to a final form but are transformed while in use. In this process, acts of construction may not be distinguishable from other categories of agricultural inputs" (Doolittle 1984:125).

In central Mexico and highland Guatemala, Wilken showed how many thousands of hectares of sloping land, originally marginal for farming, have been terraced in different ways. All this has been done "in small increments. Schedules and completion dates are rare, and construction may continue for years, even generations" (Wilken 1987:127). He called this a process rather than project approach, but the term *incremental change* is better. Incremental change takes place in many contexts, not always obvious. One aspect emerged from work among the Kalimantan people of Indonesia, who also developed the complex agroforests discussed in chapter 8 (Padoch, Harwell, and Susanto 1998:3). A number of fields that seemed to offer evidence of the "worst sins attributed to traditional agriculture and its practitioners" were in fact being prepared for conversion to wet rice. Shortening of a shifting cultivation fallow, with poor returns to labor, eliminated woody regrowth. This was not bad practice; it was deliberate. Preliminary conversion to grassland before preparation for wet rice was a means of saving labor. Long-term change was going on and was being planned. What look like backward or even destructive practices may have a constructive long-term purpose.

By creating capital incrementally, farmers can make investments that would be beyond their means in the short term. Investments in land improvement, in the form of work that is not directly productive, demand saving from direct production, but farmers are investing their own labor. Changing a farming system is an investment, one that requires the innovators to bear the risk of failure. How much farmers save in this way is little known, but it has been suggested that farmers investing in new production systems may save up to a third of their total incomes (Tiffen 1996). This possibility is surprising, but an early writer on development remarked that "we . . . could not save if we had the level of income typical of low-income societies, but to assume that this is true of the people of those societies is gratuitous. In even the lowest-income peasant societies the level of income is not so low that all of it is used for the necessities of life" (Hagen 1962:18). Farmers' purposes in investing in their land certainly include enhancing production and particularly reducing risk. Managing biophysical diversity is an essential means to this end; the more active are the dynamics of manageable biophysical processes, the greater is the need for investment. This view also is brought forward from chapter 9.

Where incremental change takes place over many decades and where this happens over more than just a few fields, it can be analyzed by an observer only when change is already far advanced or complete. It also helps to seek further guidance from the past—not the very long past discussed in chapter 4, but more recent periods on which there is documentation. Reviewing historical examples of systemwide change will help correct a short-term view of agricultural or agrarian transformation based too heavily on the uniqueness of the experience of the past half-century. It will also shed light on the relative roles of landowners, farmers, and national or regional authorities in times before there were ministries of agriculture or formal extension services and before modern agricultural science evolved. I use three examples, the first of supposed premodern agricultural revolution in the early Muslim southern and western Mediterranean, the second in sixteenth- to eighteenth-century England, and the third in seventeenth- to twentieth-century Japan. They tell different stories.

■ Two Completed Experiments

CHANGE IN THE MEDITERRANEAN REGION IN THE EARLY MUSLIM PERIOD

Great changes took place in all the areas conquered by Muslims during the first four centuries of their era in the southern Mediterranean and elsewhere in the Islamic world. Land use complexity underwent a major surge between the eighth and eleventh centuries. Watson (1974, 1983, 1995) described what happened as an agricultural revolution, taking place in association with social, demographic, and urban revolutions that created the demand for and facilitated the diffusion of innovations. Using a wide range of medieval Arabic sources, Watson regarded individualization of land tenure as a critical element: "the early Islamic owner of farm land appears to have exercised not only the full right of alienation but also the full right to determine land use" (Watson 1983:113).

Mediterranean agriculture had been well elaborated earlier, in the Roman period. It already included field cultivation of grains and legumes and production of green vegetables and condiments in kitchen or market gardens (Butzer 1996). There were orchard crops, not only grapevines and olive groves, but also a range of fruit trees. Other than draft oxen, livestock were mainly sheep, goats, or pigs. Intercropping was common and there was also segregation of fields and management practices (White 1970; Renfrew 1972; Halstead 1992; Sauer 1993; Lirb 1993; Pleket 1993). During the course of the Roman period several new crops and fruits were introduced. There is no universal agreement on the dynamism of Mediterranean agriculture in Roman times. Butzer (1996) seems to accord with those who have regarded it as already developed and diverse but changing only slowly. Complexity already had been demonstrated more than a century before 1 A.D. but increased substantially over the next two centuries when specifically agronomic writings appeared, especially those by Varro and Columella (Lirb 1993; Pleket 1993). There was only limited understanding of the use of manure and crop rotations for fertility management, and most field crops were grown only in the wetter winter season.

Agriculture declined in quality after the collapse of the Roman empire. Capital investments in irrigation, in particular, were ill maintained, although the extent to which there was an actual agricultural recession in certain parts of the former Roman domain is now in question (Brunner 1995). Whatever happened in the post-Roman Mediterranean region, there was a new surge of farming innovation in the aftermath of Islamic conquest. With the advantage of a trading and, for a time, political system that extended without impediment from India to Spain, Muslim farmers introduced and acclimatized a number of additional crops, especially crops derived from south, southeast, and even east Asia. Two or three of the introduced crops may have been present earlier, even in Spain, which was the northwestern limit of Islamic colonization (Hernández Bermejo and García Sánchez 1998).

Rotation systems and intercropping were elaborated, cropping during the dry summer was extended enormously, and major improvements and extensions of irrigation made possible the close occupation of areas unavailable in Roman times, some of them at higher population densities than exist today. The expanded use of fodder legumes and manures integrated livestock more closely with arable farming. A great increase in the range of leafy and other plant wastes used in mulches and as compost and the addition of sand, gravel, marl, and lime, were all used to sustain soil productivity. Watson (1983:123–24) wrote,

> The introduction of such summer crops [rice, cotton, sugar cane, colocasia, eggplants, watermelons, sorghum, indigo, henna, safflower, and green gram] on a wide scale radically altered the rhythm of the agricultural year as land and labor which had previously lain idle were made productive. More than this, the opening of a significant summer season was one of several factors—perhaps the principal one—permitting systems of rotation which made much more intensive use of the land.

Development of summer cropping largely but not wholly eliminated use of the plowed summer fallow. The purpose of this, as in other regions where it is still practiced, is mainly to conserve moisture by eliminating evapotranspiration through grass and weeds. Relay cropping, intercropping, and new rotations were all practices that seem to have been imported, like most of the new crops, originally from India. The system was more intensive in its use of land, capital, and labor than that of classical times. It probably attained its greatest diversity in the Muslim regions of Iberia (Hernández Bermejo and García Sánchez 1998).

In subsequent writing, Watson continued to insist that the transformation was farmer driven. He wrote (1995:69) that "the vectors of the Islamic agricultural revolution seem for the most part to have been peasants, and to a lesser extent landowners, who learned to grow the new crops in their homelands in the eastern reaches of the Islamic world and then, in the course of migrations, took the agricultural revolution, by successive stages, westward." Yet rulers and big landholders also played a role. Early agricultural science, which was strongly developed in the medieval Islamic world, contributed to the spread of information. Although few technical innovations were made in irrigation methods, the elaboration and extension of their use seem to have demanded at least a measure of authoritarian support. Irrigation technologies included the levered bucket, or *shadouf,* and the vertical wheel to which buckets were attached. This was the *noria,* known in India in more modern times as the Persian wheel. Some very large *noria* were constructed to raise water from deep wells or lift it from rivers into canals at higher level. Most were powered by oxen, but some were powered by water. Some still survive in the Near East. Both were ancient Near Eastern technologies, but the *noria* was greatly elaborated in Islamic times and diffused much more widely as far west as modern Portugal (Watson 1983). Hourani (1991) emphasized the early emergence of large landholders, and Watson (1983) indicated that there may have been deliberate experimental farming in royal gardens. Whether or not what happened can be described properly as a farmer-driven agricultural revolution, it was a set of major and rapid changes that persisted for about three centuries. Some of the changes survived a subsequent period of decline.

Enough of these innovations survived in Spain for a number of introduced crops and practices, some of which are largely gone from the Iberian peninsula today, to be taken to the New World in the sixteenth century (Hernández Bermejo and García Sánchez 1998). Although some crops and methods reached Italy, very little reached western Europe. Much more change followed introduction of New World crops in the sixteenth and seventeenth centuries. Gómez-Ibáñez (1975) described a fairly comprehensive agricultural revolution in the western Pyrenees of France at the end of the seventeenth century. Maize, introduced earlier as a minor crop, formed no part of a rigid three-course system in which wheat was both winter crop and spring crop. Maize was widely adopted after a famine in 1690–96 in which the older support crops of millet, turnips, and beets failed. First it took the place of spring wheat, but this led to replacement of a three-course rotation by a two-course rotation, further leading to cultivation of legumes and pulses, to sustain fertility, and of fodder crops, which enlarged the livestock component. Manuring and use of mineral fertilizers to improve soil all followed by the end of the eighteenth century. The revolution necessitated total reorganization of an open-field management system that had been adhered to for a long time to minimize risk. Other changes followed in the nineteenth century.

EARLY AGRICULTURAL REVOLUTION IN NORTHWESTERN EUROPE

In traditional historiography, an agricultural revolution took place in northwestern Europe and especially in Britain between about 1750 and 1880, and its central characteristics were enclosure of open fields, elimination of fallow and its replacement by standardized crop rotations, improved tools, and the beginnings of mineral fertilization. This belief depended on a great deal of technical writing published between 1750 and 1880, leading to a view that a small number of innovators developed new systems that were widely adopted (Chambers and Mingay 1966). It also relied on substantial increases in both production and yield during these years, firmly established in modern research (Overton 1996).

Yet a great deal had already happened before 1750. Characteristically, the literature on the countries of western Europe tells no integrated story. The radical changes that began earlier in Flanders and the Netherlands anticipated much of what happened in England in the sixteenth and seventeenth centuries. Dutch influence on English practices has been argued by a number of writers (e.g., Smith 1967), and certain new crops and clover appeared first in the regions opposite the Netherlands (Overton 1985; Thirsk 1997). In a work that was controversial and remains so among historians, Kerridge (1967) detailed change across a wide range of English and Welsh agricultural systems from the early sixteenth century onward. He argued that this period, before 1750, saw the real agricultural revolution in England. Determinedly nationalist, Kerridge maintained that not much beyond a few crop introductions came from abroad. It was the work of the English farmers themselves, he insisted.

Remnants of forms close to shifting cultivation in northern and western areas vanished. Most widely, a division between continuous (but fallowed) arable land on one hand and permanent pasture on the other was replaced by ley farming, the alternation of arable land with pasture and meadow. The consequences included productivity improvements, greater diversification, and lower production costs. The system "married the livestock to the soil and extracted the greatest possible cereal and animal produce from the farm, whilst continually improving its fertility" (Kerridge 1967:202). New crops and improved pasture species were introduced or bred to supply larger areas. Horses replaced oxen. The use of manure, wastes, and mineral improvements to the soil steadily increased. There were some wholly new inventions, among which the most notable were the irrigated or "floated" water meadows. Irrigated meadows were developed much earlier in southern Europe, but Kerridge presented strong evidence that the English invention was independent. These meadows served multiple purposes, among which Kerridge (1967) placed particular emphasis on their use for lambs and ewes, which, after pasturing on their fresh growth, were hastened to the arable fields and there folded (penned) to provide manure for the grain crops.

Improved livestock were bred as their production became more important, and most of the modern English breeds of sheep and cattle evolved by the early nineteenth century. The common fields, with their strips, regular rotation, and strict management rules did not wholly vanish, but management was increasingly reorganized to permit more individual treatment of plots. Thirsk (1997) treated what happened in this period not as an agricultural revolution but as a major surge of an agricultural alternative to mainstream cereal-based farming. She introduced evidence to suggest that increasingly individual operation of fields was the basis of the growth of ley farming from small medieval beginnings to widespread practice by the eighteenth century. Enclosure of the common fields and common lands proceeded apace so that only a remnant, perhaps a quarter, remained for legal enclosure after 1750. The use of plowed fallow remained where it was necessary to keep down weeds and prepare the land for a demanding crop, but it diminished greatly. Green fallows enduring up to 4 or more years, and supporting livestock, took its place.

Whether or not the change was as native grown as Kerridge argued, an affluent and educated class, familiar with developments abroad, was strongly instrumental. "The gentry were the influential pioneers of new crops and new systems" (Thirsk 1984:xxiii). Kerridge showed that the early innovators were in almost all cases the substantial proprietors of land that was already enclosed. Once an introduction was proven, "the generality of capitalist farmers hastened to adopt" (Kerridge 1967:326). The tenants, especially the small tenants with most to lose from a failed experiment, usually followed last. Nonetheless, some introductions became very widespread in a few years. The fact that the period from around 1650 to 1750 was a period of low agricultural prices barely surfaced in Kerridge's account, but it is basic to the regional reinterpretation of the record in a collection edited by Thirsk (1984). Innovation was part of a struggle for survival in hard times, and the increases in output were not well rewarded until a new period of demographic and economic growth began in the late eighteenth century. Only then, with a mass demand for basic foodstuffs in what was becoming an integrated national market, was the full harvest of ingenuity reaped (Thirsk 1984). Thereafter, a new period of higher prices and increased demand for cereal crops led to a return to mainstream cereal farming. This was the context in which labor-saving changes in machinery, improved tools, and imported mineral fertilizers formed the core of new developments in the nineteenth century.

TWO COMPLETED EXPERIMENTS IN CONTEXT

In both the medieval Muslim world and early modern England, what is described as agricultural revolution took place over periods of around two centuries, which included shorter periods during which particular innovations were clustered. This long period of change was paralleled in other regions. In eastern France the main force in the eighteenth century was expansion of cereal cultivation, especially wheat, at the expense of the common pastures on which the peasants relied to feed the livestock needed to manure their own arable strips. Only on large estates was there any harbinger of a limited agricultural revolution that took place mainly in the next century (Vogt 1953; Sutton 1977; Clout 1977; Boehler 1983).

Yet in Alsace, where Vogt (1983) minutely examined the local historical record, big changes took place during the seventeenth and eighteenth centuries. Without any major reform in land tenure, and without abandoning the open-field rotations, peasants replaced rye with wheat and oats with barley and introduced mainly leguminous fodder crops and beans into the fallow, and some planted clover and lucerne (alfalfa). Farmers also experimented with a range of mainly commercial crops, which included rape, hemp, cannabis, madder, tobacco, and hops. Some of these changes, especially use of new crops and fodder plants grown in what had been the fallow period, can be traced back into the seventeenth century, at least as early as in England. Their recognition undermines the concept of a later agricultural revolution and adds to it a long and varied series of changes that constituted what Vogt (1983:232) described as a "pré-révolution complexe."

We noted earlier that crop changes alone do not make a revolution. Also involved are new ways of organizing production, necessitating investment in both social and physical capital. Individualization of land tenure and of farmers' decision making, important as almost a precondition in both the Mediterranean and English cases, were substantial social investments. Physical capital formation included greatly enlarged and reequipped irrigation systems in the Mediterranean and walls, hedges, irrigated meadows, and soil improvements in England. In southwest France incorporation of maize led to change in the whole agricultural system. Changes in labor input and in labor relations also were necessarily involved, and in the English case included the progressive ending of collaborative work arrangements in the common fields. With the emergence of great landed estates, tenant farming and farming with wage labor evolved to dominate English agricultural organization in the nineteenth century. In no case was transformation truly revolutionary, and a large part of it took place incrementally.

The growing importance of the market was present as a background force in early modern Alsace, where its strong influence was limited to areas close to the towns (Boehler 1983). In England, the agricultural market still lacked an integrated price structure until after 1750, and marketing depended on the proximity of towns, navigable rivers, and coastal ports. To find cases in which industry and marketing seem to have been major forces in early social and agricultural transformation, it would be more productive to go the Netherlands and Flanders. Rather than do this, I cross the world to a country in which agriculture was, and even now is, more comparable with the conditions in the developing tropics and subtropics. The example also will more readily distinguish the technical from the social aspects of revolutionary agricultural change.

■ Agricultural and Social Change in Japan, 1700–1950

TRANSFORMATION IN A COUNTRY WITHOUT SIGNIFICANT FOREIGN TRADE

Not until Japan was unified under the Tokugawa shogunate at the end of the sixteenth century, when internal warfare came to an end, did town growth and rural marketing become general through most of the country. In the 1600s, towns and cities grew rapidly, possibly at a greater rate than in contemporary Europe. There was a proliferation of castle towns into which the warrior class *(samurai)* was concentrated to keep them under the firm control of the feudal lords. Thereafter, population growth almost ceased for a century and a half; growth did not resume until the mid-nineteenth century.[2] In the absence of any significant foreign trade, all the grains, tubers, timber, fish, and fibers needed to support the urban populations and permit interregional specialization of production came from rural communities. Except where most pre-1600 towns had been, in the Kinai area around the eastern end of the Inland Sea, villages had hitherto produced little or nothing for sale.

At first, rice reached the market mainly through taxation, taking 30 to 40 percent, sometimes more, of village rice production. Tax was levied by the feudal lords of each small region on village communities as a whole. Village heads, appointed by the lords and their *samurai* administrators, were responsible for collecting this tax from all landholders. Cadastral surveys were undertaken to determine the productivity of each unit of land as the basis for rice tax assessment. The tax rice was distributed in the castle towns and, together with surplus production from the largest households, also sold. Interregional trade evolved. Although the tax was heavy, the basis of assessment rarely was revised, so that the more productive farmers were able to retain a larger share of their output as time advanced (Smith 1959, 1988).[3]

Village lands in most parts of the country contained irrigated rice fields, which were the core of the farming system. Before about 1700, irrigation almost everywhere was done on a simple gravity basis, without canals, feeding down from field to field so that those with the lowest land were most insecure in their supply of water. Villagers also cultivated upland fields that, at least into and through the eighteenth century, were more important for the subsistence of ordinary peasants. These fields produced wheat, millet, sorghum, barley, buckwheat, soya beans, vegetables, and root crops, as well as upland rice. Peasants ate these foods at least as often as they ate rice, and when they ate rice it was usually mixed with barley and sometimes wheat to eke out the supply (Ohnuki-Tierney 1993). Many peasants, especially in the mountainous inland areas where only small plots were irrigated, ate very little rice; this continued until after 1950 (Moon 1989). From the nineteenth century, crop diversity was decreased sharply, with millet, wheat, and even barley and soya beans progressively dropped. Some of this crop diversity has been restored, but only under a new government policy introduced in the 1960s. Meanwhile, the rural economy became diversified in different ways, especially through the growth of silk production and through widening opportunities in industry and commerce. This diversification of opportunity was fundamental in determining the pattern of rural change.

Already by the end of the seventeenth century, although population had ceased to grow, rural employment in industrial and commercial occupations was increasing. Some of this employment arose in the towns, but a large proportion became available in the villages, with much work being put out to rural families. Cotton manufacture, little known before the seventeenth century, expanded rapidly, as did other industries. By the eighteenth century, the growth of rural industrial production, at costs below those of urban industry, began to constrain the latter, leading to widespread reduction in the population of the castle towns and smaller cities. As monopolies and other restrictions were increasingly disregarded, trade also became decentralized. Surveys in the early nineteenth century, brought together by Smith (1988), show from about one-fifth to more than four-fifths of the incomes of rural communities being derived from nonagricultural employment. This was the context in which a major transformation began to take place in the seventeenth and eighteenth centuries.

AGRICULTURAL IMPROVEMENT

In terms of the landscape transformation it generated, what happened between 1600 and 1900 in Japan was more dramatic than the changes that took place during the same period in England. Major investments in canal irrigation and field drainage were central. Employing a new engineering profession, feudal lords and great landlords who grew rich on their taxes and rents, supported by merchants, later undertook much larger works in the lower valleys. By doing so they increased their own revenues. Big areas, formerly uncultivated, were brought into production. Some of these lower valley and seashore works involved engineering on a major scale and the conscripted labor of thousands of workers. Many new villages were created. An example, near the Inland Sea in western Japan, was described by Beardsley, Hall, and Ward (1959).

Through well-organized village communities closely structured around their few leading families, resources were mobilized greatly to enlarge the small areas of irrigated land, convert dry fields into irrigated fields, and introduce many technical improvements in water management (Smith 1959). Where fields could be drained and where there was sufficient winter warmth, cultivation of a dry winter crop was possible, and drainage as well as irrigation techniques became widespread in the eighteenth and nineteenth centuries (Francks 1983). Upland fields continued to grow the dry crops needed for local consumption, as well as tea, and mulberries to support silkworms.

Even if fields were fallowed each winter, as they continued to be in the colder north, permanently cultivated fields had always needed to be fertilized in Japan since earliest historic times. Very large quantities of grass, ash, and manure from surrounding uplands were annually carried down to the fields to be spread on them. What was called guest soil *(kyakudo),* collected from the uplands, riverbeds, and ditches, was also spread on the land and especially on newly reclaimed irrigated land (Francks 1983). This use of forest litter and grass had significant ecological consequences. Deprived of recycled nutrients, large areas of Japanese forest became impoverished in species, consisting largely of those able to make heavy use of mycorrhiza. In turn, these mycorrhiza provided important supplies of edible mushrooms. Modern use of purchased fertilizers, now principally inorganic, has ended the use of organic matter from the forest floor. Some recovery in forest biodiversity is reported, but wild mushrooms are in short supply.

In the seventeenth and eighteenth centuries, commercial fertilizers, at first consisting of dried fish, oil cake, and matured night soil, began to be bagged and traded over the more accessible parts of the country, contributing greatly to higher crop yields. This heavy use of fertilizer placed a premium on crop varieties that could make effective use of the supplement, adding to grain production rather than vegetative growth. Improved, mainly short-stem varieties of rice and wheat were selected by farmers and, over time, diffused to other areas. These improvements facilitated multiple cropping.

New ways of farming came together in the late nineteenth century under the term *Meiji Noho* (Meiji agricultural ways), although many had originated much earlier (Francks 1983). Important elements were the widespread adoption of higher-yielding varieties of rice, great expansion of double cropping, increased use of commercial fertilizer, and both irrigation and deep plowing to make the fertilizer useful, involving use of draft animals and cultivation of fodder crops. There was also a large range of secondary improvements, especially in rice cultivation methods. In an era before there was any regional or national system for diffusion of new methods, "the vast majority of Meiji Noho techniques were developed by individual farmers out of their own experience and experiments" (Francks 1983:57).

Rice varieties remained essentially local, but good strains from the point of view of yield, quick maturation, or taste had even been diffused during the Tokugawa period despite feudal restrictions that were hard to enforce. The first specific high-yielding rice to win widespread adoption was developed by a farmer near Osaka in 1877. Because of this early premium placed on short-stem, higher-yielding varieties that would hold heavier seed panicles without falling over ("lodging"), some of the first wheat and rice strains used in international experiment station breeding in the early twentieth century came from Japan (Dalrymple 1976, 1986).

SOCIAL ASPECTS OF CHANGE IN EARLY MODERN JAPAN

Fully commercial nationwide rice marketing began in the nineteenth century. A decree of 1873 abolished taxes in rice and replaced them with a national cash tax at 3 percent of the value of individually registered rice land; the land was resurveyed for the purpose by the mid-1880s. This change increased the sale of produce from all parts of the country, including areas until then largely untouched by the market.[4] Agricultural improvements added greatly to production, but, to a much greater degree than in pre–nineteenth-century Europe, they also added to labor requirements. They did so in a context in which labor for farm work was growing increasingly scarce. Not only did irrigated rice demand more labor than the dry rice it displaced, but the principal cash crops—cotton, tobacco, sugar cane, and indigo—and the silk industry were all labor-intensive. They also demanded greater skill and knowledge on the part of the workforce (Smith 1959). The engineering works needed to improve irrigation systems and create drainage to permit winter cropping used labor in quantity.

Concentrated land holdings of big extended families were inherited from medieval times. At their core was a nuclear family, the possessor of the land rights. Around this was a circle of affinal and cognatic relatives, and around that a circle of unrelated bond servants of different types. In the seventeenth century, holdings of up to 5 ha were not uncommon, and these required the labor of up to 20 full-time workers (Smith 1959). As nonfarm opportunities increased, dependent and bonded labor became harder to obtain. Increasingly, large holdings were subdivided. An already old practice among large land-holding families of forming branch families but allocating them little land grew more widespread right through into the mid-twentieth century. This subdivision ensured that the new client families would still provide labor to the core family without being directly dependent on the central household for support.

Smith (1959) showed how this form of subdivision provided capital and patronlike assistance to the branch families. Freed bond servants might be added to the kinlike group of households thus formed, and this became more common by the nineteenth century. Households that could not meet their debts would sell part or all of their land to the successful core families, also becoming dependent households. By the late nineteenth century, with the disappearance or abolition of labor services, all such groups had adopted forms of landlord–tenant relationship. Until the late nineteenth century, the title of land for the whole kinlike group would be held by the core family, which was therefore the unit that paid rice tax for the whole kinlike collective of farms.

Land in downstream villages with inferior access to irrigation water sometimes was sold to upstream landowners to meet debts, the former owners becoming tenants of landlords in other communities or living away from the region. While the core families did less and less of their own farming, as tied labor ceased to be available to them, they or their successors continued to amass land (Smith 1959, 1988). Some small families and former tenants were very successful, often in businesses such as sake brewing, and became major landowners. Some old core families declined, and they fell into a subordinate relationship. Throughout the Tokugawa period to 1868 and later, crop share rents were increasingly converted into money rents, but never wholly.[5]

Intravillage cooperation was essential for irrigation management when water was fed from field to field rather than through canals. It declined as farming became a more individual enterprise, but irrigation districts formed their own associations to defend or enhance the rights of their members. There has been remarkable continuity. Cooperation between farmers linked through old kinlike relationships, in hamlets and villages, and through local associations continued in a number of areas of activity in the 1980s. Moore (1990) traced its presence in a northern region that today produces high-quality rice, commanding the highest prices, for the national market. The medieval distinction between the few core families and others did not perish quickly.

A Japanese agricultural writer, Okura Nagatsune, who wrote in 1822, saw the principal object of progress as being "to reduce the people's labor." This was achieved by substituting more efficient methods by which to increase yields, reduce costs, or require less strength and skill than the methods they supplanted (Smith 1988). This involved not only inventions but also great attention to time management in the contexts of a severe and growing shortage of farm labor, a major increase in nonfarm employment, and, until after 1850, a static population. Important as a longer-term objective was improvement in the productive capacity of the land. Okura and other agricultural writers insisted that the key to survival was seizure of the opportunities that the time offered.

THE JAPANESE EXPERIENCE IN PERSPECTIVE

The great improvements in technology and production that took place between the seventeenth and twentieth centuries could as properly be deemed an agricultural revolution as the contemporary developments in England. But in social terms, the consequence was to make farms smaller rather than larger, as the big household units were broken up into branch farms and tenancies (Smith 1959, 1988). In 1986 the average farm size in Japan was only 1.2 ha (Moore 1990). For the small farmers, and especially the landless or near-landless tenants, there was little change in status or standard of living until well into the twentieth century. They continued to eke out small quantities of rice by mixing it amply with barley, used a variety of collected semiwild foods, depended on casual labor opportunities serving the more affluent, and lived in great poverty.

A distinctive novel about the people of a village north of Tokyo, written in 1911 by a first son of one of its leading households, graphically describes the condition of the landless small farmers in the early twentieth century (Nagatsuka 1989). The small family at the core of the story rented all its patches of land, even its house site. Its members depended heavily on support of the leading household and other neighbors and on such paid work as could be found. The farmer was almost constantly in debt. He farmed only one small patch of irrigated land and scattered blocks of dry land. There was no starvation, but family poverty was extreme. Even though the site of this village is now almost immersed within Tokyo's northern outskirts, the nearest industrial employment opportunities in 1911 were a long day's walk away. For peasants like those described, there had been no social or agricultural revolution.

In many parts of Japan there had been more progress in the conversion of Japanese village communities from pure farming communities to communities in which a large part of income was derived from nonagricultural work. The number of full-time farmers in Japan is now very small. The spread of industrial employment into the countryside, which began in the eighteenth century, became deliberate official policy in the second half of the twentieth century. By 1985, the average Japanese farming family earned only 15 percent of its income from farming (Moore 1990). Because of wage income and price supports, rural incomes have stood up well by comparison with urban incomes. The real social revolution in rural Japan has coincided with a new and modern agricultural revolution based on chemicalization, standardized cropping, mechanization, and major new innovations in land and water management, almost all since the 1950s. I end the story here at its beginning. The basic differences between what happened in Japan and what happened in Europe and North America were that in Japan farms, and with them the scale of mechanization, remained small. Commercial management of the land in company farms made little penetration. Up to the time of the modern so-called Green Revolution, there is a nearly continuous and seamless story from the sixteenth to the twentieth century in Japan.

■ Japan and Java

THE COMPARATIVE ANALYSIS OF CLIFFORD GEERTZ

Java has been mentioned several times in this book, and this crowded island with more than 100 million people will be introduced again in discussing the Green Revolution in chapter 12. Writing on change in Java from the 1830s to the 1930s, Clifford Geertz (1963) introduced a comparison with Japan. He asked why the rice-based agriculture of Java had fallen into the "involutionary" trap that he famously (or infamously) described rather than attaining higher productivity and income, as in Japan. The pressures on people and the land generated by population growth in Java were described graphically by Geertz (1963), with his famous constructs of "agricultural involution" and "shared poverty." Almost every aspect of Geertz's construction has been demolished by subsequent scholarship. A good summary of the main arguments was written by White (1983). Others have shown his case material to be unrepresentative of Java as a whole. Despite all this criticism, Geertz's insights retain a strong influence.

Geertz's starting point for his comparison was close comparability between the rice yields achieved in the two countries in the mid-nineteenth century. His explanation lay in the contrasted histories, for whereas Japanese farmers remained integrated within the national economic system as industrialization proceeded, Javanese farmers remained marginal to a national economy built around foreign-owned commerce and industry. Geertz's comparison seems to give comparable weight to the tax-in-rice system of the Tokugawa era in Japan and the mandatory deliveries of selected cash crops under the "culture system" introduced in Java by Governor van den Bosch in the 1830s. They were in fact very different. There was also nothing comparable with Japanese rural industrialization in Java. Also important is the absence in Japan of anything but a brief experience of quasicolonial rule, in the immediate aftermath of World War II, and this produced the major land reform of 1946–49 rather than the exploitation that was the more normal colonial experience. A selective review of what happened in Java will help illustrate how colonialism could greatly interfere with any farm-based evolution of change.

EVOLUTION OF THE COLONIAL SYSTEM IN JAVA

In western Java precolonial rulers demanded labor service and a poll tax, but the cultivators were very free to make their own decisions. The Dutch East India Company sought tropical produce, which at first they bought on the market, but they soon found this unsatisfactory because supplies could not be ensured. After about 1700 they began to seek guarantees of supply from the west Java rulers (Hoadley 1994). This had a major effect on land use, tenure, and labor relations. It did so decisively after coffee, a wholly new crop with ecological needs distinct from those formerly grown, was added to Dutch demands in west Java in the early eighteenth century. To guarantee production, hillside terracing was necessary, which took corvée labor. In turn it required the concentration of a rural population that, under a then-low population density, had practiced land rotation farming with only a small amount of wet rice. Concentration meant creating fixed villages and wet rice fields to support them. Controlled, but supported by the company, rulers appropriated exclusive rights to the hillside terraces made for coffee and to the labor needed to keep them in production. These changes transformed almost all aspects of society in west Java. A hierarchy emerged, involving the Dutch Resident, the local ruler, and appointed village chiefs. The same basic structure later spread to other parts of Java and persisted as the means of both rule and exploitation.[6]

Under the culture system introduced in the 1830s, taxes in money were replaced with the requirement that one-fifth of the land be devoted to the state's crops, or that one-fifth of labor time be available for corvée work. The state's principal crops were coffee, sugar, and indigo, and the first two of these were grown on plantations as well. Rapid population growth quickly filled up the remaining land. After the end of the culture system in the 1870s and the encouragement of mainly Dutch private enterprise, the control thus gained over village land led in central and eastern Java to the contracting of peasant land on a rotating basis to the capital-intensive sugar factories. This put increasing pressure on the rice land. Remarkably, the authoritarian system for supplying the sugar factories with cane grown by villagers, on a regulated share of their fields, has continued well beyond the end of the colonial period. In east and central Java, where sugar production is concentrated, it continued in the mid-1990s, keeping large areas of the best irrigated rice land under commercial sugar production (Manning and Jayasuriya 1997).

Not from the eighteenth century to the mid-twentieth century were Javanese farmers free of the demands of first the Dutch East India Company, then of the colonial government and the Dutch entrepreneurs whom that government supported. Authoritarian demands on them have continued through the period of the so-called Green Revolution, described in chapter 12. The nature of the demands imposed conditions on the Javanese farmers that were qualitatively different from those experienced by Japanese farmers, though different only in degree from what happened in many other colonized countries. During none of this long time can it be said that Javanese farmers, as a whole, were in a position to make the comprehensive innovations that generate an agricultural revolution. Much of the decision making in land use, together with many investments in irrigation and drainage, belonged to the authorities and the entrepreneurs whose cash economy was the main beneficiary. This does not mean that farmers were unable to make local modifications to their land use practices. Nor does it mean that the foundation of the extensive terracing that characterizes so much of Javanese agriculture was not laid by generations of farmers before the eighteenth century. But it does mean that those who drove agricultural change in Java during the past three centuries were not the farmers. The fate of Java before modern times was the product of what was probably the most profound invasion of commercial demand ever laid on any tropical developing country.

■ Conclusions

These historical examples make clear that, at least in the absence of heavy colonial exploitation, systemwide agricultural change in the past has mainly been farmer driven, but rarely by the small farmers alone. There have always been leaders and followers, and in every case major change has been part of a wider set of social and economic changes. Where farmers have been able to do so, they have innovated, but they have always adapted their practices and social arrangements to accommodate changing external conditions. All the cases discussed in the next two chapters are in countries that have experienced colonialism. All the adaptations discussed there arose within and after the colonial period, and mostly within the last half-century. This chapter's excursion into an older history will help put these more modern transformations into perspective.

Notes

1. Paul Richards's (1985) use of the term *indigenous agricultural revolution* is ambiguous. On one hand, he regarded the development of intercropping by west African farmers through their own unaided experimentation as progress toward an agricultural revolution. On the other, he was simply calling for farmers' experimentation to become the basis for a better-informed scientific intervention. He was not using the term to describe any accomplished set of events.

2. For one central Japanese village, demographic trends were analyzed by Smith, Eng, and Lundy (1977). Deliberate family planning, including infanticide, was clearly involved. There is only the most limited evidence of any sustained high mortality.

3. There is a large literature by Japanese historians, but little is translated. I therefore rely on English-language researchers and interpreters, especially T. C. Smith, who provided extensive citation of his Japanese sources.

4. In the twentieth century, rice marketing was brought into the hands of farmer cooperatives, linked since the 1960s into a major national cooperative organization that manages a large part of farmers' dealings with the rest of the national economy.

5. Rent in kind has even resurfaced in modern times as an unexpected consequence of the 1946–49 land reform. After this reform, with its allocation of titles to former tenants and legal restrictions on the size of family holdings, new tenancies rarely were registered. They have arisen nonetheless, and rents have sometimes been paid in kind, even with as much as a half-share of the rice crop (Moore 1990).

6. Central and eastern Java have had very different histories and social systems from the mainly Sundanese western part of the island. The core of Java is in central Java, where medieval kingdoms arose already based on wet-rice cultivation. In east Java, land tenure was formerly (and even into modern times still notionally) communal, with land belonging to a village as a whole.

Chapter 11

Farmer-Driven Transformation in Modern Times

■ A Focus on Spontaneous Change

DOING IT THEMSELVES IN THE TWENTIETH CENTURY

Twentieth-century farmers have been shown at many points in this book as adapting to new conditions, in some cases changing their farming systems in the process. In only a few instances has it been wholly possible to separate farmers' own agrotechnical innovations from those that have drawn on ideas and methods imported from elsewhere. Before coming to farmers' adaptations in the context of dominant external forces, in the Green Revolution and in very recent years, I want to spend some time on cases in which there has been major change but where external intervention has been minimal or where local initiative can be separated from the effects of external intervention. We have seen a number of such cases already in this book, but those introduced in this chapter offer material of a more specific kind.

Three cases are used. In two of them, new tools, crops, and practices have been either imported or adapted, but entirely under the control of the farmers themselves. The market has been a factor in both, to a limited degree in the first case, much more strongly in the second. The third case requires separation of the effects of strong external intervention and can be interpreted in different ways. These three are all in Africa, where so little that is positive is ever supposed to happen. They carry forward, from the historical examples treated in chapter 10, the finding that farmers can introduce a great deal of productive change by themselves, sometimes by the incremental methods introduced in chapter 3, sometimes in a true burst of innovation. There is an important theoretical context to interpreting change, and it provides a framework for discussion.

INTENSIFICATION: THE CONTRIBUTION OF ESTER BOSERUP

Farmers who introduce new combinations of production to increase productivity often are said to be intensifying their systems (chapter 10). This term is also used in other ways when there is a simple increase in inputs, whether labor, materials, or fertilizers. Shifting cultivators who shorten the fallow period in a dangerous manner sometimes are also described as intensifying. This confusion of meanings is unhelpful, but it follows from an important theoretical advance that I have kept at bay ever since chapters 3 and 4, where it was first mentioned. This is the theory of agricultural change proposed by Danish economist Ester Boserup (1965, 1970, 1981). For more than 30 years, most discussion of intensification has revolved around her influential historical model. This is the chapter where it can best be presented and later discussed.

Reasoning from experience in Africa and south Asia, Boserup saw an increase in population density as the driving variable that could trigger agricultural change. She began her model with the simplest of systems, in which population growth would lead to pressure on resources. Higher outputs could be obtained and sustained only by successively more intensive agrotechnologies, all demanding higher labor input that would permit more permanent use of the land. Ultimately, it would lead to the most intensive of possible systems: multiple cropping. In the model, the inevitable downside of intensification was diminishing returns for each marginal input of labor, so that although total output enlarged, the average return for each unit of input or new skill would decline and efficiency would be sacrificed for production. Boserup's full intensification sequence has been set out in chapter 4.

Although criticized, that historical model became almost orthodoxy by the 1980s. It is reductionist in its emphasis on one main force impelling change: population growth. It is also a unilinear model, even if we allow for an opposite process of disintensification when population pressures on resources are relaxed. Many writers have used Boserup's model by assuming, without firm evidence, that a shifting cultivation stage everywhere preceded the adoption of any more intensive practices. It was shown in chapter 4 that this sequence is questionable. In much post-Boserup literature, the fact that everything hangs on a supposed stepwise transition between least intensive and most intensive forms of land use has tended to impose a deterministic template on a highly diverse set of actual histories. There has also been a more theoretical debate, with successively new layers of the issue being explored. I have contributed to this post-Boserup debate (Brookfield 1972b, 1984, 1986). Among others who have reviewed or enlarged Boserup's work, note should be taken particularly of Spooner (1972), Datoo (1978), Grigg (1979), and especially Robinson and Schutjer (1984), who elegantly linked Boserup's reconstruction to the more dismal prognostications of the neo-Malthusians. Comments and variations have been advanced by many others, including Cowgill (1975), Turner, Hanham, and Portararo (1977), Turner and Doolittle (1978), and Turner, Hyden, and Kates (1993).

DEFINING INTENSIFICATION

In my own first main contribution, I proposed what has been regarded by some as a rigorous definition of intensification. I wrote,

> Intensification of production describes the addition of inputs up to the economic margin, and is logically linked to the concept of efficiency through consideration of marginal or average productivity obtained by such added inputs. In regard to land, or to any natural resource complex, intensification must be measured by inputs only of capital, labor and skills against constant land. The primary purpose of intensification is the substitution of these inputs for land, so as to gain more production from a given area, use it more frequently, and hence make possible a greater concentration of production. (Brookfield 1972b:31)

The emphasis was on inputs measured against a constant area of land. Thus extension onto new land would become *extensification,* a term commonly used, rather than *intensification.* There are problems with the term *extensification.* There is no problem if agriculturally unused land is being brought into production, but if improvements facilitate enlargement of an arable area within land that was used less productively before, this is not really expansion onto new land. Contrary to many writers (e.g., Goldman and Smith 1995), I would not regard this as extensification. Rather, it is the application of new technology, through investment, on land previously less effectively used.

The experience of reading for and writing this book has given me new perspective on my own earlier work, as well as on that of others. In late 1998, with a large part of this book already in draft, I was asked to speak in a workshop on intensification (Brookfield 1998). I offered a revision of views that, together with the exchanges it generated at the workshop, is reflected in this chapter. Most of the post-Boserup literature, including my own contributions, has concentrated on labor input and has treated skills as agrotechnical skills. In my revision of ideas, I sought to redress the balance by giving greater emphasis to capital investment by small farmers and to farmers' skills in organization. As Mary Tiffen (1996) remarked, Boserup gave emphasis to labor because in her day the prevailing emphasis was on capital. She sought to show that many investments would have the effect of increasing, rather than reducing, labor demand. A consequence of Boserup's persuasiveness, together with changes in fashion among economists, has been huge neglect of farmers' own investment in much of the subsequent literature.

The other aspect I emphasized in the 1998 workshop paper was the importance of management skills, applied to the organization of labor and resources (Brookfield 1998). In the Green Revolution programs to be discussed in chapter 12, management of the farm itself and its workforce was central to achieving rapid production increases, yet this was one area in which farmers received minimal guidance. They made such changes as they found necessary by themselves. I now introduce two more cases that have been held back until this point. The first gives stress to capital investment, the other to labor management. Both exhibit, better than any other examples in this book, the central importance of the theme of organizational diversity, proposed in chapter 3.

In a village in a poor part of Mali, itself one of the world's poorest countries, there has been substantial innovation and investment. The security of the principal investment has depended on the ability of the people to retain control of their own resources in the presence of a growing number of outsiders. Population density is low and the market economy has been only a minor element (Toulmin 1992). The second case, from Nigeria, directly entered the post-Boserup debate, much of which was reviewed around this example by Netting (1993) and Stone (1996). The Mali example is free of any mention of Boserup. Absence of this template helped its author to isolate some very significant elements in the pattern of spontaneous change. After these two, I more briefly introduce material from a third example, that of Machakos in Kenya. This case has already assumed a classic place in the literature.

■ Management and Investment in a Sahel Village

A VILLAGE IN MALI

The Mali example draws solely on work of an economist who spent 2 years in the Bambara village of Kala in the early 1980s and revisited several times (Toulmin 1992). An updating summary is provided as a case study by Scoones and Toulmin (1999). In this account of what is evidently the same community, Kala is called Dalonguebougou. Material in this chapter is drawn from throughout Toulmin's 1992 book. The study stressed the importance of capital investment in agricultural change and its relation to farm household organization, a field that is only weakly treated in most of the literature. Kala lies 60 km north of the old Bambara capital of Ségou on the Niger River. Bambara have been settled on and north of the Niger in western Mali since the seventeenth century, when they displaced earlier people and took over land for cultivation from migratory Fulani and Maure herding groups, turning the latter into clients. In the mid-nineteenth century, the Bambara state was partly conquered by the Islamic Tukolor people, a situation ended when the French captured the region in 1890. It has a highly variable rainfall, with a mean during the 1980s under 400 mm, significantly less than in the decades before 1960. Population density was less than $10/km^2$ in the early 1980s. Intercropped with cowpeas, two millets occupied more than 90 percent of cultivated land. Though unidentified, they probably were short- and medium-maturing varieties of pearl millet *(Pennisetum glaucum;* National Research Council 1996). Diversity lay in the management of land, water, and livestock and in the livelihoods of its 29 households, which ranged in number of people from less than 5 to almost 60. In 1981, 86 percent of the village population was in complex households containing more than one married man, in some cases several. The mean size of these households was 23.8 people.

Two types of fields had very distinct cropping and management practices. In a radius of 7–8 km from the village were large bush fields, averaging more than 20 ha in area, each held and worked by an entire household. Not properly shifting fields, their boundaries advanced on one side while they retreated on the other. In the 1980s, 20–40 percent of the land was new in each year. These fields grew early-planted long-cycle millet (120 days), with cowpeas and some groundnuts, Bambara groundnuts, calabash, and a variety of hibiscus. Second were the much smaller fields around and within 1.5 km of the village, averaging 4–6 ha, most of which were permanently cultivated. About three-quarters were cultivated by households, the balance by individuals and contracted Fulani herders. They grew mainly late-sown, short-cycle millet (60–80 days) with cowpeas and small areas of maize, sorghum, groundnuts, vegetables, and tobacco. Between the two field types were a few smaller bush fields.

Between the 1950s and the 1980s there had been a major increase in production on the village fields, with some expansion of area but without reduction in the bush fields. Sale of millet, mainly to itinerant traders, had increased. During the same period the role of cattle as the main store of wealth declined. Use of cattle and grain (and, until the twentieth century, cowrie shells) as means of exchange was progressively but not completely displaced by money. The large household continued as a common institution and was not much eroded by splitting into nuclear families. The village population remained almost wholly illiterate, and up to the end of the 1980s there was no schooling; there remained no local medical services. Although there were a few village stores, itinerant traders provided most hardware, cloth, and spices for cash and millet. Most cash was obtained from visiting grain traders and from seasonal earnings of young men working in town or elsewhere as migrant laborers. A high proportion of the money drawn out of the village economy was, and had been since the end of the nineteenth century, in the form of tax.

THE PATTERN OF INNOVATION

Toulmin's main concern was with measuring all flows within the village economy and with comparing the capability in almost all matters of the large, medium, and small households. In the process she analyzed patterns of investment and provided information on their history. Kala is an old village and in the 1930s suffered from repeated locust plagues that greatly reduced grain production. Assets were sold to pay tax, and cattle numbers diminished severely. Many men migrated all the way to Senegal to earn money. Grain was requisitioned for a colonial state reserve and, in the 1940s, for support of the military.

Government was remote and remained so in the 1980s. The colonial government had concentrated its efforts in more promising parts of Mali, but its demands on Kala were heavy. In addition to tax and grain, they included corvée labor and conscription in both world wars. Government sustained the power of the chiefs, and this ended only after independence in 1960, when legislation was enacted giving all citizens free access to land and water. Even through the 1990s, the Bambara of Kala managed to confine this free access to their own people, denying it to quasi-resident Fulani and Maure, who remained in a client relationship, accessing land only through cattle-herding contracts with Bambara households. This successful restriction has had profound impact on the course of change, as did the ending of chiefly control, especially in the important matter of water.

Kala village and its inner fields lie in a sandstone basin, with groundwater encountered at about 15 m in the center and 25 m near the edges. At the time of independence there was only a single large village well, under control of the chief. Later, and more rapidly after the mid-1970s, households dug their own smaller wells. All wells were dug by hand, and all water was drawn up by hand. The 43 private wells in use in 1983 were not just for domestic use and minor irrigation. Most importantly, they were used to attract annually migrating Fulani herders who, in return for water, would corral their cattle on designated fields of the well-owning household or one of its members. There were also resident Fulani, allowed to reside, keep cattle, and cultivate small areas on contractual arrangements in which they herded and maintained the cattle of the Bambara people of Kala. They had no rights beyond those stipulated in these unwritten contracts. Investment in wells greatly increased the supply of dung to the village fields, both making production more secure and facilitating expansion in the area of the village fields.

This form of investment was much more easily affordable by the larger households, with their bigger labor forces, than by smaller households that had to pay a well-digger. By 1983, only 5 of the 29 households, all among the smallest and poorest, were without their own wells. Some had two or three. Here, as elsewhere, the fertility gain from manuring lasts 3 or 4 years, although manuring above the level needed exposes the crops to risk in unseasonably dry weather. A well-manured field could yield 2000 kg/ha of millet, although the average in the 1980s was half this. Unmanured bush fields, by contrast, yielded only a little over 200 kg/ha. Despite this major contrast in yields per hectare, returns to labor were much more closely comparable between the two field types because of the large labor demand associated directly or indirectly with what was essentially the maintenance of fertility in the village fields.

The second main area of investment was in plows and ox-teams. Before the 1950s all tillage and ridging were done by manual labor using hoes. Heavy plows were introduced to the region in the 1930s, but none were bought by Kala farmers until the 1950s, during a period in which groundnuts were becoming an important cash crop. The groundnut period was of major importance in financing the first wave of capital investment in Kala farming, in plows and plow teams, and the first private wells. As rainfall declined progressively after about 1960 through the 1980s, groundnut yields declined so that by the mid-1970s they had ceased to be a major cash crop. The same trend is recorded in the closely settled zone around Kano in northern Nigeria (Adams and Mortimore 1997), but the causes were more complex: first a severe decline in the price, then drought, then major infestation with rosette disease. Farmers in northern Nigeria have retained their seed lines, and resistant varieties were introduced in the 1990s, but without widespread adoption up to the middle of the decade (Mortimore 1998).

Village blacksmiths very soon developed a lighter ridging plow, which could be drawn by two oxen. Later, a still-lighter weeding plow was developed for use between ridges and crop rows. By the early 1980s only one household in the village had neither a plow nor an ox. Most households had two plow teams, and the largest operated four or five during the weeding season. These innovations greatly reduced manual labor requirements and facilitated enlargement of village fields while sustaining the area of the bush fields. Each year, the plow teams were used first on the bush fields, where long-cycle millet was sown a month ahead of planting on the village fields. Preferably the land was ridged before planting, but often this was not possible, and in parts of the bush fields the land was ridged after planting. It was always ridged before planting in the village fields. Oxen were bought young from Fulani herders, who performed the castrating operation, and they needed about a year of training for work. The oxen were sold for meat at ages between 8 and 10. The largest maintenance cost was in providing labor to water the animals during the 7 or 8 months when they had to be brought to the village wells to drink.

Breeding cattle were the third main form of investment, of much lower value, and kept mainly as a store of wealth, a holdout from the past. Their milk was valued, as was that of goats and sheep. Milk sometimes was of critical importance in sustaining young children, who were otherwise fed after weaning on the same millet porridge that provided the main adult diet. By the 1980s, breeding cattle provided no products that could not be obtained from other sources. A fourth form of investment, not measured by Toulmin, was in donkey-drawn carts, also an innovation since the 1950s. These were used not only for carting manure and for trade but also to carry drums of water out to the bush fields so that much longer periods of work could be sustained on these outlying fields. A fifth form of investment, also not measured by Toulmin, was in the small quantity of goods necessary to engage in trade. Using cost/benefit analysis on 1980–81 data, with a well-considered set of assumptions, Toulmin calculated that oxen plow teams were the most universally rewarding investment. Wells were of high value to the large households but much less so to the small households without oxen. Breeding cattle were the least rewarding investment.

THE SIGNIFICANCE OF LARGE HOUSEHOLDS

The investments came together to make large multifamily households viable, notwithstanding the inevitable strains that arose within them. Large households were at less demographic risk than small households, and the consumer–producer balance was better sustained through time, while the household could more readily afford the absence of some of its members seeking employment or engaged in nonfarm work such as hunting. Most importantly, the large household was better able to amass and deploy the resources needed for capital investment. Within the household, some livestock might be held individually, and women and some men maintained small private gardens, including tiny, diverse irrigated plots. The ability to provide this individual incentive also was greater in large households than in small. Little agrobiodiversity was reported within Kala agriculture. Management diversity was more highly developed, but the main diversity lay in organization. Larger households had more opportunities to engage in a wider range of income-earning activities, which included not only migrating for work but also trading and performing services in the village. Incomes from migrant labor, which returned to the household, were of particular importance in funding the payment of taxes. The Kala household thus was a corporation that offered considerable benefits to its members. Probably for this reason and despite growth in household size, only a very small number of household divisions were recorded in the period between the early 1960s and early 1980s.

The larger households also had political strength, and their assembled resources, used to pay officials in the district center or even the national capital, were critical in resisting efforts by long-resident Fulani clients to obtain secure title to land or dig wells of their own. Were the national law to be applied successfully, much greater pressure would be placed quickly on grazing resources and water. Management of the water resource as a commons was already downgraded by the burst of private well construction that took place after the mid-1970s; the need to restrict access to water by outsiders, other than as clients of the Bambara wellholders, was increased by the enterprise of the resident villagers. Eleven wells were dug in 1980 and 1981, then 16 more in 1983. After the severe drought of 1984 and an extremely poor harvest obtained in that year, the pace of investment slowed, and only 3 more wells were dug by 1988. Toulmin did not mention any fall in the water table, but some wells filled only slowly, so that herders were asked to move their cattle elsewhere.

This small example is one of large technical innovation, coupled with modification of land and labor relationships. The economy is subject to the vagaries of rainfall. As described in the 1980s, Kala was still isolated and, although increasingly involved in the market, remained dependent on its own resources and organization of those resources for almost all its livelihood. As Toulmin (1992:273) concluded, farmers have been "continually reacting to changing climatic and economic circumstances by altering patterns of production and investment." Substantial increases in production were achieved, and there was general improvement in well-being, even for the poor, who were aided through times of stress by use of a number of leveling mechanisms in society, reviewed by Toulmin in an appendix. Patron–client relationships were avoided within the Bambara community, largely because such arrangements already existed through the inferior client status of the resident Fulani. Land was not a constraint at Kala except in the form of wet-season grazing resources, already heavily used by the growing number of Fulani visitors in the 1980s, and Toulmin regarded this an impending threat.

By the 1990s the pressure of new settlement on surrounding land was increasing, but the Bambara people of this and nearby villages remained able to "maintain customary powers over access to water, grazing, and land for cultivation" (Scoones and Toulmin 1999:128).

■ Management and Migration Among the Kofyar of Northern Nigeria

THE KOFYAR BEFORE THEIR MIGRATION

Some of the best material on dynamism in an agricultural system to be found anywhere is in the work of a small group of authors, principally the late Robert Netting and Glenn Stone, who, through a period of more than 30 years, studied a tribal people in northern Nigeria (Netting 1965, 1968, 1993; Netting, Stone, and Stone 1993; Netting and Stone 1996; Stone 1996, 1997, 1998; Stone, Johnson-Stone, and Netting 1984; Stone, Netting, and Stone 1990; Stone, Stone, and Netting 1995). During that time, the Kofyar people have largely given up an old and complex system of resource management in a mountainous environment to develop a new system on the nearby plains. The Kofyar are among a number of small tribes or peoples occupying rugged uplands among the savanna lowlands of northern Nigeria. Whether or not they had to flee into their defensible natural strongholds, they remained safe from external raids and demands only while in the hills. The strong Jukun state arose on the Benue plains to the south during the seventeenth century. The raiding Islamic Fulani people spread south on both sides of the Jos plateau after establishing control over the Hausa kingdoms to the north in the early nineteenth century. There was dim memory of slaves being taken from Kofyar and their immediate neighbors who lived on the edge of the plains, but the Kofyar area was inaccessible to mounted warriors. Only with firearms were the British able to impose colonial control on Kofyar in the twentieth century, ultimately only in 1931 (Netting 1968; Stone 1996). There was substantial local warfare among the Kofyar and with their immediate neighbors, and in the early twentieth century this led to the destruction of several villages.

When Netting began research on the Kofyar in 1960, they lived in a restricted environment on the southern edge of the Jos plateau and in the immediately adjacent pediment. Average population density was more than $100/km^2$, ranging from less than 50/km² in some hill communities to more than 400/km² on the closely cultivated land below the hills (Netting 1993; Stone 1996; Stone, Johnson-Stone, and Netting 1984).

The Kofyar system of the mid-twentieth century included management of steep slopes by stone-walled terracing, although the terraces were used only on a rotational cycle and were no longer central to the land management system. The terraces were first built by generations of which people in the 1960s had no memory, and it seems probable that the complex agricultural system had evolved over a period since at least the seventeenth century and possibly earlier. Close to the groups of compounds that formed loose villagelike aggregations, the system included a range of tillage methods adapted to site conditions and the needs of specific crops. These included raised beds, ridging, rotational ridging over mulch laid between former ridges, tied ridges (called waffle gardens in the reports) to retain water in the resulting basins and check erosion, and tillage to different depths for different crops. The system included burning of trash and cleared vegetation to create ash that, together with household ash, was applied as manure to specific crop plants as well as being dug into the soil. Most importantly, it also included use of manure from corralled goats and other small livestock, composted with the bedding grass. Almost all households had goats and chickens, and there were some sheep and pigs and a few dwarf cattle. At the beginning of each farming year, late in the dry season, the compost was applied heavily on the inner village fields. Rubbish was piled and burned in the fields, to add to the compost. It was used as seedbed material for millet.

By these and other means, inner village fields were cultivated permanently under a large number of field and tree crops. Cropping was dominated by an intercropped association of early-planted pearl millet *(Pennisetum glaucum)*, sorghum, cowpeas, groundnuts, pumpkin, okra, gourds *(Lagenaria vulgaris)*, and some maize (Netting 1968). Also important was a late-planted variety of *Pennisetum* millet. Fonio *(Digitaria exilis)* was of minor importance here but was and remains a major crop for the people living on the northern part of the plateau; Kofyar broadcast fonio seed, whereas most crops were either seeded individually or transplanted from seedbeds. Other crops listed in a large range included sesame *(Sesamum indicum)* and taro (both *Colocasia* and *Xanthosoma* spp.), here individually replanted from among old corms rather than being a volunteer crop as in Ghana (chapter 9). There were at least two types of yam, also the sweet potato and the Bambara groundnut *(Vigna subterranea)*. Tree crops, other than banana, included mango, papaya, and the oil palm, which does well, even this far north, in the wetter climate of the plateau edge. It is noteworthy that this crop complex contained several introductions from Asia and the New World, as well as African domesticates, and new crop varieties were still entering the area in the mid-twentieth century.

Most crops were subject to distinct cultivation and manuring treatments. As thus practiced on the fields close to the villages, this was a highly elaborated system that would bear the adjective *intensive.* It included crop rotation. The system involved capital investment in the form of soil improvement, but it did not apply to all land. Only the inner Kofyar fields were cultivated permanently. Outlying village fields and the terraces received little manure and grew mainly late millet and fonio in rotation with groundnuts, together with some tubers. The groundnuts received some ash. These outer fields were rested for a few years from time to time. More unambiguous use of shifting cultivation practices was confined to the furthest out-lying fields, including some already on the lowlands.

THE MOVE ONTO THE PLAINS

Netting (1968) was describing a conservationist farming system. In his 2 years in the field, he saw little erosion to report but instead stressed its prevention and the development and maintenance of soil fertility. He also showed how the organ-ization of labor and rights to resources were essential to maintenance of the system. Already, change was in progress. Temporary farms on the plains began to be opened even in the 1930s, and by 1950 settlement extended beyond the thin soils of the pediment near the escarpment onto much better soils overlying sedimentary rocks further south. Land was either unoccupied or could be had for a nominal payment to local chiefs whose people were not using it. By the 1950s, good land was being used for cash cropping, especially of yams for the national urban mar-kets, using shifting cultivation methods within large farm blocks. The settlers cleared the savanna forest, mainly by burning individual trees, and rapidly occu-pied the land at fairly low population density.

Netting and Stone (1996:53) remark wryly that migrating Kofyar felled the for-est in a manner "betraying no expressed ethic of noble savage conservationism." Boserup-like, Kofyar seemed to be responding to land abundance by adopting a system of far lower intensity. Netting (1968) reported farmers as neglecting old practices because the land was good, and the care applied in the hills was no longer necessary. But this was not a casual shifting cultivation, and even in the early days of settlement crops were planted "in a thoroughly worked field, the grains going into parallel ridges and the yams into knee-high heaps" (Stone 1996:89, citing Netting's 1961 notes).

By the 1980s the population density in the new settlement areas had reached 100/km^2, and current cultivation occupied 50–80 percent of all land. Stone (1996) and Netting and Stone (1996) described how a new household farming system came into being, with the mainly commercial yams being grown in rotation with an intercrop of sorghum, *Pennisetum* millet, cowpeas, and groundnuts. A number of new sorghum varieties were grown. A greater variety of crops was grown in kitchen gardens near the houses, though without the former range of spices. Other crops occupied appropriate niches, including taro under trees and rice by streams and in swampy areas. Manioc was introduced in the 1980s and became established as a minor part of the farming system. Because manioc remains in the ground into the dry season, it had to be fenced against the free-ranging goats, which were turned onto the individually held arable land in this season. Until the new agricultural year began, the arable land became a common-property grazing resource (Netting and Stone 1996). As well as pigs, goats were reintroduced, tethered during the growing season and free-ranging in the dry season but penned at night so that their composted manure became available for each new crop.

At first the plains settlers retained their hill farms, but by the late 1960s many were moving their homes onto the lower ground. This continued until the 1980s, by which time most Kofyar lived on the plains, and the hill villages had become small, the homes mainly of old people. The old village fields in the hills "were cropped extensively by the few neighboring families still present, growing unfertilized groundnuts or *acha* [fonio]" (Netting, Stone, and Stone 1993:242). Despite appearances in the mid-1980s, hill communities had to some extent revived by 1994, now obtaining significant infrastructure support from the state in the shape of roads, schools, and water supply. Kofyar retained a permanent territory in the hills, something they did not have on the plains, giving them political leverage (Stone 1998).

LABOR MANAGEMENT

Particularly important was labor management. In the hills, farm families were small. On the plains, where labor rather than land was the constraining element, larger farm households evolved quickly, with the mean size almost doubling in 10 years (Stone, Johnson-Stone, and Netting 1984). Stone, Netting, and Stone (1990) analyzed labor scheduling in depth in a sample of 15 plains households in 1984–85. The households averaged 4.1 adult workers. Work was classified by task, through a period of 50 weeks, and shown against the calendar of the agricultural year, in which the tightest scheduling involved planting and harvest of the quick-growing millet. Depending on the rainfall, the peak labor demand of ridging the land might conflict with the equally pressing need to plant millet early, as at Kala, necessitating adaptations in sequencing.

The heavy labor of preparing mounds for yams, best performed when there was still rain to soften the soil, was done in the period after the millet harvest and before the sorghum harvest. This also had the effect of weeding the still-standing sorghum and preparing the ground for catch crops of sesame. Including groundnuts, "work on five crops had thus been arranged into five separate 'shifts'" (Stone 1996:109). Tasks were scheduled to diminish the effect of bottlenecks. Such periods of stress were managed by drawing on two forms of cooperative labor, traditional from the homeland but of less importance there. Hired labor was little used. Stone (1996) showed how the organization of people into communities, and their distribution of settlement, facilitated the mobilization of cooperative labor. Slack time was filled with work on crops that were less demanding in their timing.

Few jobs were gender specific, and family members worked almost interchangeably through a great range of tasks. The system as a whole was scheduled to achieve efficient use of a total working time that was longer than had been normal in the hills and much greater than that of most shifting cultivator populations, calculated on the basis of comparative figures not always well controlled for area. Kofyar annual adult labor inputs into fieldwork and processing in 1984–85 were measured at an average of more than 1500 hours per adult. Mean farm size, in two samples, was 5.4–5.7 ha in 1984, with 4.3–4.5 ha under cultivation. This would yield more than 350 hours per adult per hectare on average (Netting, Stone, and Stone 1993; Stone 1996; Stone, Netting, and Stone 1990).

EXPLAINING CHANGE

Netting and his colleagues saw the reintensification of Kofyar farming on the Benue plains mainly as a Boserup-like response to the rapidly increasing population density. Yet the primary basis of the farmers' economy had also changed from a mainly subsistence orientation to a commercial one, supplemented by subsistence crops. Yam cultivation, overwhelmingly for sale, absorbed more than a third of total labor inputs in the 1980s (Stone 1996). Heavy new demands were placed on the organization of farm work, within a field distribution different from the more fragmented distribution in the homeland. Invention of a new system, drawing on the old but not identical with it, was necessary for maintaining the new commercial economy as well as the growing population. New intercrop mixes, new social relations, and an expansion of women's specific roles in agriculture were all elements. As the settlement frontier filled up, it also became important to establish more firmly the control of land. In the 1980s, land was again growing short, and conflict with Tiv neighbors was emerging (Stone 1997). By this time, there was much use of inorganic fertilizer, then cheaply available. The record does not tell what happened in the later 1980s and 1990s, when fertilizer became more costly and harder to obtain.

In Netting's final work in the early 1990s, the theme became the agrodiversity of the system (Netting and Stone 1996). A common supposition that a decline in biodiversity follows adoption of more intensive practices was questioned. On the hill farms still maintained by their owning households at a low intensity of use, different kinds of beans and some upland rice had been introduced. The upland farms also became a source of seed stock for the lowlands. On the plains, the seminatural biodiversity of a forested landscape that had not been much used since before a period of wars in the seventeenth to nineteenth centuries had largely gone. However, a lot had also been conserved by being protected for its utility.

A large number of wild plants remained or became established on vacant land, by paths, and on field breaks. Reporting a single informant in 1994, Netting and Stone (1996) listed and reported uses for 31 trees or parts of trees, 7 wild tubers, 5 wild legumes, 11 wild plants used as vegetables or for seasoning (3 with craft use), 7 wild grains (some with several varieties), and 12 varieties of a wild yam. Tree crops, still confined to the hills in the 1960s, now became important on the plains farms. As field borders changed position and as specific microenvironments were brought into use, individual farms became "shifting mosaics of land-use and crop types" (Netting and Stone 1996:63). Although most households bought some of their food by the 1990s, none depended only on store-bought or market supplies. The resilience of agrodiversity is by no means the least important among a great range of findings from this remarkable piece of long-term research.

A NEARBY COMPARISON

The northern Jos plateau provides some interesting points of comparison with the Kofyar. Farming technology, under population densities between 150 and 300/km^2, was in many ways comparable until the 1970s. It lacked the stalling of goats but included deep compost pits where mixtures of ash, domestic wastes, cow manure, sawdust, crop residues, and grass were collected, burned, then covered with green grass to be dug up when needed (Phillips-Howard 1994). This compost, applied to the land, improved the soil. In the early 1990s, most cow manure was provided by the herds of migrant Fulani pastoralists, who were given access to the crop stubble and fallow areas. The unimproved soils on the ancient landforms of the plateau, which is partly mantled by a colluvium that developed more than a million years ago (Morgan 1979), are of strikingly low mineral fertility (Phillips-Howard 1994). Severe land degradation, described in the 1950s (Grove 1952), included gullies that concentrated surface flow to make it usable for irrigation on the valley terraces.

As also in Kofyar, this system was greatly modified in the 1980s by the availability of subsidized inorganic fertilizers, permitting increased crops of maize, potatoes, and other vegetables for sale in a nearby urban market in place of the traditional fonio, millet, and sorghum. After the mid-1980s, distribution of fertilizers became sporadic and unreliable, and by the early 1990s the entry of middlemen made it hard for small farmers to get supplies except from a black market on which the price was many times higher than the official price. Some farmers experimented with ingenious mixtures of the scarce inorganic fertilizer with the old organic mixes. Some mixed animal and poultry manure with town refuse and ash. Use of these composted mixtures improved the soil even on the mining spoil of old opencast tin mines, making possible intensive vegetable production (Alexander 1996).

■ Interference and Invention in Machakos, Kenya

INTERPRETATIONS OF THE AKAMBA

The third case has already become famous as one of the success stories of modern Africa. The subhumid to semiarid spur of hills and their associated pediments that runs southeast from Nairobi, paralleling the road to Mombasa, is the home of the Akamba people. Here, despite a sevenfold population increase, farming systems have moved since 1900 "from extensive cattle and goat rearing, accompanied by shifting cultivation . . . to permanent terraced fields with ox-ploughing, intensive tree-crops and horticulture, and relatively integrated crop and livestock enterprises" (Tiffen, Mortimore, and Gichuki 1994:226). What is more, the region has been transformed from one in environmental crisis to a state of good and improving management. This is not a case of people who have adapted and invented in isolation or neglect, certainly not since the 1930s. External interference began early when large areas were taken from Akamba for European settlers or put into a reserve of "Crown land" that the people were not able to recover until after national independence in 1960. Interference has continued throughout.

Taking a jaundiced view of the perceptions of government, both colonial and independent, and their specialist advisers, Rochelau, Steinberg, and Benjamin (1995) showed how an overlapping series of perceived crises produced their own problem-specific solutions. Perceived crises have been constructed successively around bad livestock management and overgrazing, soil erosion, underproduction and underdevelopment, deforestation, fuelwood shortage, and then loss of biodiversity. These external constructions have overlain real crises of epidemic disease in the 1890s and two famines since 1980. Capacity to cope sometimes has been undermined by the solutions imposed during previous perceived crises. Tiffen, Mortimore, and Gichuki (1994), writing for the World Bank, also were critical of some of the interventions but emphasized those that were more positive. They also noted many spontaneous changes made by the Akamba themselves.

Since 1930 the volume of agricultural output has increased markedly, especially on a per-hectare basis, despite large-scale population movement into areas of lower production potential. Use of an improved set of maize varieties, developed on a government research station, has been a valuable aid to higher and more reliable yields. Ability to cope with very great climatic variability has been enhanced, for the population as a whole if not for all its members (Tiffen, Mortimore, and Gichuki 1994). How this happened is not fully explained by Tiffen, Mortimore, and Gichuki (1994), nor by other writers, but at least for the Machakos hills and adjacent lowlands, its outline can be pieced together. Perhaps the best account of how farmer innovations were combined with technology transfer was provided by Rochelau et al. (1989). Other writers on the Akamba include Porter (1970) and Silberfein (1989).

THE PATTERN OF CHANGE IN MACHAKOS

Historically, cattle were the wealth of the Akamba. Alienation of large areas previously used for seasonal and dry-year grazing concentrated cattle onto a much smaller area, where severe loss of pasture and then of soil became apparent early in the twentieth century. The first major intervention was an effort to secure culling of family herds, which encountered such stiff resistance that it was abandoned in

1938. Arable land historically had only casual management, but as population and commercial possibilities grew it came under increasing pressure. Mixed farming systems were advocated from an early stage, and experimental terracing was introduced before the end of the 1930s. After 1945 the government strategy involved resettlement and compulsory terracing, using a so-called narrow-based or contour ditch design. This sort of terracing was promoted until the 1960s, but the terraces were hard to maintain and were often neglected. They were given up in favor of the thrown-up or *fanya juu* terraces. Although the design may have been borrowed from another area, its introduction in the late 1930s seems to have been an Akamba initiative. Only in the 1960s did this home-grown form of terracing acquire official blessing under the somewhat inappropriate title *bench terrace*.

Fanya juu terraces, which usually have a gentle forward slope, were as important for water as for soil conservation. Initially water conservation probably was more important. They were widely adopted from the late 1940s onward in all areas where people were able to seize the opportunities of vegetable production for the Nairobi market. Wider than the narrow-based terraces, *fanya juu* terraces also were associated with the spontaneous adoption of the oxen-drawn plow. Heavy plows were bought in small numbers from European settlers, but they were replaced in the 1950s by a lighter two-beast design of local manufacture that could be used on terraces.

Ox-plowing was one change in the role of livestock. More important was the use of cattle manure, already practiced in the 1930s, spreading much more rapidly by 1960. In the more crowded areas it was associated with a shift from pasture to stall feeding, now common in all areas where individual farms are small and no common pasture remains. Though supplemented by inorganic fertilizers, the nitrogen and phosphorus supplied by manure are critically important on soils deficient in these elements. Composting, promoted by nongovernment organizations rather than the authorities, has been used mainly by farmers without livestock; its demand on plant materials competed with the needs of cattle, goats, and sheep. By the 1980s livestock, though highly valued within the production system, were no longer a main store of wealth. Livestock were fewer in number in 1990 than they were in the 1950s, partly because of droughts since 1980 but more because of competition for land and fodder.

This is by no means all of the story, which also includes adoption of tree crops so extensive as to have rewooded the Machakos hills. This was almost wholly an Akamba innovation. Also important was change in labor organization and land tenure. Since precolonial times Akamba have maintained individual title to land, but formal registration of title under a national plan initiated in the late colonial period gave greater fixity to individual rights, creating a class distinction between land rich and land poor. The effect of individualization has become very visible in the more crowded parts of the Machakos hills, where excellently terraced, well-treed and well-managed holdings stand side by side with other, usually smaller holdings on which management is of a visibly lower quality.

In terms of both production and the health of the land, the success story cannot be questioned. Erosion has not been controlled fully in the drier rangelands but has been reduced greatly in all arable areas. *Fanya juu* terracing came to be practiced in dry and gently sloping areas of new settlement, being taken up much earlier in the life of a new farm than used to be the case. But soil fertility remained a problem. Data assembled by Tiffen, Mortimore, and Gichuki (1994) still showed low levels of soil nutrients in cultivated areas and only small improvement under grazed fallow. They did not dwell on this problem but reported lack of firm evidence that yields of staple crops were necessarily higher on terraced than on unterraced land; manuring was more important (Tiffen, Mortimore, and Gichuki 1994). After walking over some impressive *fanya juu* constructed over a 10-year period in a lowland area adjacent to the Machakos hills in 1997 and talking to farmers, my People, Land Management and Environmental Change project colleague R. M. Kiome, who guided me through Machakos, remarked that "the erosion problem is solved, but the problem of fertility still remains."

■ Intensification, Revolution, and Agrarian Transformation: A Review

KALA, KOFYAR, MACHAKOS, AND INTENSIFICATION

In all three cases discussed in this chapter there have been rapid increases in production coupled with population growth and expanding commercialization. All could be said to have intensified. At Kala and Machakos, but not in Kofyar, there was important technical innovation, including at Kala investment in the fixed landesque capital of wells. In both Kala and Kofyar, the organization of land and labor through time emerges as of central importance, and in Kofyar it has been "in some ways more important" than total labor inputs (Stone 1996:111). In both Kala and Kofyar, control over resources has been a critical element in successful farm management.

Investment and skillful organization can be the means to make labor inputs more efficient, whether the total labor input is reduced or increased. The Kala wells facilitated production increase by attracting larger supplies of manure. This has been used, with aid of the further labor-saving introduction of oxen-drawn plows, to increase yield on the village fields and permit enlargement of their area. The Akamba of Machakos invested in plows and terracing and also in trees. Investment, except in farm and settlement systems, has been of much less importance in Kofyar, where organization has been everything. Yet risk was not eliminated. So tight was the scheduling that "the loss of one adult to illness during a key part of the season may endanger an entire crop" (Stone 1996:114). The value of the risk-reducing adoption of larger household sizes and of expanding old practices for use of cooperative labor is clear.

The question of population pressure is ambivalent. In the case of Machakos, it has undoubtedly enforced use of land-efficient practices. On the other hand, it has created means of interaction among the people, and Tiffen, Mortimore, and Gichuki (1994) argued persuasively that it has provided the labor for improvements, increased the number of "idea-generators," and provided the mass of people needed for investment to be worthwhile. The market has been of major importance. On the basis of comparative data from India and Africa, Goldman and Smith (1995) concluded that successful agricultural transformation arises most readily from the seizing of commercial opportunities and that population growth alone cannot be shown to produce any comparable result. They were writing of modern change only, but the same broad conclusion would apply to the historical cases discussed in chapter 10.

Whether population growth caused people to develop their production systems in the way that they did or whether it merely facilitated their seizure of a particular set of opportunities is a question that cannot be answered readily. In a situation in which so many new opportunities were being created, one might well be forgiven for suggesting that it is not worth asking. Farmers pay close attention to market signals and change their land use accordingly. It is certainly easier to see this happen than to see quick response to perception of decline in the marginal product of additional labor, on which the Boserup trigger relies. We need to take account of other variables that have been uncovered in the long post-Boserup debate. They include production to meet social demands (Brookfield 1972b), innovation to reduce risk, and exploitation of a subordinate class or gender by those with the power to appropriate the surplus from those whom they oppress. So many variables are involved that to isolate only one or two as most important is at best unhelpful and at worst dangerous.

FROM INTENSIFICATION TO AN UNDERSTANDING OF COMPLEX REALITIES

At this point I am going to propose, as I have done already in the 1998 workshop paper (Brookfield 1998), that there are better ways of thinking about agricultural change than pursuing any further the mess that intensification theory has become. We can talk about intensive practices, but the term *intensification* has been so misused that it is best to avoid it. Wherever possible, this has already been done throughout the book. For understanding transformation, it is perhaps more useful to go back to the three elements involved in my early definition of *intensification,* quoted earlier, and to consider them separately. They are very different from one another.

Labor inputs have dominated the debate, but it has also been shown that simply to intensify use of labor inputs, without qualitatively new ways of combining the factors of production, would lead only to stagnation. However, in the context of marginalization, this is what Geertz's (1963) "involutionary" argument was all about. In the different context of a west Indian island, I have shown how first slavery, then an abundance of cheap labor, discouraged land-owning sugar planters from adopting technological innovations that were both well known and easily available. Thus highly inefficient methods of production and management persisted well into the twentieth century (Brookfield 1984).

Capital and skills can be combined with labor in very different ways. Yet this also is not simple because capital and skills are not simple categories. Working capital is used along with labor and technical skills. Fixed capital, or landesque capital, must be created before it can be used. Whether it is created quickly or incrementally, it can endure a long time, as this book has amply demonstrated. Landesque capital, used by today's cultivators, is not a current input but an inheritance, requiring only maintenance.

ORGANIZATIONAL SKILLS

Technical skills, on which the post-Boserup debate has concentrated, are of limited value without organizational skills, and use of the latter alone can achieve a great deal. Organizational skills have been given prominence in this chapter, partly because they have been insufficiently stressed in the past. These skills are the property of the farmers themselves, and for a book about small farmers, this is a fact of central importance. The organization of farming operations is not just the socio-economic aspect of the management of biophysical diversity and of a range of crops, noncrop plants, livestock, and other fauna. It is the element that is most strongly tested whenever new forces come into play. Sensitivity and resilience in the more physically obvious parts of the system often are causally related to the sensitivity and adaptability of farmers' organization of their working resources. The importance of how people organize their land, workforces, and resources has come through very strongly in the last two chapters, especially in the Japanese story told in chapter 10 and in Kala and Kofyar in this chapter.

There are other important dimensions. Agricultural change can be triggered by a new crop, a new variety, or a new production method. It can be enforced by a ruling class, a government, or a political party, or it can arise through the seemingly uncoordinated efforts of many thousands of individuals. The roles of the farmer and the authorities have varied through time, and today they vary substantially between places. The role of the agricultural scientist is substantial in modern times, but there have been many scientific treatises on agriculture in the past 2000 years, and some have had an influence on events. Farmers' own knowledge, so much despised by a very recent generation of specialists on agricultural development, has played a major role in all the transformations reviewed in this book. Sometimes its role has been only to help farmers decide which instructions and advice to follow and which to reject or quietly disregard. In other times and places, farmers' knowledge, acquired from experimenters among them, has been the real foundation of change.

Both the Kala and Kofyar cases, and to a much lesser extent Machakos, also demonstrate the importance of gaining and retaining access to resources, an aspect of major importance only touched on in earlier chapters. Chapter 1 drew attention to the importance of controlling access to resources among the Chimbu of Papua New Guinea. During the 1960s, there was a debate among anthropologists, and some geographers, studying Papua New Guinea societies. Disagreement arose from contrasted findings among the Chimbu (Brookfield and Brown 1963) and the Enga about 150 km to their west (Meggitt 1965). Meggitt argued that wherever the rules of land tenure rely on agnatic descent, increasing population pressure will lead groups to insist more firmly on that principle. Population densities in the two areas were comparable, both high. In Enga clans, only 10 percent of residents were nonagnates; in Chimbu there were 21 percent. This argument waxed for several years, becoming quite intense at times, before it came to be realized that longer-term data showed a different picture and that we were dealing with two different behavioral responses, each with its own rationale.

If the Bambara of Kala had not succeeded in keeping their Fulani neighbors in a client status, without rights to land or water, they could not have changed their own resource-constrained system as they did. Had the Kofyar not been able to charm land out of the chiefs on the plains and forestall the aggressive northward migration of the Tiv, they would still have been confined to their mountainous homeland and its thin-soil pediment slopes. Akamba were able to reclaim only a part of the land they had lost to alienation, but that amendment was important to them in making possible rapid adaptation to commercial and population pressures after 1960. For Kala and Kofyar there has been a need to shore up vulnerable negotiating positions, and in neither was future security certain at the point where the stories came to an end. An important political dimension of transformation is demonstrated here better than in most other cases discussed in this book.

THE TIME ELEMENT

Finally, there is the question of time, with which chapter 10 began. Until the Kala, Kofyar, and Machakos cases were presented, it might have seemed that most farmer-driven innovation had to proceed incrementally. Yet these cases exhibit the sort of innovation surge, involving more than one innovation at a time, that rapidly transforms the nature of systems. It has been possible to show, quite securely except in the case of Machakos, that the rapid changes arose from initiatives taken locally. I go back to an earlier argument that innovation must be given much more weight in discussions of agricultural change (Brookfield 1984). We now turn to innovations that clearly came from outside the local system. What has happened in the last 50 years may be historically unique, but that should not blind us to the fact that farmers themselves can make far-reaching changes in the way they do things. That is the basic lesson of these last two chapters.

Chapter 12

The Green Revolution

■ Science and Public Policy as the Drivers of Change

THE RISE OF EXTERNAL INTERVENTION

This chapter reviews the Green Revolution in wheat and rice in two Asian countries. The experience of farmers with new higher-yielding varieties of other crops in other regions has been touched on elsewhere in the book, but it is best here to concentrate on areas where a great deal happened in a short time. The story must be viewed in the context not only of agrotechnical change but also of state policies and the remarkable link between plant science and policies that evolved in this period. Such a link continues in new ways, as we shall see in part IV.

External intervention, which forces farmers to grow crops needed by the state-led economy or work for wages, is not new. It was already a strong feature in the colonial period. We briefly examined the extreme case of nineteenth-century Java in chapter 10. Writing mainly of west Africa, Michael Watts (1983:247) succinctly described the "colonial triad of taxation, export commodity production and monetization" used to enforce integration of local economies into the global economy.[1] The elements of value to the new authorities were labor and produce and, from the yield of these, revenue. Taxation made farmers produce cash crops or hire their labor to local or remote enterprises. The resultant monetization created demand for new goods and services to entrench the new economic system.

Independence did not bring this to an end because the postcolonial states continued to need production, labor, and revenue and saw it as necessary to coerce farmers into providing what the national economies needed. Few of the independent states have abandoned the instruments developed under colonialism or the elaborate local administrative systems devised to make them operative. This background was shared in most regions heavily affected by colonialism, and they included the larger countries of south and southeast Asia. Not everything was repressive, and in India extensive canal irrigation systems were built in the mid-nineteenth century. The British purpose was to create the conditions for a more commercial agriculture, anticipating the Green Revolution by a century (Whitcombe 1972).

State intervention blossomed anew in the mid-twentieth century in the hands of newly independent governments determined to develop their national economies rapidly. It was joined by new scientific initiatives, using the best known breeding methods to create new crop plants that would yield far more bountifully. In this chapter, drawing more heavily than any other on Indian and Indonesian material, the linkage of science and government becomes a dominant theme. The new wave of intervention and top-down innovation coincided with a period of widespread concern about future world food supplies and with the rise of wholly new approaches to rural modernization. In its most classic areas, this is how the Green Revolution has to be understood.

GREEN REVOLUTION?

More commonly, the Green Revolution is seen in simpler terms. As conventionally and still often defined, it involved the distribution and acceptance of seed of high-yielding varieties, especially of wheat and rice, from research stations to developing country farmers, supported by and initially almost wholly supplied from international research centers located in Mexico and the Philippines. These are Centro Internacional de Majoramiento de Maiz y Trigo (CIMMYT, or International Centre for Maize and Wheat Improvement) in Mexico, and the International Rice Research Institute (IRRI) in the Philippines. CIMMYT began with Rockefeller Foundation support in 1943 and became an international center in 1963. IRRI was founded in 1960 and was fully established in 1962. By that definition, the revolution began in the mid-1960s and was largely complete by about 1985 when the high-yielding or modern (post-1960) varieties (MVS) had been generally adopted in all the areas for which they were best suited.[2]

Such a definition is inadequate. A great deal has been involved beyond simple transfer of germplasm and appropriate technology. Even the technical aspects of the revolution were not so new. The scientific basis has been built on long experience. Both selection of pure lines and hybridization were already used in formal experimentation with a range of crops before the end of the nineteenth century. A major early success was with maize, a vigorous hybrid being developed by farmers in the U.S. cornbelt in the nineteenth century. They used controlled pollination to develop superior varieties. In turn, this led to seed company breeding of commercial hybrids, first grown in 1933 and already occupying more than half the U.S. cornbelt land by the 1940s (Sauer 1993). This experience sowed an important new set of ideas, coming together in Mexico with the breeding of high-yielding wheat under Rockefeller Foundation auspices in 1943.

Formal plant breeding on experimental stations had already begun in some developing countries between 1860 and 1910. Improved varieties (IVS) already were being distributed from these stations to farmers during the 1920–40 period. In rice, major progress was first made by the Japanese in Taiwan and to a lesser degree by the Dutch in Indonesia (Hayami 1973; Carr and Meyers 1973; Fox 1991). The first IRRI MV rice crossed a Taiwanese with an Indonesian variety, both the product of long-period breeding efforts. The first Mexican MV wheats drew on four decades of plant breeding in the United States and two in Mexico. Use of manufactured inorganic fertilizers was already being encouraged in the 40 years before 1965. Improved tools and mechanization also began earlier in several developing countries. Even the most ecologically debatable part of the innovation complex, the use of chemical pesticides, began in the 1940s. To define a Green Revolution only in terms of the events since 1960 shows a remarkable disdain for history. The period 1960–85 saw a major burst of innovations, especially in the development of heavy-tillering, short-stemmed varieties of wheat and rice, within a much longer chain of events.

There is diminishing agreement over what the term *Green Revolution* includes (Jirström 1996). There is now a growing tendency to write of a post–Green Revolution period since 1985 in which a widening range of problems has emerged. Among examples are books by J. R. Anderson (1994) and Pingali, Hossain, and Gerpacio (1997). However, this new term seems to imply that increases in cereal output and profitability were all that the years of revolution were about. The complex of changes involved is far from being at an end. It has moved with scarcely a pause into a new gene revolution discussed in chapter 13.

QUESTIONS THAT IMPINGE ON DIVERSITY

The Green Revolution has a very large literature, both technical and social. One major question has concerned the differential impact on large and small farmers and on the landless. The first group was expected by many commentators to reap most of the benefit and the second and third to be further impoverished. The issue has been largely resolved by a substantial body of empirical evidence at community level in the most thoroughly transformed areas, and mainly from the 1980s, showing substantial benefit to poorer farmers and to landless laborers who depend on wage employment. Lipton and Longhurst (1989) and Rigg (1989) provided good reviews. I shall not discuss the topic. Certain technical questions are of closer relevance, and they have social as well as ecological dimensions. They include the following:

- Although bred to have wide environmental tolerance, the MVS are best suited to a narrow range of environments, especially environments in which water supply can be controlled tightly so that the uptake of commercial fertilizers can be optimized. Later expansion of MVS into rain-fed areas has been substantial but with less yield benefit (Byerlee 1996). The spatial impact of the Green Revolution has therefore been very differential, leading to substantial diversity at an interregional level and even at a local level. There remain wide differences in the rice yields obtained both between and within Asian countries (Chang 1993; Fox 1993a). Although yields on experimental farms may rise as high as 11 tons/ha, the best regional farm yields were below 6 tons/ha in the early 1990s, and many have remained below this level.

- Local varieties (LVS) and the IVS of the period before the 1960s have been displaced by the MVS, and many varieties have been lost. In addition, production of leguminous crops and the so-called coarse grains, the nutritious millets and sorghum, has declined. Intercropping has been replaced widely by monoculture and organic methods of fertility maintenance by manufactured chemicals. Production has come to be derived from a narrowing genetic base, increasing risk from disease and pests. This has taken place despite the fact that for more than a century a narrow genetic range has been known to increase risk of disease and pest infestation. However, Lenné and Wood (1999:453) drew attention to the widening number of landraces in the background of successive generations of MVS in both wheat and rice; the number increased from 5 to 50 and from 4 to 46, respectively, between 1960 and 1990. Multiple stress resistances have been incorporated.

- Breeding has focused not only on high grain yield, responsivity to fertilizers, and resistance to lodging (best attained by the short-stemmed or semidwarf varieties), but also on growing and maturing periods that are insensitive to daylight length, offering much greater flexibility in planting time.[3] Most MVS also have a shorter growing period than the IVS they have replaced. This increased flexibility has facilitated great expansion of double cropping and even triple cropping, reducing the time and space that can be devoted to other crops.

- The conditions needed by the MVS are exacting. Although soil quality can be managed to a considerable degree by fertilizer application, hydrological conditions are another matter. The Green Revolution took off first and most strongly where modern irrigation systems already were in place.

In the early years, new germplasm reached farmers by various channels, many of them informal. The information that came with the seeds was imperfect. In the mid-1960s all this changed. Governments built up national and regional organizations to multiply seeds and provide comprehensive recommendations for their cultivation. New information dissemination systems were developed, and with the emergence of packages a strong element of direction was introduced. Fertilizers and pesticides were made available at heavily subsidized prices, and government-supported credit in cash or kind to small farmers often was provided only if the whole package was accepted. Coercion soon became part of the set of changes, and it included coercion even of governments to impose the new regime on their farmers and their own agricultural research systems.[4]

Although the packages included instructions on how much fertilizer should be applied and when, they were not always well received, and farmers had problems with them. Until the packages were improved, farmers had to learn for themselves what quantities of fertilizer, initially only nitrogen fertilizer (urea), would suit the growing conditions of their own fields. Many farmers were unconvinced that the expense of buying fertilizers was justified. Also, poorer farmers lacked the resources needed until credit became more generally available. Fertilizer application rates far below the recommended levels initially were common.[5]

In several regions, chemical fertilizers displaced use of more labor-demanding organic manures, despite the value of the latter for soil constitution on nitrogen-rich land. The deleterious consequences of reliance on chemical fertilizers alone were widely disregarded. There was a tendency to irrigate to excess where sufficient water was available, and overapplication of fertilizer led to serious losses.

In some rice-growing regions the existing practice was to plant rice in seedbeds then transplant into the field, whereas in others rice was sown directly in the field, or broadcast. In the century before the 1960s, transplantation had become a much more widespread practice. Subsequent trends moved toward further use of seedbeds or away from it, leading to weed infestation problems where broadcasting was used (Jirström 1996). At least partly because much less new knowledge was needed to grow the MVS of wheat, progress in the first years was more rapid than with rice.

The areas in which the Green Revolution had its most dramatic successes were mostly those in which there had been the greatest agricultural change in the previous century and in those that enjoyed good irrigation conditions and a sunny season to maximize photosynthesis. These were the areas in which national Green Revolution programs were most strongly developed (Lipton and Longhurst 1989). Where the environmental and infrastructural criteria are not met, the MVS often yield little better than the old IVS and LVS.

Generalities abound in the literature, but to understand agricultural transformation while it is still in progress demands a perspective that takes account of people. Farmer (1977) and Bayliss-Smith and Wanmali (1984) made very successful use of village studies in combination with regional and general papers in two important edited volumes on the Green Revolution in India and Sri Lanka. For the last time in this book, I turn again to case material, limiting the selection to studies from India and Java and in the process adding some detail on agrodiversity in regions of high population density.

■ North and South India

TWO CASES FROM NORTH INDIA: SHAHIDPUR AND KARIMPUR

In an unusual paper comparing change in India with change in Nigeria, Goldman and Smith (1995) relied heavily on the interpretation of a Green Revolution success in the Punjab village of Shahidpur by Murray Leaf (1983, 1984, 1987). Leaf's work is often cited because it does record success, and I use it later in this chapter. It describes change in precisely the classic Green Revolution period, the 1960s to 1980, in one of the main initial areas, on the well-irrigated plains of the Indian Punjab. The village lies not far from a key institution, the Punjab Agricultural University at Ludhiana, which was the first source of wheat MV distribution in India and already a base of agrarian research before the mid-1970s (J. Harriss 1977a). Before I get to it, I first draw on a different sort of village study from the plains of Uttar Pradesh, on the edges of the Green Revolution heartland, where ethnographic observation began in the 1920s and ended in the 1970s just as rapid change was getting under way. Both are communities in which rice remained secondary to wheat and in which the so-called miracle seeds of the 1960s gained very rapid acceptance. The two studies can be used in a complementary manner because the time overlap in the material presented occupies only a few years.

KARIMPUR, UTTAR PRADESH, FROM THE 1920S TO THE 1940S

The village of Karimpur, on the extensive interfluve *(doab)* between the Ganga and Jamuna rivers northeast of Agra, was studied by missionary–ethnographer Charlotte Wiser between 1925 and 1931 and then again, on the basis of long residence, in the 1960s and early 1970s (Wiser and Wiser 1971; Wiser 1955, 1978).[6]

Her material rested on long personal knowledge of a few individuals and their families across the range of social strata. As described in the 1920s, Karimpur had about 750 people, and there were twice as many in the 1970s. Agriculture depended on the summer rain and winter irrigation. The nearest distributaries of the Ganga canal system, constructed between the 1840s and the 1870s, lay a few kilometers north, but seepage from them had raised the water table, with mixed consequences. The main low-rain season supply came from shallow wells from which water was drawn in leather buckets by oxen or buffaloes and by people. They were masonry lined; throughout this region, older earth-sided wells had mostly collapsed before 1900 because of subsoil saturation (Whitcombe 1972).

In the 1920s, some areas were too saline for cropping, and waterlogging was a serious problem in parts of Karimpur lands, as it was widely in the whole region. Seepage from both ancient minor irrigation and the large nineteenth-century canal systems, together with obstruction by railway and road embankments, led to flooding of the soil with more water than could be drained horizontally or vertically. In the late 1960s, areas less badly affected than the Mainpuri district in which Karimpur lies were described in the following terms: "dark green Mexican [MV] wheat fields stand out against ragged patches of grey-white, waterlogged and saline soil" (Whitcombe 1973:187). Differences in land quality were distinct, for this "featureless" plain has a complex microgeomorphology. The distinctions creating farming spaces within Karimpur land depended in part on microtopography in relation to the high water table, in part on human use, and in part on the proportion of wind-blown sand, very high in some areas (Wiser 1955). A small area around the village received heavy manure from buffalo and cattle, and a larger area was lightly manured. Karimpur people classified the land by its relative proportions of sand and clay and its closeness to the water table.

Most holdings were subdivided into as many as 60 or 70 tiny plots. Involvement in trade was limited mainly to acquiring sufficient money to buy necessities. There were two main production seasons, of which the winter, dependent on the daily labor of irrigation, was the more important. The principal winter crop was wheat, harvested and often grown mixed with barley or one of the leguminous grams (mainly chickpeas, *Cicer arietinum;* pigeon pea, *Cajanus indicus;* and *Phaseolus* spp.). Most was for daily consumption as bread, also made of millet, barley, and maize. For the poorest there was little wheat. Winter vegetables, and mustard and sesame for oil, also were important.

Summer crops were principally maize, millet, sorghum, and a summer range of legumes. Rice, grown in the wettest ground, usually was planted together with millet so that if the rains failed and the rice with them, the drought-resistant millet would still yield a crop. A little sugar cane was grown for home consumption. Crop mixing combined the values of insurance, saving land, sustaining fertility, and making optimal use of moisture. Trees were of major importance; most yielded fruits, berries, or nuts, and the leaves of some were eaten. Among 20 spices in common use, some such as cardamom were grown in the fields among grain crops and some were grown in courtyards. Others were purchased. Storage of grains, pulses, and even greens was critical to life, and there was a lean and hard time in the hot, dry premonsoon months when harvest of the winter crops was over and very little was growing. Almost all crops were grown by intercropping, and few were grown in monocultural stands. The main exceptions were sugar cane and some of the wheat, the latter sold to pay rent to agents of the absentee landlords.[7]

GOVERNMENT-INDUCED CHANGES AT KARIMPUR AFTER 1947

One of the earliest steps taken by the state governments in newly independent India was abolition of the *zamindari* system of land holding by absentee landlords, who charged the peasants rent but had paid land tax during British times. By the mid- to late 1950s, the former tenants in Uttar Pradesh owned their land and could decide how to use it. They could buy it, sell it, or rent it to sharecropping tenants. Much land changed hands, and by the 1970s the proportion of land cultivated by the dominant Brahmin caste at Karimpur had fallen from four-fifths to about one-half. Many members of the lower castes acquired pieces of land, thus enhancing family security, even as traditional occupations such as carpentry and oil pressing were lost to urban competition (Wiser 1978).

In the 1950s a Community Development Program was initiated, grouping villages together and setting up cooperatives. These community development programs were initiated in 1952 in several parts of India. Each covered about a hundred villages, later called a block, which became the principal unit for agricultural extension, with village-level workers appointed through them. After 1957, a system of village, block, and district councils *(panchayat)* took over many of their functions. The system of cooperative societies at first worked poorly, but credit and marketing were increasingly handled through them after 1960, backed ultimately by the Reserve Bank of India. In 1960 came the system of Intensive Agricultural Development Programs (IADPS), set up in the best agricultural areas of each major state, supported by one of the new agricultural universities. Their activities included developing packages of seeds, fertilizers, and cultural recommendations, supported by infrastructural improvement and technical staff. The IADP innovations, which contained many elements that were later included in most national Green Revolution designs, anticipated the MVS by several years (Pearse 1980). After the mid-1960s, when MVS began to become available, these IADP areas became the core areas of change under a new agricultural strategy.

The reaction of Karimpur farmers to each of the stages they experienced (not the IADP stage, of which they formed no part), was minutely illustrated in Wiser's material (Wiser and Wiser 1971; Wiser 1978). Important innovations were the seed stores, one of which Karimpur received. They offered seed to farmers on loan, to be repaid from the crop. Cotton was an initial introduction at Karimpur, followed by high-yielding wheat and improved varieties of sugar cane, potatoes, and other crops. Composting in pits was augmented by introduction of inorganic fertilizer. New plows were introduced. New ideas, seeds, fertilizers, and tools were at first made available to Karimpur farmers singly rather than being presented as a whole package. They were taken up only after successful experimentation by leading farmers with enough land and enough food in store to risk a trial. Not all innovations were adopted. The farmers themselves, who traveled more and more widely as transport and communication systems improved, introduced some new germplasm and ideas. Farmers might ask the village seed store to acquire new varieties that they had seen in other villages or had seen or heard about in the market of a nearby town.

Government intervened directly and increasingly in irrigation. The first deep tube wells at Karimpur were sunk in the 1950s; their strong flow of deep groundwater was fed into irrigation channels on a rationed basis and provided means to get rid of saline accumulations. Swampy land was later drained, a great bonus to farmers who had little but waterlogged land. In the 1960s came a provincial Land Consolidation Program, achieved in Karimpur between 1966 and the early 1970s. Negotiations and conflicts led to some being very satisfied and others very disappointed, but almost all arable land was reallocated so that farmers had one or two large fields rather than many scattered plots. Those who got good farms through this process were better able to invest in new machinery and in private tube wells, 10 of which were sunk in 1970. By 1975 there were 15 privately owned pumps.

Much remained unchanged, and for the larger farmers multifamily household organization endured. Division of households was not uncommon, jealousies and disagreements over land often being involved. Wiser (1978) provided intimate detail. Family members leaving the village mostly remained in a supportive relationship or might return to farm. A higher proportion of lower-caste people left permanently, although employment opportunities improved. Education facilities were greatly improved, presaging further changes. By 1970, MVS with a shorter growing season, which permitted farmers to grow a catch crop of soya beans, were being distributed from the Uttar Pradesh state agricultural university and no longer only from Ludhiana. The first tractors and the first mechanically operated threshing machine arrived in the village at the end of the period of report (Wiser and Wiser 1971).

SHAHIDPUR, PUNJAB, 1965–78

The Sikh village of Shahidpur, in the north central Punjab, was first studied by Leaf in 1965 (Leaf 1972). At that time, sugar, maize, cotton, and wheat were already sold in quantity by village farmers. Several members from different social groups worked away from the village, on salaried work or in the armed forces, something that had also begun to happen in Karimpur in the late 1950s. Very importantly, the Punjab canal irrigation system, begun in the 1830s, was reconstructed after the separation of Pakistan from India in 1947, and new distributary canals were constructed in the 1950s. One branch was opened to pass through the village lands in 1962, creating immediate improvements. The new Mexican varieties of wheat reached Shahidpur in 1968, and the consequences of this and related introductions are the main topic of Leaf's second (1984) report, based on work in 1978 and subsequent visits. The tone is enthusiastic throughout.

Shahidpur lies south of the Sutlej River, on plains with a uniform brown loam soil with a fairly high clay content that can lead to pan formation below the level of disturbance. Land was valued according to its water supply, and the amount of compost and human and animal waste it received. The best land was along the canal in 1965, the second best was near wells, and some land had no water supply and could yield only summer crops. The new canals had a dramatic effect on the groundwater environment. All unlined, they permitted much of their flow to drain into the subsoil. The groundwater situation in this part of the Punjab probably had been deteriorating for centuries (Spate and Learmonth 1967) toward desiccation. In the 1950s the water table at Shahidpur was 30 m underground, but after construction of the new canals it rose to 10 m in 1965 and to 7 m in 1978.

This made well construction cheaper. There were 26 in 1965 on 178 ha, the number increasing to 52 by 1975. In 1965 most wells were still worked by the Persian wheel, driven by oxen or camels.[8] By 1978 all were driven by pump, powered by diesel or electricity, whether they were old masonry wells or modern tube wells. All parts of the village land were by then able to receive irrigation water, and farmers could manage it individually. Continuation of the water table rise could lead to waterlogging and salinization, but it was checked at least in part by drawing out large quantities of underground water through the tube wells.[9]

It seems likely that this rapid change in the biophysical environment helped make Shahidpur a dramatic example of Green Revolution success. The range of winter and summer crops grown in Shahidpur in 1965 was much the same as in Karimpur in the 1920s–1970 period, but with important differences in emphasis. Sugar cane, still grown only for domestic use in Karimpur up to the mid-1970s, was a major cash crop in Shahidpur in 1965, requiring more water and labor than any other crop. It was sold to a government sugar factory or locally processed for consumption. Cane was left in the ground to grow ratoon crops for 4 years. Cotton also was an important cash crop. After 1965, there were big changes in cropping and related practices at Shahidpur, analyzed in detail by Leaf (1984).

Comparing his data obtained in 1978 with that from 1965, Leaf found that among winter crops there had been a large increase in the area sown to wheat, now exclusively in monocrop and no longer mixed with gram. Everything else, including gram, oilseeds, and fodder crops, had declined. Among summer crops, the area under maize had doubled, whereas cotton had become only a minor crop. The most important innovation was a large area under summer-grown rice. Sugar cane, grown the year round, had diminished in area and in production. Most of the wheat was MVS developed in India, but IVS still were important. Some of the maize was of a composite variety from which the farmers could continue to use their own seed, but most was hybrid, requiring constant resupply of seed from the plant-breeding department at the agricultural university.

Introduction of rice as a major crop was a direct product of improvements in irrigation, coupled with low electricity prices for pumping; farmers could maximize the use of irrigation time by using the available water on this crop. Leaf estimated that available water increased by four times between 1965 and 1978. Improved indigenous varieties of rice were grown, including the low-yielding but high-priced *basmati,* which has enjoyed a wide international market. Despite all this change, all food consumed in the village still was produced there at the end of Leaf's period of report.

By the end of the 1970s, a rice–wheat rotation was evolving and was giving the farmers very high returns. This rotation has since become very common in the northern Punjab, but it is problematic. Soil must be puddled to grow irrigated rice. It then develops a pan below the limit of disturbance, and this pan can severely inhibit the growth of a following dry crop. A more deeply cutting plow is needed to break up the hardpan. In turn, the water-holding capacity of the soil is then reduced for the next rice crop. Leaf (1984) noted this problem at Shahidpur, and it has become widespread in the Punjab (Pingali, Hossain, and Gerpacio 1997). There probably is no solution other than to change to a farming system without annually alternating impact on the sensitive physical properties of a clay-rich soil.

Despite replacement of working livestock by machinery, livestock numbers increased for reasons of preferred subsistence. The meat of goats and sheep, formerly a rare luxury, became more common in the diet. Most importantly, many of the smaller farmers took up dairying as their speciality, and milk production increased greatly. Because ghee (clarified butter) was the preferred form of oil, the need for oilseeds declined at Shahidpur. There was also an increase in the making and use of compost, in addition to the chemical fertilizers, which by 1978 were applied at the recommended rates published by the agricultural university. Compost can best be absorbed by moist soils rich in nitrogen, and advantage was taken of the combination of chemical fertilizers and improved irrigation. There were additional labor requirements, and these spread the benefit in the form of higher-wage employment. Some small farmers sold their land at high prices to devote themselves entirely to farm labor and other work, some specialized in livestock, and a few invested in machinery that they could rent out once it had performed the small tasks needed on their own land. Nonagricultural employment, both in the village and in nearby towns, became much more generally available.

Fundamental to these changes was a great improvement in arrangements for marketing and credit, especially through cooperatives, which, after the early years, enjoyed strong government support. Leaf discussed these new arrangements in detail, together with changes in social and political structure within the village. He concluded that social and cultural changes were fully as important as technological changes. He perhaps gave less weight than might be warranted to the fundamental change in groundwater, which made everything else possible. For example, it was the basis of increasing use of vegetables and potatoes as intercrops on land that could be kept permanently watered. Within a benign political and economic environment, farmers at all levels had adapted quickly to opportunities; they retained diversity in their systems of production and income earning even as the old crop diversity that once provided their insurance was reduced.

RICE IN NORTH ARCOT DISTRICT, SOUTH INDIA: WHAT WAS MOST IMPORTANT?

Villages in the inland North Arcot district of Tamil Nadu were studied by a Cambridge/Madras team in 1973–74 (Farmer 1977) and then by an International Food Policy Research Institute (IFPRI) team with members of Tamil Nadu Agricultural University in 1982–84 (Hazell and Ramasamy 1991). Both teams were studying the Green Revolution in rice. Both were seeking generalized results, so a large part of the village data are aggregated or referred to in a comparative context.[10]

This is an area very different from the north Indian plains. Low hills and valleys are developed on deeply weathered basement crystalline rocks, covered with a depth of regolith and colluvium. It is a subhumid region with a short late-year wet season, a very dry season, and a season with intermediate and very variable rainfall associated with the southwest monsoon. Regionally, groundnuts were the dominant crop, and millet and sorghum were more important than rice. Millet and sorghum also benefit from irrigation, but wherever there was sufficient water, rice probably was the main crop long before the Green Revolution.

Hydrological conditions varied greatly from village to village, and they had not remained constant through time. The village of Dusi, one of those sampled by the Cambridge/Madras team, was supported by a very large tank (reservoir) fed by a diversion system completed in 1857. This tank adequately fed double cropping of rice early in the twentieth century, but by the 1940s and 1950s increasing demand and an uncertain supply of water made only a single harvest secure (Dupuis 1960).[11] In 1973–74 Dusi was described again as having reliable water from its large tank, which it shared with 17 other communities. All 18 had cooperated in 1971 to improve the long channel system diverting water into the tank from the seasonal Palar River (Chambers 1977).

Over most of North Arcot, irrigation water was derived mainly from wells sunk into the regolith overlying the impervious basement rocks. In the past water was lifted in leather buckets by oxen. Wells increased greatly in number in the 1950s and diesel pump sets were substituted for oxen. After rural electrification, which began in 1957, these in turn were progressively replaced by more efficient electric pumps, used in about half of all wells by 1973–74 (B. Harriss 1977). By that time, 72 percent of all rice in the area was irrigated by wells, with the undesired effect that a drop in dry season groundwater levels could already be measured (Bandara 1977).

The dates are important for understanding the course of change because in 1973–74 only 13 percent of all rice in the North Arcot survey area was MVS (Chinnappa 1977). Most of the rest was IVS that had been developed and distributed during the previous century. The most widespread variety had been released in 1952 and, with a shorter growing season than the early MVS, it remained attractive in a region not well endowed with water. The share of MVS quickly rose later in the 1970s to reach 93 percent in 1983–84 (Ramasamy, Hazell, and Aiyasamy 1991). In the early 1970s, the Cambridge/Madras team concluded that possession of a pump set, giving cultivators holding any size of land area independent control over irrigation water, was at that stage the most important factor in MV adoption. In all this change, the successful use of MVS seems to have depended on a complex of earlier changes, among which rural electrification in the late 1950s had critical importance.

SOME OBSERVATIONS ON THE INDIAN CASES

Only a small part of the huge Indian material on the Green Revolution experience is reviewed here, but enough is presented to show how the pace of externally driven change gave farmers a wide range of new opportunities. This is even more strikingly demonstrated in a comparative study of a group of villages in the Meerut district, in northwest Uttar Pradesh, by Sharma and Poleman (1993). There, MVS came along with major changes in farming practice, electrification, and truly revolutionary changes in marketing, employment, and investment opportunities. What is remarkable is the degree to which the Indian farmers retained flexibility. Although plentiful irrigation prevented the major droughts of 1980–81 and 1982–83 from having any severe effect on crop production in the Punjab plains, the swift response in North Arcot was a drastic reduction in the area planted under rice, concentrating on the fields best supplied with water. Areas under groundnuts, sorghum, and millet were increased.

The Indian material highlights the importance of water and the unintended consequences of its management. Even in the early 1980s adjustments in land use practice, land and labor arrangements, livelihood, and social organization were still in progress. Some of the economic and ecological problems today described as part of a post–Green Revolution were not yet as serious as they became in the late 1980s and 1990s. The oil price shocks of the 1970s, with their steep increase in the cost of chemical and mechanical inputs, were weathered mainly because the received prices for agricultural produce remained rewarding in comparison with higher input costs. The heavy subsidies to rural development that had been introduced in south Asia in the 1950s remained largely in place. To pursue events into the 1990s, I turn to the island of Java in Indonesia (see figure 6.1), where some different dimensions of change are very well exemplified. In that country, and especially on that island, the drive to increase grain production exceeded in scale, and in apparent success, that encountered anywhere else.

■ Farmers and the State in Java

GREEN REVOLUTION UNDER MASS GUIDANCE

Between the mid-1960s and the mid-1980s, Indonesian rice production doubled, and the country moved from being the world's largest rice importer to a state of near self-sufficiency. Production increased by two and a half times between 1970 and 1996. The island of Java has consistently contributed almost two-thirds of Indonesia's total production. This development was the direct consequence of government policy and its thoroughgoing and very regimented implementation, especially in Java. The period of success coincides with the New Order regime that in 1967 emerged out of the chaos and bloodshed of the 1965 coup and countercoup. Dominated by the army, Soeharto's New Order regime replaced the discredited but popular regime of Indonesia's founding president, Soekarno. One priority of the new government became achieving a large increase in food production.

Not everything began with the New Order. Dutch-initiated work on rice breeding had been in progress since 1905, and large irrigation works were undertaken before the end of the colonial regime and were then further extended after 1950. A pilot rice intensification program was initiated under Soekarno in 1963. With important changes in 1965, it formed the foundation of the nationwide campaign that began in 1967. This program was in lowland west Java in the 260,800-ha command zone of the large multipurpose Jatiluhur dam on the Citarum River, which was then under construction and was completed in 1967. I return to this area later.

In a manner indicative of its underlying philosophy, the new government program was called mass guidance (*bimbimgan massal,* or BIMAS). It was reconstituted several times under different titles, each stage designed to overcome problems in the previous version. Fine tuning has been almost constant. At one early stage, the provision of seeds and fertilizer, on credit to the farmers to be repaid from the crop, was contracted to multinational firms. Later, Indonesia built up its own oil-based chemical industry, with the capacity to export fertilizers beyond its own needs. A whole agricultural support system was created, including an administrative bureaucracy to direct national policy at local level as well as a strong scientific system. An entire rural banking and credit system was created after initial disasters, and its credit activities have been closely linked to the changing packages of BIMAS and its successors (Pearse 1980; Hardjono 1983; Fox 1991, 1993a, 1993b).

A striking characteristic of the whole Indonesian program has been the close direction of farmers' activities. For several years at the start of the program, farmers received specific instructions on what to plant and what inputs to use. These instructions could be disregarded safely only in fields well away from observation by local and BIMAS officials. Farmers have had to adjust to reversals in direction, especially the 1986 decision to halt the use of a wide range of pesticides, described later. Also in the 1980s, the high cost of fertilizer subsidies to the national economy led to their reduction and to an end of instructions to use ever more fertilizer. In the 1990s, there was new encouragement of diversification. In the main, the government orders were accepted in the early years, and the partial relaxation of tight control did not lead to significant slackening in the rate of growth. With constraints, farmers have learned to make individual decisions on how to apply the packages. Although at no stage have they been encouraged to do so by the authorities, they have also continued to experiment.

INSECT PESTS IN A RICE BOWL: NORTHERN WEST JAVA

By 1974–75 more than half of Indonesia's wet-rice land was planted with four closely related MVS. Much of the rest was still planted in LVS and IVS, especially of the distinctive *javanica* race on the volcanic upland soils of Java and Bali. In 1974 the first of a series of outbreaks of a hitherto minor pest took place. The pest was the brown planthopper *(Nilaparvata lugens);* it lodges in the stalks, and the larvae suck the juices of the plant. Between 1974 and 1986, this pest devastated not only the new MVS but also many of the local rices, including *javanica* varieties, which were largely dropped in consequence. It attacked rice in other countries as well, especially the Philippines. The response was to increase pesticide applications and develop new strains less attractive to the planthopper. This response continued through successive outbreaks into the mid-1980s, through a period in which there was temporary success with an IRRI-bred variety.

The planthopper breeds very rapidly and spreads very quickly. In the 1970s, rice came to be grown continuously, three crops in a year in almost a relay pattern, with the new crop in the ground in some fields before harvest was over in others. In a valley close to Bandung, I once saw, within about 1 ha, some standing crop, some harvested rice being threshed, a new seedbed sown broadcast that day, and another field being prepared for crop by a buffalo-drawn plow. Managing a farmer's resources to achieve this dovetailing of activities is very demanding of skills in planning use of the farm workforce.

In these conditions the insects were able quickly to shift their population structure to adapt to new cultivars. The effective span of a resistant rice variety was reduced to 3–5 years (Chang 1993). Pesticides, in Indonesia subsidized at 85 percent of their price at the peak in 1983, were applied regularly and heavily (Fox 1991) but greatly reduced the natural enemies of the planthopper without bringing the latter under effective control. By the mid-1980s, enough scientific knowledge about the pest had been built up to propose alternative methods. When a new and major outbreak threatened in 1986, the president of Indonesia agreed to accept scientific advice and in a single decree banned the use of 57 varieties of insecticides and authorized introduction of integrated pest management (IPM) in their place. Also involved, although not in very precise terms, was to be a progressive end to the continuous cultivation of rice and its replacement by rotations. By the mid-1990s, the IPM program was successful in achieving a major reduction in pesticide use in Indonesia as a whole, while rice production continued to increase.

It took time to organize training of farmers in IPM methods, which involved training in skills rather than the simple adoption of a package. By contrast to everything that had gone before, this was a program in human resource development, making farmers experts in IPM and able to create a more sustainable agriculture. Although the basic assumption at the outset was that farmers knew nothing and had to be taught, this was more true of the agricultural extension officers. The chain from policymakers and scientists through the field staff to the village officials and only then to the farmers was not easily broken.

At the end of the 1989–90 wet season, anthropologist Yunita Winarto (1996b) went to the village of Marga Tani in the rice bowl of northern west Java. The area is wholly irrigated from the Jatiluhur dam, and much of it was brought into sustained rice production in the 1960s. Her purpose was to study how the IPM program was operating among the farmers. Shortly before her arrival there had been a new and unexpected outbreak of another pest, the white rice stem borer *(Scirpophaga innotata)*. This pest was already known and had a recommended insecticide, to be used for preventive spraying. The unprecedented scale of this outbreak was a calamity to the farmers. It represented a major challenge to the IPM program. Winarto began by attending the school without walls in which selected farmers learned the principles of ecology largely in the field, then she followed up developments over four planting seasons, ending with the wet season of 1991–92. Only two crops were taken annually in the Jatiluhur irrigation area by that time, with a 2-month fallow period at the end of the dry season during which other work was done and water was not supplied to the fields.

Farmers, she found, did not need lessons in ecology as such. What they did learn was to express their knowledge in new ways and to incorporate aspects they had not considered before. In particular, they had very limited previous understanding of the behavior and life cycle of the insect pests and their predators, especially the latter.[12]

Farmers soon learned to observe pest behavior in the fields, and a minority among them also undertook not only sustained observation but also experiments, learning new knowledge they had not acquired in the school. Use of pesticides, which they called medicines, did indeed diminish as they used the recommended chemical as a control method rather than as preventive. On their own initiative, they also learned about the importance of rotating rice varieties.

The schooled farmers were at first a minority. Among the majority, some continued to accept the whole credit package or even purchased additional pesticide from among the banned list, still available in the shops. In the second year, a number of farmers tried to reject the pesticide part of the credit package. The Rural Extension Centre officers responsible for administering the program refused to allow this. These problems were accentuated for farmers in the Jatiluhur command area by the fact that this rice bowl was one area selected for a program of "super special intensification" *(Supra Intensifikasi Khusus)* initiated in 1987 (Sawit and Manwan 1991). In 1991, a minority of farmers rejected the package altogether and thereafter experimented with different crop management methods. Widely, the farmers used kerosene mixed with fertilizer as a substitute for insecticide. They reintroduced a number of nonrecommended rice varieties, some of them old *lokal* varieties little used in 20 years.

Some sought information from the scientists rather than the agricultural authorities and found that they could talk to them more readily than to the extension workers. "The methods of experimentation and observation that the scientists used to acquire knowledge were perceived by farmers to be similar to their own ways of learning" (Winarto 1996a:19). By 1992, farmers were gaining confidence in their abilities to manage the ecological problems, going "back to the old condition of growing rice . . . without medicines" (Winarto 1996b:315). Their problems with the Rural Extension Centre authorities still continued and were not resolved at the time when Winarto left the field.

■ Back to Diversity

JAVANESE FARMERS MAKING THEIR OWN DECISIONS

Among the reasons why the Marga Tani farmers found it hard to resist the package and its low-interest credit was the sharp rise in the price of farm inputs, especially fertilizer and pesticides, that was taking place at the end of the 1980s. This was happening in company with a stagnant price for the rice that they mainly sold off the farm. Discussing a paper by T. T. Chang at a 1991 conference in Yogyakarta, central Java, leading Indonesian agricultural scientist Mubyarto (1993) criticized the manner in which urban consumers and the government had forgotten rice farmers once self-sufficiency was achieved. Although what he said contradicted official figures (Tabor 1992), he reported that the terms of trade for rice farmers in Java had declined by 13 percent between 1976 and 1988. As in other countries, the comparative profitability of rice cultivation and its alternatives changed as subsidies were reduced. Farmers' yields had ceased to improve dramatically, and in some countries and regions were showing a small deterioration.

The widespread adoption of triple cropping, made possible by the MVs, had reduced the dry season cultivation of a cluster of crops known in Java as *palawija:* maize, manioc, sweet potatoes, groundnuts, soya beans, and mung beans. Although these crops still were grown on unirrigated land, there was no longer space for them on land that had secure year-round irrigation water. In villages away from the lowland rice bowls, they survived only as crops of choice by farmers who could grow sufficient rice for their needs in two seasons or had other reasons for reducing their input costs and efforts during a part of each year. After the devastation wrought by the planthoppers in the late 1970s, pressure on farmers to grow more and more rice relaxed to some degree, and a wider range of choices began to be used.

Joan Hardjono's (1987) study of an upland village, Sukahaji, south of Bandung provides detail. In this remote area, government control was lighter than in the plains of northern Java. As early as 1981, the year on which her data were based, MV rice was less profitable than tastier *lokal* rice, any of the *palawija* crops, or vegetables, which very few farmers grew for the urban market. It was mainly risk avoidance that led almost all farmers to ensure their own household supply of rice, using the MVs to get at least two and usually three crops in each year.[13]

As the Green Revolution began to ebb, farmers' own initiative increasingly surfaced from beneath the falling flood of government intervention. Even in heavily regulated Java, farmers refused to accept yet another command initiative in 1995–96: the replacement of powdered urea with tablets. Experimental results had shown some yield improvement but had failed to show that significant additional labor was needed. The farmers knew better (Manning and Jayasuriya 1997). Farmers' decisions have been important at all stages except in introduction of the MVs themselves. At Sukhaji in upland west Java, the complex system described for 1981 by Hardjono (1987) had been developed over 80 years before MVS were used successfully there in 1978, and in 1979 farmers already were starting again to diversify their crops. The first MVs had been introduced in 1974, but in this upland area they yielded less well than the old LVs, so they were rejected initially (Hardjono 1987).

A stage in diversification beyond that described by Hardjono is recounted for an upland village near Malang in east Java by Cederroth (1995), without Hardjono's excellent quantification. Here, irrigated land that can bear three high-yielding crops of MV rice was increasingly under higher-value vegetable crops by 1986–87. Cederroth (1995) estimated that 40–50 percent of irrigated land was planted with vegetables during the wet season and 75–80 percent during the dry season. Low-risk cabbages and cucumbers, beans, and maize were grown together with higher-risk tomatoes and chilies. All these generated higher incomes than rice. In the case of successful tomatoes and chilies sold in the wet season when they were scarce, income was many times higher.

Here and in Sukahaji, intercropping was the norm with vegetables and was central both to risk spreading and to the high productivity of the system. At Sukahaji, where commercial vegetable production had begun only 2 years before Hardjono's 1981 fieldwork, this production was as demanding in terms of labor, fertilizer, and manure as any rice cultivation. Soil preparation, including the digging in of manure from stall-fed sheep, had to be very thorough, and mounding and drainage were of major importance. Timely harvesting, except for the root crops, was critical for getting good prices. There were no slack weeks as occur even with MV rice (Hardjono 1987).

DIVERSIFICATION IN PLACE OF RICE MONOCULTURE

Diversification is increasingly a recommended strategy for the regions of rice monoculture. It includes creating fish ponds for aquaculture, already an element at Sukahaji in 1981, and integration of livestock production (Pingali, Hossain, and Gerpacio 1997). In the areas most strongly affected by the new system in India, technical change and urbanization were leading to diversification of the rural economy by the end of the 1980s (Sharma and Poleman 1993). On-farm diversification has the labor advantage that many of the activities demand high manual inputs, so that rural employment opportunities were retained even while there was major growth in nonfarm employment in all the more developed areas.

At Sukahaji, some distance from the nearest large town, more than half of the occupations of both men and women were nonagricultural in 1981–82 (Hardjono 1987). In three villages close to Malang in east Java, studies some years apart in the 1980s found more than 80 percent of adults and more than 40 percent of all villagers involved in nonfarm work. This included trading and some significant entrepreneurial activities as well as wage employment (Nibbering and Schrevel 1982; Cederroth 1995). For landless people in north India, off-farm employment had become the principal source of income by 1988 (Sharma and Poleman 1993).

■ Conclusions

From the point of view of this exploration, there are four conclusions to draw from this partial examination of the Green Revolution experience. Diversity in the biophysical environment has never been fully overridden; agrobiodiversity has been very greatly reduced but has shown some recent signs of recovery. Management diversity has arisen again as farmers have learned to adapt to new conditions and work around rather than under the authorities. As for organizational diversity, all its dimensions have been changed radically. This is the area in which a transformation begun in the 1950s clearly continues.

Notes

1. More harshly, Polanyi (1944:164, 179) earlier wrote that even in modern times "white men may still occasionally practice . . . the smashing up of social structures in order to extract the element of labor from them . . . [and hence dissolve] the body economic into its elements so that each element can fit that part of the system where it is most useful."

2. By the beginning of the 1990s, the proportion of total rice land planted under MVs was 66 percent in India, 77 percent in Indonesia, 89 percent in the Philippines, and 100 percent in South Korea (IRRI 1995).

3. Grains that come into flower at a certain time of year are described as being photoperiodic; those that take approximately the same time from sowing to harvest whenever they are planted are nonphotoperiodic. The early MV wheats developed in Mexico under different climatic and daylight regimes were insensitive to photoperiod and adapted readily to a similar range of climatic conditions in northern India and Pakistan. In the case of rice, there are three main Asian varieties of *Oryza sativa*. The Indonesian *javanica* variety and the east Asian *japonica* variety are both nonphotoperiodic, whereas the south Asian *indica* varieties are photoperiodic but also tolerate less favorable growing conditions. The Asian Green Revolution in rice involved cross-breeding to obtain nonphotoperiodic, quick maturing strains capable of being grown across a range of latitudes regardless of season (Dalrymple 1976, 1986; Fox 1991; Jirström 1996).

4. Even Consultative Group on International Agricultural Research (CGIAR) centers were pressing for displacement of some very good improved varieties by their own fertilizer-responsive MVS. A particular case in India, with rice in the earlier years of the Green Revolution, is recounted in some detail by Juma (1989). It led to destruction of a promising program that might have competed strongly with that of IRRI.

5. In western Java in the 1960s, where urea came together with the new rice varieties in the form of credit that had to be repaid, farmers sometimes threw the urea bags into streams in the early years. By the 1990s, the same farmers sought the then wider range of chemicals, no longer heavily subsidized, even if they had to borrow from the moneylenders at high rates of interest to acquire them (Winarto 1996b).

6. Wiser's monograph (1955) was originally published, with the same title, in 1937 as a Bulletin of the Bureau of Statistics and Economic Research of the United Provinces of then-British India, now Uttar Pradesh.

7. Before the extension of irrigation in the nineteenth century, most of the people of the Uttar Pradesh plains depended mainly on the rain-fed summer crops, particularly millet and sorghum, with barley, peas, and gram as winter crops. Wheat spread much more widely in the later nineteenth century, and sugar cane was a new introduction (Whitcombe 1972).

8. The Persian wheel is the *noria* of the medieval Arab world (chapter 10). There were Persian wheels also at Karimpur in the 1960s, but it appears that they were a modern innovation there. In the Punjab they were an ancient introduction.

9. In the adjacent Haryana state to the southeast, before 1965 the southern part of Punjab

state, the water table rose widely to only 3 m below the surface, with severe deterioration recorded between 1955 and 1977 (Ghassemi, Jakeman, and Nix 1995).

10. The main exception is J. Harriss's (1977b) material on the village of Randam, but his principal concern was with the social impact of agricultural change.

11. Dupuis relied on Slater (1918) and Thomas and Ramakrishnan (1940) for his historical material. Dusi was included in the 1973 and 1982 samples because the historical material was available, but not much use was made of it in either book.

12. This is perhaps surprising, given the detailed knowledge of the ecology of the variegated grasshopper shown by Nigerian farmers, a knowledge they used to control the pest (Richards 1985). Perhaps the Javanese farmers had simply forgotten because the previous major outbreak of the white rice stem borer took place 53 years earlier, in 1937 (Winarto 1995). Perhaps more importantly, they had been brainwashed by a mass of technical instruction received during the 25 years before the 1990 outbreak.

13. The Sukahaji area may have been used by shifting cultivators in the nineteenth century, as was most of the volcanic upland region south of Bandung. Ancestors of the modern villagers moved into a little-occupied area around 1900. They grew an annual crop of rain-fed rice in bunded fields, rotated with *palawija* (dry-field) crops in the drier season. Terracing was incremental until after a weir and some main canals were constructed, around 1930, on a nearby river upstream of Sukahaji. Further reticulation through a system of small canals, leading water down into rebuilt and carefully designed terraces, was carried out rapidly by the farmers themselves. Double cropping was then practiced, leading to triple cropping after MVS were introduced. During this period, population had continued to increase rapidly. In this area, without fully controlled irrigation, sometimes there was a shortage of water for part of the village land during the dry season.

Part IV

The Future of Agrodiversity

Chapter 13

Recent Trends in Agriculture

■ Economy and Ecology Hand in Hand

INTRODUCING PART IV

In this final part of the book, the story of modern change is carried forward from the Green Revolution into recent and current trends in agriculture and its sciences. The purpose is to prepare the way for a concluding evaluation of agrodiversity as a continuing strategy. Four main areas are discussed. One is a set of developments sparked by fears of genetic erosion, and the second is a group of revolutionary changes in plant breeding technology. Alongside these are the emerging science of agrodiversity itself, and especially agrobiodiversity, and the influence of modern biological management developments in the North on scientific thinking. The issues are complex, and the present chapter deals with them mainly in general terms. Chapter 14 goes more deeply into scientific issues. I begin with an account of events.

FARMERS IN TROUBLE

We saw for Nigeria in chapter 11 and Indonesia in chapter 12 how the cost of chemical farm inputs began to increase rapidly in the 1980s. The same happened all over the world. Countries that subsidized their provision to farmers found their financial burdens soaring. Long before financial crises began to affect one developing country's economy after another in the mid-1990s, many countries had greatly reduced subsidies for farm inputs, and some had eliminated them altogether. There was another set of economic problems, closely linked with the growth of massive grain surpluses in the North. In the names of free trade and economic restructuring, developing countries receiving aid came under pressure to cut back on their own protection to food producers, where this existed. Farm prices failed to rise, and cheap imported foods competed with the produce of national farmers, who had to pay higher prices for everything they bought. Farmers who had adopted monoculture of commercial crops suffered severely when the price of these crops declined. Monoculturally grown crops also were more susceptible to attack by insect pests, necessitating costly pesticide applications. With few resources, many small farmers who had adopted cash crop monoculture soon fell into debt. This continued in the 1990s. Indian cotton production was revolutionized in the 1980s after introduction of high-yielding varieties of American origin. Grown monoculturally by small farmers, these need frequent seed replacement, fertilizers, and pesticides, all bought commercially because these varieties are not covered by the old Indian seed store system (chapter 12). By 1998, combined misfortunes left many small farmers in debt for sums larger than the value of a good year's crop.[1]

These problems generally have been viewed as in some way different from those that afflicted many farmers in the North. Yet in a globalized economy there is no longer any clear separation between South and North. There are many linkages. To discuss the future prospects for agrodiversity, we must put the situation in the developing countries into a perspective in which the focus of the most dynamic new ideas, methods, and scientific innovation—both by industry and by farmers—has been in the North. I now abandon the exclusive focus, in discussing modern events, on farming in developing countries. What is now happening in the North has big implications for farmers in the South. Since the 1980s, the drivers of the most vigorous initiatives in plant breeding have been multinational corporations based mainly in the United States and Europe. They have their eyes on farmers in the South as buyers of their products.

EVENTS IN THE NORTH

Recent events in the North have parallels with those in the South. During the years of the so-called Green Revolution in Asia and elsewhere, discussed in chapter 12, even greater changes took place in the North, with results as dramatic as the best Green Revolution results in the South. By the 1980s, problems were emerging. Massive overproduction of several food crops and products had led to the creation of grain mountains, especially of wheat. Their existence depressed prices and led to subsidization of exports to the rest of the world, whether in trade or as aid. Worldwide, almost all farm crops sold on the international market lost a large part of their value between the early 1980s and the early 1990s (Pretty 1995). Since then, there has been some improvement. In developed countries, many farmers who had invested in expensive new machinery and had taken loans to develop their enterprises became deeply indebted. There were widespread financial failures, and most of those who failed were smaller farmers.

To the cost–price squeeze on farmers was added a more insidious set of problems. As in the South, either yields stagnated or more fertilizer was needed to sustain them. Insect pests multiplied and outbreaks increased, demanding heavier and heavier use of pesticides. Their use has consistently created problems by wiping out a range of nontarget insect and bird species and some soil organisms, including many potentially useful in pest control. Public attention first focused on the damage to wildlife, and then on the dangers to human health from use of agricultural chemicals and the pollution of waterways by these chemicals (Carson 1962).[2] Rising concern led to bans on the more persistent toxic chemicals, the organochlorines, and to growing pressure on farmers to change their ways. The pressures came mainly from a group of environmental movements that arose in the 1970s among the urban middle classes of the North. Soon these latter were "in effect telling farmers what is in their and the nation's best interest to do" (Lowitt 1985:324). Many Northern farmers became unhappy about the consequences of modern agriculture for the environment.

STAGNATION IN YIELDS

There are many reasons why crop yields in so many regions have ceased to grow. Rosset and Altieri (1997:285, 288) state that stagnation has resulted from the "steady erosion of the productive base of agriculture through unsustainable practices." They contrast "a healthy, biologically rich soil with ample organic matter and a diversity of micro-organisms" with "a dead, sterile, chemically poisoned soil with little organic matter." This is not all, as we will see in chapter 14, but they are largely right. The agricultural use of nitrogen fertilizer tripled worldwide between 1960 and 1979. Among fertilizers, nitrogen has increased in proportion to potassium and phosphorus to reach about 60 percent of the whole (Socolow 1999). A lot of this has exceeded the potential uptake by crop plants or has failed to be absorbed in the plants. A great deal is lost into the soil and waters and, in the case of nitrogen, into the atmosphere. It has led to acidification of the soil, demanding more fertilizer inputs. Excess nitrogen can accelerate microbial decomposition, destroying organic matter faster than plant residues can accumulate it.

Increasing loss of topsoil by erosion is also well documented. Topsoil contains a high proportion of soil organic matter (chapters 5 and 9). By the 1980s, this occurred not only in the well-documented cases of the United States and the former Soviet Union (Blaikie and Brookfield 1987) but also in countries that had not experienced soil erosion problems in modern times, such as England and northern France (Boardman and Bell 1992). It was estimated that erosion rates in northern France had attained levels not reached since the land was a subarctic desert about 15,000 years before (van Vliet-Lanoë et al. 1992). The identified cause was mechanized land surface management, involving deep plowing, and the introduction of cropping systems that left the land bare for significant parts of the year. Problems of soil compaction by heavy machinery are widely recognized. Soil compaction also results from the loss of organic matter. Some chemicals cannot be degraded by the carbon cycle and persist in the soil and food chain. Their use fails to contribute to soil structure and capacity to retain water in the way that former rotational practices did before the modern period. Soil organic matter, though very persistent in humus form, has slowly been mined.

ALTERNATIVE APPROACHES

Two sets of responses to what some perceive as a crisis have emerged among Northern farmers and scientists. Supporters of what Pretty (1995) has described as modernist agriculture essentially seek to revitalize the drive of the 1950–80 revolution with the aid of a dazzling new set of technologies. Others seek a return to ways that are biologically more sustainable. I will use Pretty's (1995) term *modernist agriculture* in this chapter, in preference to *mainstream* or *conventional*, because the mainstream has never been constant and what is often now called conventional farming applies only to the past half-century. On the other hand, I shall use the increasingly common term *alternative agriculture* to group together the various biologically friendly strategies. This is in preference to the emphasis on sustainable agriculture, common in much of the literature, because some modern practices might also become more sustainable. Modernist agriculture has adopted conservationist changes, especially in the matter of plowing, and there is a large gray area between the two extremes. Initially, I present the two approaches as opposites, starkly so in their more extreme manifestations. Both are complex and they are linked through a new valuation placed on genetic diversity.

■ The Background: Genetic Erosion and Conservation

GENETIC EROSION

One central concern in the second half of the twentieth century has been with the major loss of cultivated crop varieties, already mentioned in chapter 12. A great many have been replaced by uniform higher-yielding varieties of a smaller number of crops produced by modern plant breeding. The modern varieties have a narrow genetic base. To many people, some scientists included, the losses may be regrettable but are necessary because of population growth and the need for modern higher-technology agriculture to produce more food. Since the early 1980s, it has been known that more than 90 percent of worldwide nutritional needs are met by some 30 crops, so that future global food security depended on this range, and mainly on only a small part of it. For example, Harlan (1995) bemoaned the concentration on a small number of species and saw the risks that it involved, but he viewed the loss of variety as inevitable. There are suggestions that the genetic erosion argument has been overstated, and the topic is discussed more fully in box

13.1. Loss of habitat diversity, which is very widespread in modern times, may be an even more serious reality if it threatens the integrity of whole dynamic ecosystems. With genetic erosion often goes the accumulated knowledge of farmers about varieties and their cultivation, which is an essential part of agrobiodiversity (Thrupp 1998; Cromwell 1999; see also chapter 3).

Box 13.1
The Background and Reality of Genetic Erosion

Most of the world's genetic diversity is in the tropics and subtropics; the temperate countries are not genetically well endowed. The agricultural development of Europe and North America has been based heavily on both products and genetic material imported from other regions, principally the Near East and central America. Juma (1989) calculated that 65.6 percent of the genetic resources of food crops were derived from west central Asia and Latin America and that Latin America alone has contributed 34.4 percent of the genetic sources of industrial crops. Very tiny contributions have come from the Euro-Siberian, North American, and Australian regions.

Scientific crop breeding began in the temperate countries, and in some tropical colonies of those countries. Modern high-yielding varieties of preferred crops rely for improvement, and especially for resistance to pests and diseases, on genetic material imported from the areas of greater diversity. In these areas, in the Near East especially, the diversity of landraces and their weedy relatives is reported to be severely diminished or confined increasingly to isolated areas. It is further threatened by the widespread withdrawal of public institutions from crop breeding and the modern dominance of the seed industry by a small number of multinational companies also involved in producing chemical inputs to crop production.

Although the situation is alarming, some cautions are in order. Genetic diversity still may be more robust than many commentators have suggested. Prescott-Allen and Prescott-Allen (1990) searched available national data on food plants only, limiting themselves to those that supply the top 90 percent of each country's food supply by weight. This yielded an incomplete list of 103 species used for food only, without including home garden crops not also found elsewhere. Several species contain widely cultivated varieties that are distinct enough to be subspecies. Some crops that are cited frequently in the literature are not included in national lists. In his "select roster," J. D. Sauer (1993) discussed the named varieties of some 200 food and fruit crops. The Food and Agriculture Organization of the United Nations (FAO 1996) provided some information on about a third of these but also noted that of 270,000 known species of higher plants, about 7000 are used in agriculture and horticulture. In addition to cultivated crops is the hidden harvest of crops that are collected but not cultivated (Scoones, Melnyk, and Pretty 1992). Wood and Lenné (1997) further commented that, very often, farmers' practice is to plant new varieties together with their old varieties so that there may be an enrichment of agrobiodiversity rather than a loss. Moreover, genes may survive in cross-bred varieties, so "it is quite possible to have a considerable loss of landraces with no genetic erosion whatever" (Wood and Lenné 1997:113). Erosion of the whole genome takes place only when crops are totally replaced, and genetic substitution or replacement takes place when one variety of a crop replaces another. Even genetic enhancement or enrichment can be the outcome (Qualset et al. 1997).

Landrace erosion is not a new phenomenon. It followed naturally from adoption of preferred new lines, and a great number of former crops have vanished altogether or their cultivated forms have vanished. Mass selection of good seed may be as old as all agriculture, and germplasm that was not selected has been lost. Farmers have always discarded varieties that are no longer of interest to them. Hybridization has been used purposively with a wide range of crops since the seventeenth century after scientific discovery of sexual reproduction in plants. The ancient discovery of grafting made this form of hybridization important much earlier among tree crops. Once national markets were created by improved communications, seed merchants began to seek more productive and uniform varieties, and many named varieties of both crops and livestock arose in western Europe in the eighteenth century. They were already eroding local races by early in the nineteenth century.

EX SITU AND *IN SITU* GENETIC CONSERVATION

In the early 1970s, after two decades of rising alarm, major attempts were initiated to rescue the endangered varieties of seed by collection and long-term storage. Within the network of International Agricultural Research Centers (IARCS, constituting the core of the Consultative Group on International Agricultural Research, CGIAR), a new body was set up in 1974 to coordinate germplasm storage internationally, including national storage as well as in the IARC facilities. This International Board for Plant Genetic Resources had fairly dramatic success in its first decade, and in the 1980s its main functions were taken over by a newly formed United Nations Food and Agriculture Organization (FAO) Commission on Plant Genetic Resources for Food and Agriculture.[3] Collection of germplasm extended through most countries of the world.

The conditions of *ex situ* conservation are demanding. Seeds must be kept dry for even short-term preservation by farmers, and long-term conservation demands that they be very dry and very cold. The first very-long-term storage at about −20° was undertaken at St. Petersburg (Leningrad) in the 1920s, under the direction of Vavilov (Juma 1989). Deterioration of the stock after a few years is common, and seed must be planted out and replaced. By the mid-1990s more than 6 million accessions were stored worldwide *ex situ*, including proportions of wild relatives of the main crop plants that range from 10 to 60 percent. They are held under a wide range of quality control conditions, often with inadequate documentation and in many cases without the necessary ability to regenerate the collections every few years (FAO 1996, 1998). Taking account of duplication, FAO (1996) estimated that only 1 to 2 million of the accessions are unique. A high proportion of the world's seed storage has substandard conditions, and there is concern about the viability of many collections. Most of the securely preserved stock of seeds is held either in the IARCS or in large, well-equipped stores in countries of the North. In recent years there have also been developments in the preservation of plant tissues of clonally propagated crops *in vitro* at ultralow temperatures (Withers and Engelmann 1998).

There is an important political aspect. Under the conditions of their establishment, germplasm from the internationally backed genebanks is freely available. A high proportion of the wheat grown in North America, Europe, and Australia derives from varieties originally developed by Centro Internacional de Majoramiento de Maiz y Trigo (CIMMYT), the IARC located in Mexico. Germplasm of a range of crops, further improved by developed country breeders, has been sold to developing countries, and in modern times it has very commonly been protected by either legally enforceable plant breeders' rights or by patents. It is often pointed out that great profits are being made out of genetic material imported at little or no cost, the sources of which are now under threat by competition from the modern crops. This seems to violate agreements such as the Convention on Biological Diversity, which, as stated many times in the documentation of the convention, require "fair and equitable sharing of the benefits of biodiversity." This long and complex story is told in many places, with fairly uniform concern but different degrees of indignation. The most comprehensive accounts are by Mooney (1983), Fowler et al. (1988), Juma (1989), Fowler and Mooney (1990), and Mooney (1996). Certain gaps are filled in by Querol (1993) and Shiva (1993). Other recent discussions are in FAO (1996, 1998), Gliessman (1998), Cromwell (1999), and the Web sites of FAO's Commission on Genetic Resources for Food and Agriculture and of the Rural Advancement Foundation International (RAFI).

The alternative or complementary approach is to favor *in situ* conservation on farmed land. Except by newly developed *in vitro* methods, *ex situ* conservation is difficult or impossible in the case of clonally propagated species such as tubers and recalcitrant seeds that will not survive in the genebank conditions. Except for some conservation of live plants on experiment stations, most *in situ* management of crop species and varieties is in the hands of farmers, who replant seed saved by themselves or acquired from nearby. This is mainly done by developing country farmers, but a minority of Northern farmers do the same.

No country has a record of what is managed in this way, but there is certainly greater diversity, especially intraspecific diversity, managed on farms than is held in the *ex situ* genebanks. *Ex situ* conservation is expensive, it is subject to high risk from electric power failures, and the samples are so small that seed is not at all readily accessible to farmers (Cromwell 1999). Critics also have maintained that *ex situ* conservation destroys evolutionary dynamism and support the view that local or landrace varieties are adapted to environmental conditions. Until lately, little use has been made of the germplasm of wild relatives held in the stores. Critics have increasingly urged that much greater efforts be made to support *in situ* conservation (FAO 1996, 1998; Maxted, Ford-Lloyd, and Hawkes 1997; Jarvis and Hodgkin 1998; Brush 1999).

There has been significant research on how to achieve this goal, but most of it has concerned how the material should be selected, not the modalities of generating farmer cooperation. Most work has been sponsored internationally, and only a few national agricultural research services have participated. There has been one major pilot project in Ethiopia, and this work is now leading into an international program undertaken by the International Plant Genetic Resources Institute (IPGRI; Jarvis and Hodgkin 1998; Brush 1999). There is also a 12-country mainly nongovernment organization–based Community Biodiversity Development and Conservation Programme that has related objectives (Visser 1998). Experience is being assembled, but the problem of facilitating the necessary participation of farmers remains mainly the subject of suggestion (Cromwell 1999). FAO (1998) reported eloquently—through its silence—on the dearth of effective *in situ* conservation activities before these new initiatives were begun.

Wood and Lenné (1997) took issue with the emerging paradigm that *in situ* management is better than *ex situ* conservation. Both involve storage as well as management, and storage under stable conditions is hard to achieve on a farm. They argued that little evidence supports the widespread belief that varieties grown on farm often benefit by crossing in of genes from wild relatives. Local adaptation is also questioned because it is more likely that maladaptation progressively takes place in the absence of importing new germplasm from other areas. This view corresponds with nonequilibrium ecology findings in which the concept of local adaptation has been questioned both on empirical and theoretical grounds (Zimmerer 1994, 1996, 1998, 1999). For centuries most farmers have regarded germplasm importation from other areas as necessary. This practice has ensured that genetic diversity has sometimes been enhanced by the introduction of modern varieties. In many areas, Wood and Lenné maintained, there may already be no true local varieties remaining but many crosses between original landraces and the introduced varieties (see also box 13.1). Qualset et al. (1997) similarly showed how erosion is not the necessary consequence of introducing modern varieties.

Moreover, there is considerable debate about the purity of landraces in open-pollinated crop plants. Taking the example of maize in Mexico, Louette (1999) showed that every landrace is a metapopulation created by geneflow within the local region and by new seed brought in from other areas. She remarked, "A landrace is far from a stable, distinct, and uniform unit. Its diversity is linked to the diversity of the material sown in the area, and then related to the diversity of the introduced varieties" (Louette 1999:137). A main object of *in situ* conservation therefore should be to enhance the processes that create genetic diversity, not to protect any actual body of genetic material.

In situ conservation of traditional landraces demands that the agroecosystem also be conserved, and this can be extremely difficult and probably impossible in economically dynamic regions. No agroecosystem is a museum. One alternative, raised by Jana (1999), concerns creating biological diversity in modern agricultural systems by planting heterogenous populations under contrasted conditions. Such mass reservoirs may develop greater diversity than has been found in landrace populations.

There is increasing emphasis on the agroecosystem as the unit in which to work, including all its wild and semimanaged components and its arthropod and microbial members (Aarnink et al. 1999; Cromwell 1999). It is noted that although crop plant biodiversity may be limited in many modern systems, other aspects of biodiversity remain of major importance, just as in small farmer regions. There has been a new look at the meaning of agrobiodiversity on different scales, from the whole ecosystem down to the individual field. Agrobiodiversity, as distinct from the wild biodiversity that was the focus of all attention only a few years ago, is a major new field carried forward into the twenty-first century.

NEW USES FOR BIODIVERSITY

The developments just described have taken place at the same time as a new wave of plant breeding. The latter includes new sets of technologies and began to evolve in the 1980s. It has led to new germplasm collections to supplement what breeders could get from the genebanks. The advances in biotechnology discussed in this chapter give potential value to a much wider range of biota than was formerly the case. Bioprospecting has become an important economic activity, no longer simply a matter of scientific research. It now includes large-scale soil sample collection to be searched for useful soil biota. Some of this modern collection work has become highly controversial, especially in the context of medicinal plants, and it has been the subject of major international dispute.[4] Because detailed discussion of medicinal plants is specifically excluded from this book, this aspect of what is often called biopiracy is not developed further.

■ Innovations in Plant Breeding

BIOTECHNOLOGY: THE GENE REVOLUTION

Among the most important scientific advances in modern times was the discovery in 1953 of the physical structure of deoxyribonucleic acid (DNA), the molecular substance carrying the code that determines genetic structure in all living organisms.[5] About the same time, work on plant tissue culture, which had already advanced quite far before the 1940s, began to achieve important successes (Cocking 1986). The latter, with the development of new cloning (micropropagation) techniques, provided the foundation for the exploitation of genetic engineering. By the early 1970s, work done in universities established methods by which specific genes could be transferred from one organism to another, thus creating transgenic (or genetically modified) organisms. In the same period, a whole series of other advances was made in the area of mapping, or sampling, the genetic sequence of a wide range of plants, making possible new developments in both conventional and tissue culture plant breeding. Producing disease-free germplasm derived from the meristem of the plant, also a low-cost innovation, became of major importance in the 1980s.[6] Micropropagation of disease-free material has now been very widely used with many crop plants in many countries (Spillane 1999).

In vitro tissue culture techniques miniaturize and accelerate conventional plant-breeding methods. Genetic engineering, although supported by tissue culture, sets out to transform the plant by deliberate introduction of foreign genes, not necessarily only from plants but from any life form. Growing out of 1970s work in the pharmaceutical sector, it became important in agriculture in the 1980s. The ability to introduce and express foreign genes in plants was first described for tobacco in 1984 and by the mid-1990s had extended to more than 120 species in 35 families (Birch 1997). There have been rapid developments in technology, and a substantial number of field trials have been conducted, a growing number of them in developing countries. The earliest and still basic method is the use of a natural infection, through the plasmids of *Agrobacterium* spp., to carry the transferred gene or genes into the plant or tissue cell. Another method, used also in other forms of biotechnology, takes advantage of the opening of cell walls by enzymatic digestion so that foreign genes can be introduced. These quasinatural methods have been supplemented by use of a "gene gun" to fire the genes into the cell, and a number of other methods have been developed experimentally. Potrykus et al. (1998) offered a comprehensive but quite technical description of the technology.

Over time, work has become concentrated on the crop plants in which most commercial profit can be made, especially maize, wheat, rice, soya beans, cotton, tobacco, canola (rapeseed), potatoes, and tomatoes. Commercial sale of transgenic seed, incorporating nonplant genes, began in 1994, and since 1996 there has been rapid expansion. By 1999 the area planted already reached 40 million ha, principally in the United States but also in Argentina and Canada, with smaller areas in several other countries. Most of this was soya bean, maize, and cotton, and the crops were engineered first for resistance to herbicides and second to contain their own insecticides. The Web pages of the RAFI (1999a, 1999b, 1999c, 1999d, 2000) are an up-to-date source. The Web pages of the USDA and the Monsanto-sponsored Life Sciences Knowledge Center (1999) provide information of another type.

PRODUCTION OR PROFIT?

Concentration on cash crops has happened because the environment of plant-breeding work has undergone major changes. The new enabling technology of gene transfer had enormous monetary value, and its products were protected quickly by patent rights. In the 1980s there began a series of mergers between the agrochemical companies, pharmaceutical companies, and seed-producing companies in both North America and Europe. They quickly became multinational. These mergers coincided with ideologically driven pressure on public institutions to withdraw from plant breeding and to hand this task over to the private sector. Surviving public sector research in several Northern countries often is required to seek private funding. Public sector research in the developing countries, including that in the IARCS, has been starved for funds. The account by Juma (1989), although it presaged much of what has happened since, is already historical.

Genetic transformation is possible in two main areas: the output characteristics of yield and performance and the input characteristics, of which herbicide resistance is an example. Much enthusiasm for transgenic work has been generated by the possibility of improving output characteristics, and a great deal of research is in progress, some of it discussed in chapter 14. Yield is controlled by many genes, but the initial transgenic technology made possible the transfer of only single genes. By the mid-1990s capacity had increased to about five genes in one transfer. Therefore it has been both easier and more profitable to work first on input traits. As soon as strong commercial links were established between the seed and chemical industries in the early 1980s, it became an objective to insert resistance to the herbicides manufactured by the same companies into the seed they produced for sale. Global sales of herbicides and other pesticides rose above $30 billion in the late 1990s (Lappé and Bailey 1999).

Increased herbicide application has been associated closely with reduced tillage, the soil conservation benefits of which have become clear to many farmers. Already in 1987, reduced tillage was used on more than a quarter of all U.S. crop land (Committee on the Role of Alternative Farming Methods 1989). By the end of the 1990s, minimum tillage practices had become increasingly common. Generally, herbicides are applied to the land before the crop is planted, but they could be used much more effectively and economically with the crop already growing if it was resistant to the herbicide. At the end of the twentieth century, more than 70 percent of commercially planted transgenics had herbicide resistance. The most common herbicide-resistant technology in use operates by stimulating an enzyme that assists photosynthesis, enabling the plant to withstand at least two applications of company-produced herbicide each season. The small number of companies that own most of the technologies have been able to develop them in an order that, as Lappé and Bailey (1999) remarked, enabled the short-term goal of profit to assume more importance than the long-term goal of improved productivity and nutrition. They argued that there has been a cynical rush to lock farmers into use of proprietary chemicals before the patents on them expire.

Developing insect resistance within plants has become important because of the high toxicity of most insecticides initially used and some of those still used. Frequent spraying also is costly. Until the 1980s, plant breeders sought natural resistance among landraces for crossing into new commercial varieties. More recently, emphasis has been placed on incorporating genes expressing substances toxic to insects into the crop plant itself. The greatest success has been achieved with varieties of the soil microorganism *Bacillus thuringiensis* (Bt), long known to produce proteins that killed a range of moths and butterflies but not animals or people (Juma 1989). Among commercialized transgenic varieties used in 1998, insect resistance characterized 28 percent (James 1998).

CONTROLLING THE FARMER

Through purchases, mergers, and takeovers, the five or six largest agrochemical companies together with their subsidiaries, are also among the largest seed companies worldwide. They compete keenly for market shares but also support one another in reinforcing their common oligopoly. In addition, they have been able to exert influence on decision makers in national governments. Because transgenic work is very costly, specific control of what farmers do with seed is increasingly important for the breeders. It has become of greater and greater concern to the commercial breeders that farmers be not only discouraged but prevented from replanting or selling their own seed. Otherwise, the breeders would be unable to get a sufficient return from their large investment. Since the 1930s, most commercial maize in the world has been grown from hybrid seed, vigorous in the first generation but widely variable in subsequent generations. Farmers therefore buy seed every year. Using a combination of biotechnology and traditional methods, there has been substantial progress in developing other hybrids. In China, a hybrid rice with the highest yield known anywhere was produced in the early 1990s. Annual repurchase or supply is not necessary with nonhybrids unless new varieties are sought. Especially with transgenic soya beans, there has been mounting pressure on farmers not to replant seed, including clauses in contracts with seed dealers, legislation, and policing by company agents. Farmers reusing seed are in violation of the agreements they had to sign when buying the new transgenic material. In North America, there have been several prosecutions for violation of patent rights.

The companies probably are right in claiming that transgenic developments offer farmers in developing countries higher and more secure yields in consequence of reduced weed competition and insect damage, although few could afford to buy them annually. Several developing countries have now joined the Union for the Protection of New Plant Varieties (UPOV), which, in its 1991 revision, makes it optional for member governments to enforce rules preventing use, as distinct from sale, of farmers' seed (Correa 1999). Yet the practical possibilities of enforcement among millions of small farmers are very limited. Some developing countries and a number of civil society organizations have mounted a sustained case for farmers' rights against those of the plant breeder. The seed and chemical companies therefore took new steps of their own, developing a set of technologies that would do away with the need to police farmers. This bold but misguided initiative is discussed in chapter 14. In this chapter we look at the other side of the conflict of approaches.

■ Alternative Agriculture in the North

THE RISE OF ALTERNATIVE AGRICULTURE

A major reason for widespread adoption of minimum tillage practices by Northern farmers has been the need to reduce production costs rather than awareness of the damage done to soil organic matter. The change is ecologically benign except from the point of view that it has led to increasing dependence on herbicides. There have been other changes, particularly greater care in applying insecticides by a minority of farmers who have used integrated pest management (IPM). More farmers have shied away from the more toxic chemical preparations and are seeking biological products. However, many Northern farmers are continuing much as they have done since the 1950s, except with greater care to control their costs.

A very small minority of Northern farmers never adopted modernist methods in full, and since the 1970s a small but growing number have joined a hitherto fringe group in the comprehensive adoption of biological management. Although these approaches rarely draw overtly on developing country experience, some of them are very similar to practices reviewed earlier in this book. There are a number of different schools, three of them reviewed in more detail in box 13.2, but they can usefully be collected together under the term *biological husbandry* that was preferred by one of the founders of the modern organic school of farming (Balfour 1943, 1948, 1977). I use her expression because the more common term *organic farming* has been debased by being used in other ways. The central notion of biological husbandry is concern with the health and vigor of the soil.[7]

Box 13.2
Three Schools of Modern Alternative Agriculture in the North

Biodynamic farming. Founded in the 1920s in Germany, this was the first modern alternative, and as a small minority movement it has become completely international. It relies wholly on natural inputs, especially to create humus in the soil, and create biologically active soil. Treatments also include use of two sprays, one using organic ingredients and designed to condition the soil and the other using mineral ingredients and designed to sensitize plants to light. Also involved is timing of agricultural activity in relation to the phases of the moon and the signs of the zodiac. There is a good deal of mysticism in all aspects. Acceptance of very tight discipline is needed. In modern organic producer certification, bio-dynamic farmers qualify for registration (Podolinsky 1985; Koepf 1989).

Natural farming. This approach initially was unique to Japan and also unique in other ways. It did not grow up all at once but was developed through experiment over time by its innovator, Masanobu Fukuoka, who, in his younger days, was an agricultural research scientist. Compost and farmyard manure are not used, and there is no plowing even to shallow depth. Summer and winter grains are broadcast in the standing crop and among clover. Straw mulch is used heavily, but no green manures are cultivated. Irrigation is used only to reduce the clover and kill weeds at the time of rice germination. Vegetables also are broadcast in suitable spots on the hillsides. The system has gained popularity in other parts of Japan, in Europe, and the United States but has not been widely adopted. Yields obtained compare with those of modernist farmers in the same (high-yield) region of southern Japan, but Fukuoka did not plow his level and irrigable field between 1950 and at least 1985 (Fukuoka 1978, 1987).

Organic farming. This is now the largest of these schools, growing out of fringe movements of the 1950–80 period to embrace 0.5–12 percent of the agriculture of Northern countries. There are much smaller movements in some developing countries. Set up to meet a demand for food free of chemical additives and influences, organic farming is now subject to standardization in most regions. Minimum standards require no use of chemical pesticides or fertilizers, no artificial ingredients in food, and maintenance of soil fertility by natural means. Crop rotation is a common practice. However, natural supplements can be used and can come from far off-farm. Unlike other movements, there is now agribusiness involvement in organic farming.

There have always been alternatives to prevailing agricultural practices, commonly used by a minority of innovators, often experimentally. There have been historical periods, mostly associated with changing economic conditions, during which such innovations have been more widely adopted. Thirsk (1997) recognized four such episodes in English agricultural history since the time of the Black Death in the fourteenth century.[8] This long view is important, but the circumstances of modern alternative farming in the North are different in several ways. It responded to an opportunity. From the 1960s onward, it was supported by growing concern about the quality of food among mainly urban consumers. Almost simultaneously in most countries of the North, but most strongly in western Europe and California, alternative agriculture began to flourish in the late 1970s as a movement with an ideological basis. 'Organically sound' farmers sold to a mainly middle- to upper-income niche market, which was supplied from small farms, by younger farmers, principally in regions where specialisms had endured since the nineteenth century.

In its large literature, the single common characteristic of modern alternative agriculture is the reduction of external inputs, especially of an inorganic nature, and their replacement with management that improves the organic matter content of the soil by means of rotations, recycling, and the addition of organic inputs. There are close links with the ideals and concerns of modern environmental movements. Diversification is a major element: crop diversity and the use of ecological niches, with field boundaries aligned closely to soil types, together with crop rotation, are distinguishing marks of an alternative farm in the United States (Gardner, Jamtgaard, and Kirschenmann 1995).

Ecologically, the rationale of alternative agriculture is the enhancement of biological cycles within the farming system, as far as possible working the farm as a closed system with regard to organic matter and nutrient elements, the true measure of biological husbandry.[9] The farm builds its reserves of soil fertility, is sustainable, and acquires much greater resilience in the face of pests, diseases, drought, and other external shocks, as we shall see for agrodiversity as a whole in chapter 14 (Committee on the Role of Alternative Farming Methods 1989; Dumaresq and Greene 1997; Fernandez-Cornejo et al. 1998). The approach calls on much older styles of farming, practiced in Asia at least as much as in Europe and North America. Fukuoka (1978:117–18), whose distinctive natural farming approach involves no use of manures (see box 13.2), pointed out this dependence: "When you look over the principles of the organic farming popular in the West, you will find they hardly differ from those of the traditional Oriental agriculture practiced in China, Korea and Japan for many centuries."

Biological husbandry calls for discipline and skilled management, and in most versions it is more labor intensive than modernist farming. In its full application, it has appealed only to a minority of Northern farmers. In some countries of continental Europe, more than 10 percent of farmers meet the exacting organic registration standards, but there have been subsidies to make this possible. In Denmark, for example, a special extension service existed for a time alongside the regular extension service (Kaltoft 1999). The proportions in the United States and in Britain did not exceed 1 percent until the end of the 1990s (Niggli 1998).

REGISTERED ORGANIC FARMING AND MODERNIST AGRICULTURE

This is not all that has happened. Production of more natural produce has become profitable and, to accommodate business interests, regulation has come to be more concerned with what may or may not be used as inputs than with ecologically sustainable modes of production. Beyond the fringes of the small group of farmers who try very hard to manage their farms biologically, with a minimum of external inputs, is a larger group that is not inspired by ecological values. Their version of organic farming includes a range of mineral and organic products such as seaweed, fish meal, crusher dust, potash, bone meal, bat guano, and commercially produced compost. These are manufactured from widely separated sources and transported over long distances. Organic registration is not lost by use of these inputs, which therefore make organic production accessible even to farm businesses that produce monoculturally with substantial mechanization (Rosset and Altieri 1997; Guthman 1998). In a situation where there are no clear boundaries between alternative and modernist practices, regulation becomes a political issue.

Beginning in the 1980s, a growing number of countries have established national standards for produce allowed to be sold as organic. The standards vary widely, but an emphasis on methods of production is giving way to emphasis on permitted and prohibited inputs. There are usually long lists of what may and may not be used. In the United States several years of dispute led to the decision that products of genetic engineering and irradiation, or municipal sewage sludge, could not be included. Most countries exclude use of genetically modified seed or other inputs (Guthman 1998; "USDA abandons three contentious issues" 1998).

In addition to those who meet registration standards, many farmers now use some biological practices together with modernist methods. A growing range of biologically benign inputs is being offered commercially to essentially modernist farmers. In the late 1990s, a lot of research was going into biological preparations, including biofertilizers inoculated with bacteria to accelerate formation of organic carbon and the release of phosphorous and potassium. Other organic products are marketed for pesticide use. Among the soil conditioning products now listed in the trade catalogs are some that blend chemicals with biologicals. Some of the advertising is redolent of the literature of the biodynamics movement with its claims for a seemingly magic stimulation of soil-forming processes and plant growth (box 13.2). As Rosset and Altieri (1997) argued, alternative inputs can readily take the place of a holistic alternative agriculture, and in consequence the organic movement has been devalued as a true alternative.

In Northern countries at present, organically produced food commands a premium in prices, and this is principal among the attractions for investment. To move into the mainstream of a postmodernist farming, the innovations would need to become substantially cheaper in relation to the general level of prices. Modern market developments are heartening to those who want to see diversity restored to agriculture in the North, but only changes in outlooks and taste that extend beyond a minority of middle-class buyers would permit alternative agriculture to become a wide-ranging reform of mainstream modernist practices. It would also be necessary for publicly assisted research into organic methods to become more than the very minor aspect of total research expenditure that it currently is. At the turn of the millennium, a much larger amount of public support is being devoted to developing agriculture based on modern biotechnology.

FIRST AND THIRD WORLDS

A few developing country farmers deliberately practice Northern biological management methods, for both ecological and economic reasons (Haverkort, van der Kamp, and Waters-Bayer 1991; Bunders, Haverkort, and Hiemstra 1996). Although these individuals are noted, surprisingly few participants in the alternative movement in countries of the North realize how much of biological husbandry simply reflects common practices among small farmers in the South. This is so even though one of the acknowledged founders of the movement gained his experience, and most of his ideas, in India (Howard 1940). A few of the core intellectuals of the modern movement, such as Altieri and Gliessman, make frequent reference to developing country farming in their writings. The more common view seems to be that if learning is to be in any direction, it should be by Southern farmers from their more enlightened brethren in the North. There is apparent reluctance to accept the view that indigenous farming is truly organic. As an example, a Briefing Paper on organic farming issued by the British Soil Association adds as a last item that "products produced and processed in a traditional way by indigenous and/or traditional groups can be certified as organic, provided that the sites are subject to a normal annual inspection and production and/or processing are in accordance with the principles in the Basic Standards" (Soil Association 1999a, 1999b).

Yet there is a substantial literature by developing country authors on the merits of a diverse biological husbandry, whether or not strictly organic. From India, they include a valuable book by Pereira (1993) as well as the work done by the M. S. Swaminathan Research Foundation. Swaminathan himself, a former director-general of the International Rice Research Institute, does not call for a fully biological agriculture and believes that all available technologies, including biotechnology, should be considered, but with care and after proper testing and evaluation (M. S. Swaminathan Research Foundation 1999).

■ The Special Case of Cuba

CHANGING UNDER THE PRESSURE OF EVENTS

Farmers in Europe, North America, and Japan who wanted to adopt biological husbandry methods, or simply use some of these within a modernist framework, were able to take their time. It was generally expected that full conversion to an organic or related system would take up to 5 years, and since the 1980s a number of governments wanting to see farming become environmentally more benign have subsidized farmers to help them make this transition. Such luxuries have not been available to farmers in developing countries. For Cuba, major changes became unavoidable for compelling reasons in the space of a very short period of time. It will be useful to examine this example in concluding this chapter.

Cuba's revolution in 1959 was essentially a revolution against the dominance of its economy and politics by the United States ever since the Spanish–American War of 1898. American companies and their affiliates controlled a large part of the land, its single-crop sugar economy, and its international trade. Postrevolutionary measures to take national control aroused immediate opposition from the United States. Trade sanctions, leading ultimately to America's total embargo on trade with Cuba that continued to the end of the twentieth century, were applied in little more than a year, ahead of the bungled military intervention in 1961. Cuba's economy survived partly by rapid internal reorganization but largely because the Soviet Union and its eastern European partners provided aid and took the sugar that could no longer be sold in the United States. The initial intention in Cuba was to diversify the economy, but hard economics determined a return to sugar on the basis of 1963–64 agreements with the Soviet Union.

For the next quarter of a century, the Soviet Union and its partners in Europe supplied oil, chemicals, machinery, animal feeds, and basic human foodstuffs at close to cost and took almost all of Cuba's principal exports, principally sugar, nickel, citrus, coffee, and tobacco (Le Riverend 1967; Boorstein 1968; Blackburn 1972). By around 1980, 85 percent of Cuba's trade was with the Soviet Union and eastern Europe. Rising economic problems in the latter region led to a worsening of the terms of trade between Cuba and its partners during the 1980s, and in 1989–90 the entire trading system collapsed.

Between 1964 and 1990, Cuba had developed a modernist agriculture on the Soviet model, with heavy intervention by national planners. Although the initial plan to diversify farming was dropped, an important second strand was developed in meat and dairy production, largely dependent on imported animal feeds. During the 1970s, mechanization and heavy use of fertilizers and pesticides turned sugar production into an industrial, modernist agriculture. Although a cooperative sector emerged after 1975 to occupy some 12 percent of land by 1990, 80 percent of land remained in state-owned farms in 1990. Peasant farmers became a tiny minority. The prerevolutionary diet of most Cubans, based on rice, beans, and root crops, was modified to include meat and dairy produce, and also bread made from imported wheat.

Fortunately for Cuba, its revolution included several home-produced innovations, of which the most important were in education and health. Not only was illiteracy eliminated, but technical and scientific capacity expanded to reach levels much higher than in most other Latin American countries in the 1980s.[10] In a generally critical review of post-1990 trends, "the human capital represented by the Cuban people [was described as] the greatest accomplishment of the revolution" (Centeno and Font 1997:218).

With a national policy open to ideas, the international rise of environmental concern did not bypass Cuba. National IPM initiatives and the making of biological products to control insects and diseases began in the 1980s (Rosset and Benjamin 1994). The old drive to diversify agricultural production resurfaced, leading by 1989 to a National Food Program to increase production of root crops and vegetables and restore a measure of food self-sufficiency. This was soon overtaken by crisis and the hard measures of the "Special Period in Peacetime." It was necessary to find ways to feed the population and rebuild trade with only small supplies of petroleum, fertilizers, and animal feed and no means to import capital goods in significant quantity.

REVOLUTION TOWARD BIOLOGICAL HUSBANDRY

In less than 3 years after 1989, Cuba's petroleum imports fell by more than a half, and there were even greater losses in agricultural chemicals, machinery and spare parts, and animal feeds. Many foods, including meat and milk, became short. Although there was no starvation, there was substantial hunger. The high-input system of agriculture could no longer be sustained, and it became official policy to apply science to the land in new ways. The intention was revolutionary. Crop diversity was to replace monoculture, and organic fertilizers, biofertilizers, and biopesticides were to replace inorganics. Oxen, which had never vanished from the peasant sector, were widely used in place of tractors, and in some areas energy-intensive dry season irrigation was replaced by a more seasonal farming.

There were major organizational changes, especially after 1992, and these continued into the late 1990s despite significant economic recovery after 1995. Most of the big state farms were broken up into smaller cooperatively managed units. Necessarily, these innovations required farmer cooperation, so there was new emphasis on making better use of the peasant sector (Deere, Perez, and Gonzales 1994). Partly to ensure farmer cooperation and partly because of the sheer impossibility of sustaining control by national planners, a number of organizational innovations arose at local level, restoring diversity in this element.

Parts of this ambitious plan have enjoyed more success than others. Despite massive production of *Azotobacter* for nonsymbiotic nitrogen fixation and energetic use of worm culture to make large quantities of compost, there was insufficient experience with biofertilizers for prompt success. Inorganic fertilizers continued to be used, as on three successful cooperatives in the Havana region that North American visitors praised for the success of their organic approaches in the early 1990s (Simon 1997). Yields continued to decline because rebuilding soil organic matter takes time, and it is not easy to move quickly from monoculture to intercropping and rotations. Rehabilitation of degraded soils has advanced only slowly.

The most outstanding success has been with biological programs for control of pests and diseases, especially through more than 200 decentralized Centers for Production of Entomophages and Entomopathogens. This network was initiated in the 1980s and was expanded rapidly in the 1990s. In these centers, products of modern biotechnology are produced on a local scale by a few scientists assisted by technicians recruited from among local high school graduates. As of the late 1990s, this system was still unique in the world.

IS CUBA A MODEL?

Up to the mid-1990s many aspects of the Cuban program remained in the research stage. As economic conditions began to improve, the urgency to substitute alternative for modernist practices had already decreased, and foreign exchange was again spent to rebuild agricultural infrastructure and buy inorganic fertilizers. However, much may endure. It makes good economic sense for a country that has had to rely heavily on its own resources, and make the best use of its own strong scientific community, to sustain its self-reliance. The ecstatic enthusiasm of some North American environmentalist observers for Cuba's embrace of the organic model notwithstanding, Cuba has not been seeking a full-blown organic agriculture. The aim has been to produce a diverse range of goods at least cost in foreign exchange, not to manage the soil to produce food of high quality without significant external inputs. Adopting biological husbandry practices was a measure toward a production goal, not a primary goal in its own right.

Therefore, in some ways the pragmatic Cuban approach might yield a better model for other parts of the developing world than the purist approaches of biological husbandry in the North. In other ways, it harks back to the early debate about fertilizers in the 1940s. In a very careful critique of the writings of Howard (1940) and Balfour (1943), founders of the biological management movement, Hopkins (1948) argued that maintaining soil organic matter was an essential goal of all farming. He believed that it was self-evident that chemicals alone are inadequate maintainers of soil fertility. Only a few years later many people believed that chemicals were all that was necessary, but Hopkins's persuasive argument for balance gained renewed validity by the turn of the new century.

Notes

1. In a part of Andhra Pradesh, more than 500 such farmers committed suicide in 1998–99, in some instances by drinking the pesticide that had failed to solve their problems (Vidal 1999). This is a particularly harsh example of the risks of cash crop monoculture for small farmers.

2. Awareness was sharpened when certain major examples of health damage came to be known in the 1980s. To cite only one case, it became clear that the fall of the level of the Aral Sea was not the only consequence of modern agricultural development using irrigation to excess in the southern republics of the former Soviet Union. There were also very grave problems of public health along the irrigated valleys flowing toward the Aral Sea, and they sprang largely from chemical pollution (Glazovsky 1995; Mainguet and Letolle 1997). Although replacement of a diverse system of irrigated farming with cotton monoculture had begun in the 1890s, it took only some 30 years, after initiation of a major drive for intensified production in the late 1950s, to achieve this result.

3. The reasons for these changes are political and do not directly concern us here.

4. This is discussed at length by Juma (1989) and Mooney (1996), among many other sources. The collection network includes major botanical gardens and universities, as well as explorations and research projects sponsored directly or indirectly by pharmaceutical companies. Collecting material necessarily includes collecting information on the known uses of the medicinal plants, and this aspect has become particularly sensitive. Crop plant germplasm to provide new genetic sources is sought in the same ways. After patenting, the intellectual property of the genetic material becomes that of the company or organization that developed it.

5. For plants, the significance can be described by the following simplification of a simplifica-

tion. All plants begin life as single cells. Except for plants that do not reproduce sexually, eggs are fertilized by sperm delivered to the egg by pollen. The cell divides many times to form the tissues and organs characteristic of the species. These characteristics arise from the kinds and amounts of proteins made in the cells. Cells control how much of which kinds of protein are present by controlling which genes are functioning. Proteins are chains of amino acids, the order and length of which are unique for each kind of protein. The sequence is specified by code, made of DNA, on a chromosome in the cell's nucleus. A gene is a piece of DNA that contains the code for a specific protein (Crouch 1998).

6. The meristem is undifferentiated tissue from which new cells are formed. It is a localized zone of specialized cells capable of division and growth in which one of the daughter cells of each division remains unchanged and the other differentiates into parts of the developing plant. Virus and fungal infections are widely found to stop short of the meristem in their penetration of plants.

7. Balfour eschewed chemical inputs (while also experimenting with them, both in parallel and association with organics, for 25 years). Her method included use of compost, farmyard manure, and mulches in large quantity. Work done on her experimental farm in England was scientifically based and measured and was first responsible for the finding that levels of mineral nutrients in the soil fluctuate seasonally in response to demand. With others, she founded the Soil Association, now the certifying body for organic food in the United Kingdom.

8. The period between about 1650 and 1750, identified by Kerridge (1967) as that of the major early modern agricultural revolution, is reinterpreted by Thirsk (1997) as a period of strong diversification (chapter 10). It was an alternative agriculture that gave way to the cereal-dominated mainstream farming of the more widely recognized agricultural revolution between 1750 and about 1880. The latter was in turn succeeded by new diversification and innovation in the face of depressed cereal prices. Modernist agrotechnology continued to evolve, especially after the development of industrial fixation of nitrogen in the early twentieth century, but did not again emerge to dominance until the time of World War II (1939–45).

9. This is a different meaning from the closed and open nutrient cycling within the soil that was discussed in chapter 5. Here, the farm is the unit.

10. When the president of Cuba visited Uganda just as that country was beginning to climb out of its two decades of ruin under two of Africa's nastiest dictators, he was asked to provide staff and supplies to help improve Ugandan medical services. Instead, he offered to help the Ugandans improve their own services through education. A new university in western Uganda, founded around its medical faculty, was staffed largely by Cubans until the end of the 1990s.

Chapter 14

Science, Farmers, and Politics

■ A Check to the Seed–Chemical Juggernaut

Chapter 13 did not complete the story of modern scientific advances in crop breeding, nor did it analyze the claims for sustainability of alternative agriculture. There is a good deal more to be said before we can address the future of agrodiversity. It is also necessary to evaluate the political strengths and weaknesses of both the modernist thrust and small farmers' agrodiversity in the present-day context. In the process, it will be possible to bring together a good deal of what we know and do not know about the science of agrodiversity. Everything written in this chapter concerns small farmers in the South, although little of it is directly about them. I begin with the changes in biotechnology introduced in chapter 13 and take their progress forward.[1] I begin with the proposed introduction of a new set of technologies that have been expected, by many observers, to pose a major threat to small farmers in the South. The story is not complete because the events are not completed, but it is at least possible to gauge where the future might lead.

UNEXPECTED EVENTS IN 1999

In late 1998, the juggernaut of the seed–chemical multinationals looked unstoppable. Further rapid expansion of the area planted under transgenic crops and of genetically modified product sales were anticipated. In the United States, it was suggested that seed stores might have only transgenic soya bean and cotton seed to offer within 1 or 2 years (Lappé and Bailey 1999). Opposition to use of transgenics in Europe, India, and Brazil was expected to crumble. Centrally, it was widely argued that conventional crop breeding had reached the limits of its potential, so that the only way forward was through transgenics. Most persuasively, it was argued that without transgenics, agriculture would be unable to feed the enlarged populations expected during the coming decades. Transgenics could offer a cure-all for all other limits to crop production. These views have been very widely accepted, and acceptance has led to fairly massive support for advanced biotechnology research in countries that have sufficient scientific capability and infrastructure.

During 1999, the situation changed dramatically. The opposition of European consumers and food processors to transgenics not only hardened but also extended to the United States so that seed sales and sales of genetically modified produce declined rather than increased. Temporary bans on transgenic crops remained in place in a number of countries and had extended further, and more research and testing were required. None of this resistance was based on proof of any unintendedly harmful consequences of the new introductions, except to valued butterfly species. A great body of claims and counterclaims continued to be made. The real event that hardened at least short-term opposition to transgenics as a whole was separate, and it followed from a monumentally inept misreading of public opinion perpetrated by the companies themselves and their allies in government.

GENETIC USE RESTRICTION TECHNOLOGIES: THE TERMINATOR AND FRIENDS

Company interest in preventing farmers from replanting their own seed has already been discussed in chapter 13. A new generation of transgenic technologies, capable of disabling a wide range of plants to enforce seed or chemical purchase, emerged in 1998 after several years of very secretive development in which the U.S. Department of Agriculture (USDA) collaborated with one seed company. This new class of technology is called a Technology Protection System (TPS) in the United States and Genetic Use Restriction Technologies (GURTS) internationally. It became better known by the name *terminator* given to it by the civil society organization Rural Advancement Foundation International (RAFI) when the first unambiguous patent was issued in early 1998. At once it became a subject of major controversy, news of which sped around the world.

GURTs are a whole technology that can be applied to many plants, not simply a new set of transgenic introductions. By mid-1999, more than two dozen patents for genetically sterilized or chemically dependent seeds had been obtained or were being sought in almost all possible countries. All the major seed–chemical multinationals became involved in the research. The original terminator patent had the simple objective of making second-generation seed nonviable unless it was treated with products available only to the companies or their agents (Crouch 1998). Second-generation seed planted without treatment would not germinate.

By mid-1999, the array of patented strategies was more sophisticated. An induced disabling of the seed itself can be reversed by chemical application to the plant, probably in the form of proprietary herbicide or fertilizer. Among traits designed to be disabled unless a regulator is applied were natural plant functions that

enable them to fight disease. Potentially, positive traits for higher yield, drought resistance, and other desirable characteristics could be turned on by chemical regulation, but the technology to provide these benefits was not ready in 1999. In some
designs seed saving would be possible but at the price of total chemical dependence. Farmers would need to buy chemical preparations to make their seed useful.
With the costs shifted to the farmer, company costs would diminish and profits
could soar.

In the face of a rising barrage of criticism, the USDA responded in defense of its
terminator involvement (United States Department of Agriculture 1998). Preventing seed-saving was seen as beneficial because it would provide companies with a
stronger incentive to develop new and more useful varieties and to extend their activities into crops in which research investment hitherto offered a low rate of return, particularly in rice and wheat. The USDA argued that developing country
farmers could be expected to benefit from the resulting research by reaping more
bountiful crops.

The civil society organization RAFI stressed the consequences for small farmers
in the South of a development they expected to be commercialized early in the
twenty-first century. They went so far as to paint a future, not at all distant, in
which so many farmers might be forced or coerced into buying the new seed that
their own plant breeding would collapse (RAFI 1998, 1999a). More modestly, the
United Nations Development Programme (UNDP 1999:68–69) remarked that
"the 1.4 billion rural people relying on farm-saved seed could see their interests
marginalized . . . and the scope for producing alternative crops will most likely
decline, depleting local genetic diversity." The threat of GURTS worried farmers,
nongovernment organizations, agricultural researchers, sections of the public, and
some governments so much that it greatly enhanced opposition to all transgenics.[2]
In India it was widely believed that the terminator already was present, and opposition to transgenics assumed a violent nature. The American-based Monsanto
Corporation had to issue denials in 1998 and 1999.

Monsanto was at the core of all this. Until the end of 1999, they were still planning to acquire the terminator patent by taking over the smaller company that had
developed it together with the USDA. In the fall of 1999 Monsanto bowed to intense pressure, not least from the powerful Corn Growers' Association within the
United States, and announced that no more work would be done on the technology. A second company followed suit, but the USDA still refused to reject the
terminator (RAFI 1999c). At the end of 1999, faced with a difficult financial position, Monsanto both gave up its takeover bid and agreed to merge with a
seed–pharmaceutical company (RAFI 1999d).

Throughout the industry, the advanced biotechnology was being degraded in
importance. The multinational companies had done a very poor public relations
job by pressing forward initially with resistance to their own herbicides and then
with patenting of the terminator and its fellow technologies. Opposition was able
to draw not only on concern about risks but also on widespread hostility toward

the pressures that had been applied and toward the profit-oriented motivations. Certain companies have continued with the research, so the terminator and related technology remain as possible introductions (RAFI 2000). Other forms undoubtedly will emerge because making farmers buy seed yearly or pay for its use remains an obvious commercial objective.

Yet farmers' rights to plant their own saved seeds are enshrined in principle in some international agreements and have been defended strongly by many developing country governments, agricultural research organizations, and civil society groups. They have been under threat for some years through the Union for the Protection of New Plant Varieties (chapter 13). As late as mid-1999, trade boycotts of countries that refused to accept GURTS were threatened, within the framework of World Trade Organization rules, at an international meeting (RAFI 1999b). Nonetheless, the check to transgenics that followed swiftly may be a turning point, opening the way to a more productive and farmer-friendly use of the advances in biotechnology.

■ Progress in Wider Biotechnology Fields

TRANSGENIC ADVANCES

Herbicide resistance, a principal characteristic of the first generation of transgenic commercial releases, remained prominent in the 1999 listing of U.S. patent applications. Insect, fungal, and viral resistance were the subject of other proposals. A number of applications concerned product quality (of importance in storage and transportation) and, in a few instances, crop reproduction biology and higher yield. They also included new *Agrobacterium* plasmids (USDA 1999). Other traits reported to be under study include tolerance of abiotic stresses such as acid soils, drought, salinity, and cold, shade tolerance, and water and nutrient use efficiency (Spillane 1999). Swiss research made a very distinctive advance in engineering truly useful output traits by developing a rice that offers greatly improved iron and vitamin A content. There is a real risk that promising advances of this nature might be blocked by the strong opposition to all transgenics. There is a long way to go before the real benefits of genetic engineering that were forecast more than a decade ago—higher yields and enriched oil and protein content (Gasser and Fraley 1989)—can be achieved widely. Up to the end of the 1990s, in this important area there was more progress outside the corporate field than within it.

ADVANCES OUTSIDE THE CORPORATE FIELD

The five largest seed–chemical companies own 95 percent of gene transfer patents (UNDP 1999), but they do not own everything. Some technologies used in genetic modification within the plant domain, as distinct from modification involving nonplant genes, remain available to support private or public sector breeding without the need to seek licenses. They include plant tissue culture, molecular genetic mapping, and marker-assisted selection. Anything that has already been published is unpatentable, and a good deal of public sector work is in this category. Genome maps, currently of the simpler crop plants (rice in particular), are being made available on the Internet and by other electronic means so that researchers in any country can have access to them (Paul, Goto, and McCouch 1999). Thus it is possible for scientists in Indonesia or Colombia to participate interactively in a Cornell University project with Chinese partners in a search for yield-enhancing genes in wild rices.

This project, already some years old, has particularly interesting implications. It is often remarked that the yield potential of modern grains has remained almost stationary since the late 1980s. This is one important cause of the problems reviewed at the beginning of chapter 13. Modern plant breeding has reduced genetic diversity not just by displacement, but also by drawing on a smaller and smaller number of productive strains. Search among wild relatives was unproductive, and little pursued, until genetic linkage mapping (genome maps) became available, locating quantitatively significant traits. First with tomatoes and then more importantly with rice, it was found that wild relatives, which themselves yielded poorly, contained some genes in some locations that were superior to those in the cultivated varieties. Extracted from the wild plant and bred by molecular breeding (tissue culture), then crossed with a high-yielding Chinese hybrid, these produced new plants that showed a yield improvement of 18 percent over the latter by the late 1990s (Xiao et al. 1996; Tanksley and McCouch 1997; Coghlan 1999).

Using hybridization through tissue culture (somatic hybridization), it is possible to make wide crosses between plant species that are sexually incompatible (Tanksley and McCouch 1997; Lineberger n.d.). It is then possible to do quite a lot by using conventional methods in the field. Once the genetic markers have been connected to plant performance and the right plants for cross-breeding identified, a new variety can be achieved in only three or four plant generations using simple hand-pollination of the normally self-pollinating rice plant (Wheeler 1995; Tanksley and McCouch 1997). In this way, improved regionally suitable varieties can be created out of regionally suitable cultivated germplasm. Then the new breeding material can be multiplied and further crossed, even by farmers themselves. Using essentially natural methods in both laboratory and field, this promising new set of developments draws on a much wider range of plant genetic diversity. It could make some of the patented high technology, especially the controversial introduction of nonplant genes, largely unnecessary.

There are substantial resources outside the commercial system, some of them in developing countries, where centers for molecular biology work were set up during the 1980s, both nationally and under UN agency sponsorship (Juma 1989). Very important work has been done in China, India, and Brazil, in particular (Sasson 1998). Such centers must be networked in collaboration with other centers devoted to the spread of knowledge and participation in new discovery (Altieri n.d.). The Consultative Group on International Agricultural Research (CGIAR), with its limited budget, has proposed to undertake public research into ways of developing improved crops and to collaborate both with national centers and with other organizations that will permit findings to remain public property. One such independent nonprofit organization is the Center for the Application of Molecular Biology to International Agriculture (CAMBIA), set up privately in the early 1990s in Canberra, Australia. Its successful emergence is seen by some as a hopeful development in opposition to company dominance. Although it relies largely on the royalties obtained for its own early innovations in transgenic breeding, and on foundation support, CAMBIA has been selected to work in collaboration with the CGIAR Institute for International Tropical Agriculture (IITA). CAMBIA has the specific purpose of providing critical new technologies outside the patented web to agricultural research communities in developing as well as developed countries (Finkel 1999).

A number of initiatives are not industry-funded and therefore may escape capture before they can reach the farmers, as long as the web of patent litigation does not block them. Notwithstanding the size and power of the multinationals, an international network of small laboratories is growing, and it is collaborating with the CGIAR and the foundations. Separately, there has also been important progress in the higher-technology field of transferring asexual reproduction (apomixis) from the small range of plants that have this characteristic into others. The potential implications for farmers everywhere are monumental.

THE PROMISE OF APOMIXIS

It has long been an aim of plant breeding to develop varieties that will contain the normally rare apomixis trait. Between 300 and 400 plant species have this trait, most strongly developed in some tropical fruit trees and a number of wild grasses. It permits seed to develop without pollination and probably arose in each case by mutation from sexual ancestors (Jefferson 1994; Bellagio Apomixis Conference 1998). With apomixis, plants can be cloned through seed, and the offspring are genetically identical with the parent plants. Work in Russia in the 1960s established that apomixis could be transferred, but the possibility has not been developed further until recent years. Among several recent projects, the United States Agricultural Research Service (USDA-ARS) has worked with Russian colleagues in the 1990s. An apomictic strain of maize was patented by the partners in 1997. It relies on introducing genes from apomictic wild grasses into cultured cells of maize. Patents also have been obtained by Centro Internacional de Majoramiento de Maiz y Trigo and a French national research organization.

Apomixis potentially can allow persistence of hybrid vigor in almost every crop species, including many to which hybrid technology has not yet been applied. It would free farmers of the need to buy new hybrid seed each season. Together with the methods discussed earlier, it might permit genetic combinations between distantly related species, expanding the diversity of the genetic resources used. In theory at least, it would allow local plant breeders to adapt stable varieties to suit very specific environmental conditions, if—contrary to some views discussed in chapter 13—this is a desirable aim everywhere. Importantly, it could permit reproduction from seed of many plants now planted only vegetatively (Spillane 1999; Bicknell and Bicknell 1999).

Introducing apomixis would drastically reduce the phytosanitary problems associated with tissue culture and micropropagation and hence reduce their costs. This is splendid as long as it remains widely available to breeders with limited resources in all countries. But it also appeals to the commercial breeders, who may very well be able to obtain patents on apomictic plants as the USDA-ARS has already done, so that even if apomixis were developed separately in other countries, their products could not be sold on the global market. An advance of enormous potential value to the developing country small farmer would then be captured by international commerce. Most seriously, the benefits in restoring value to local diversity would not be achieved.

Meeting in 1998, a group of biotechnologists supported by the Rockefeller Foundation issued a declaration calling for preservation of freedom to operate in this highly promising new field (Bellagio Apomixis Conference 1998). In association with other modern developments, including the use of molecular breeding, a publicly available apomixis technology might reinforce, rather than weaken, the competitive scope for agrodiversity in farming.

■ Biosafety and Ethical Issues

BIOSAFETY, TRANSGENICS, AND THE ECOLOGY OF FARMS

Biosafety concerns about new crops, transgenic or otherwise, are a public issue that has led to the development of extensive and costly testing procedures. Biosafety issues apply to all new crops, but the greatest concern has been with transgenics, for which there is no precedent in evolution. As recently as the mid-1990s, little was known that would enable researchers to predict outcomes (Rissler and Mellon 1996). Some of the most publicized issues concern hazards to consumers eating transgenic produce, and a great deal of unsubstantiated propaganda has appeared on both sides. Evidence remains inconclusive, and I will not follow this aspect further.

Many issues concern the environment and the effect on ecosystems. Movement of transgenes into wild relatives is a virtual certainty, but there is little knowledge that would help us to even guess at outcomes. Because so many weeds are related to crop plants, there is serious concern that herbicide resistance can be transferred to the weeds themselves, creating problems of a new order. By natural selection over years of herbicide application, already a great many weeds are herbicide tolerant. There is concern that the transgenics may themselves become weeds. The terminator was even advocated as one answer to these worries because escape of traits through pollen would have no effect if they carried the sterilizing gene with them. Several commentators on the transgenic introductions have called for male sterilization on ecological grounds. One basic reason for the strong opposition to transgenic trials that has arisen in Europe is the belief that organically grown crops cannot avoid being contaminated. Opposition involved crop destruction in Britain in 1999 because transgenics were being field tested widely in that one European country. Insects and especially birds and the wind can carry pollen over vast distances, and probably there is no such thing as a safe buffer zone.

One specific case for concern is that of canola rape *(Brassica napus),* one of the initial herbicide-resistant releases. It belongs to the huge Brassica family and can breed readily with other polyploid members of that family. These include the widely grown oilseed *Brassica rapa.* Herbicide-resistant canola has been sold to farmers in North America and Argentina since 1996 after several years of trials. Rape probably originated as a weed among wheat and barley and was selected for its useful oils. Quite often it is grown in rotation with wheat and becomes a weed among the wheat. If it becomes herbicide resistant, it will be very difficult if not impossible to eradicate. Trials did not demonstrate this problem, but it is reported to have happened on farmers' land.

Concerns about the use of built-in insecticide capacity are widely recognized even by the companies themselves. It is not only possible, but highly probable, that there will be a quick buildup of insect resistance that, if it occurs in the case of *Bacillus thuringiensis,* could wipe out the utility of what has become one of the most environmentally benign insecticides. In the case of cotton, in particular, purchase of the new seed brings with it a compulsory strategy to reduce this risk, that of leaving refuge areas in which unprotected crops are planted. Feeding on the unprotected cotton, insects can multiply free from the built-in insecticide; when they breed with survivors from the treated areas, their progeny will have less chance of carrying resistance. The effectiveness of this strategy is uncertain (Salleh 1998; Lappé and Bailey 1999). It would be hard to apply on small farms. In addition, the use of built-in antipest protection may be a disincentive for farmers to use the now well-developed methods of integrated pest management (IPM). This is less likely in the developing countries where the Food and Agriculture Organization–sponsored program of farmer field schools in IPM methods is beginning to have a very large international effect.

Implications that have been discussed less widely are also important. Transgenics bred for herbicide resistance may undermine efforts to secure progress in ecologically benign alternative methods of weed control, discussed later. Given the incentive for farmers to improve yields simply by eliminating weed competition, it is likely that herbicide consumption will increase rather than diminish. Herbicide sprays already escape over large areas, affecting other crops that lack the protection.

Lappé and Bailey (1999) argued that plant growth and resistance may be affected by gene insertion in ways that are not anticipated. In plants that reseed, the new gene may cause instability in the genome, an effect made evident only a few generations down the line. Widespread crop failures might occur.

These longer-term and wider effects are hypothetical. Notwithstanding the increasingly rigorous biosafety testing procedures now required, many still feel that experiment under field conditions has been too limited in space and time. The degree of caution advocated by Rissler and Mellon (1996) has not been applied. What is now going on in North America, Argentina, China, and a growing number of other countries in which transgenic crops are already widely planted is a gigantic experiment on millions of hectares of farmers' fields. The outcome of these massive releases into the environment will be watched keenly by all concerned, but it could take 10–15 years before damaging environmental consequences become apparent (UNDP 1999).

Meanwhile, negotiations on the planned international Biosafety Protocol, envisaged in the Convention on Biological Diversity, were not concluded until early 2000, and the agreement is unlikely to be ratified soon by all the negotiating partners. A pause in the rate of expansion of transgenic developments might in itself be valuable. In addition, it might give time for the increasingly well-established science underlying agrodiversity to gain higher public visibility and become linked to the new science of plant modification. The possibility of an alliance between the science of agrodiversity and a manipulative biotechnology is not closed off.

■ Understanding the Scientific Basis of Agrodiversity

TWO TYPES OF SCIENCE

All farming depends on science, but the science that has underpinned agriculture for thousands of years and still underpins the diverse agriculture of most of the world's small farmers is based on the farm. It is a science based on trial and error, sometimes involving something like formal experimentation. Farmers' science, as expressed in the management of land and biota, takes place in the laboratory of the whole farming landscape. Understanding of interactions is central.

Agricultural science is compartmented and has never been able to take full account of interactions. To do this required the emergence of the science of ecology. It took a long time. The term *ecosystem,* now universally understood, was proposed in 1935, and the first major textbook statement on the emerging science of ecosystem ecology was that of Odum (1954). The *ecology of agriculture* soon was taken to mean the reciprocal interaction between cultivated or induced organisms and the physical habitat, but it was not until the late 1960s and 1970s that methods already developed for the analysis of wild and semi-wild ecosystems were applied to the study of agriculture. The term *agroecosystem* emerged from several sources between 1975 and 1985. The reason for this slow entry was that agriculturally managed ecosystems were for a long time thought to be unstable, because of their lower species diversity and trophic complexity, in comparison with wild systems. The old equilibrium-seeking model, introduced in chapter 3, could not therefore be applied to them (Tivy 1990). Emergence of the new nonequilibrium ecology removes that difficulty. Indeed, the pendulum has swung so far that in a redefined agrobiodiversity, agroecosystems sometimes are interpreted as being all and any ecosystems with human management (Aarnink et al. 1999). By this definition they would embrace the greater part of the whole biosphere.

Ecological science is distinctive within the scientific domain by its emphasis on interactions. Most formal science, in agriculture as well as in other fields, uses a reductionist approach in which narrow sets of phenomena are the object of study. The variables that are not being measured or subject to experimentation must be excluded or held constant. Major advances therefore take place along very narrow fronts. Biotechnology, including the applied molecular biology of transgenics, is an example of such a discipline or set of disciplines. Interactions are studied only during plot and field testing. Thus far, the study of such interactions does not range into the impact on agroecosystems as a whole.

RESILIENCE IN THE TECHNOLOGY OF AGRODIVERSITY

At a meeting of the Advisory Group of the People, Land Management and Environmental Change (PLEC) project in Paris in September 1998, Stein Bie, director of the International Service for National Agricultural Research (ISNAR), remarked that if we were all to meet again in 20 years, we would hear little about sustainability, and a great deal about resilience. Although sustainability remained the central concept at the end of the twentieth century, there are reasons for writing also of the resilience of agrodiversity. As defined by Holling (1978:11), "Resilience is a property that allows a system to absorb and utilize (or even benefit from) change." Abundant evidence of the manner in which agrodiversity has persisted and has successfully adapted through change has been presented in the preceding chapters.

To evaluate the future, it is important to establish why agrodiversity, in all its agrotechnical elements, should have remained adaptive for so long, even in the face of massive uniformitarian interventions such as the Green Revolution. Chapter 5 introduced some of the reasons why diversity in crops and management methods has advantages in managing the nutrient status of soils. In temperate lands, these values are being rediscovered by those who practice alternative, more biologically friendly agriculture (chapter 13). Ecological conditions are different in the temperate and tropical lands, and scientific knowledge of the natural processes involved in the latter is still inferior. The best sources span both sets of environments; among them Altieri (1995) and Gliessman (1998) provide thorough worldwide reviews of the scientific background of diverse agriculture. A short review of knowledge, in uncomplicated language but fairly comprehensive, is provided in a chapter on the basic ecological principles of low–external-input and sustainable agriculture (LEISA) by Reintjes, Haverkort, and Waters-Bayer (1992). In what follows, I review elements of agrodiversity technology from the standpoint of their resilience.

MANAGED AGROBIODIVERSITY: THE VALUE OF MIXING CROPS

The practice of intercropping, cultivating two or more crops and varieties of crops together on the same piece of land, with or without perennials among them, was until recent years regarded with great disfavor by agricultural authorities and most scientists. We have encountered intercropping at many points in this book, and it is one of the most widespread characteristics of agrodiverse farming systems. Scientific study of its advantages did not seriously begin until around 1960; to fewer than 200 possibly relevant titles available before that date, a literature review by Rao (1984) added more than 1700 published by 1980. The comprehensive (if unbalanced) review of the intercropping literature by Innis (1997) listed very few titles before the 1970s. It is interesting that intercropping achieves almost no place in the classic African-based review of shifting cultivation by Nye and Greenland (1960), yet it is in Africa that intercropping is most extensively practiced. The first major analysis of such systems, in northern Nigeria, was by Norman (1974). Also writing from Nigeria, Greenland (1975) urged that intercropping be a major element in the design of new and more productive farming systems for the lowland humid tropics, and Okigbo and Greenland (1976) made the first substantial attempt to provide a typology of intercropping systems. A major step forward, on an international scale, was the review of knowledge collected by Francis (1986).

Earlier, a French agronomist working in Chad invented a method, since widely used, for evaluating productivity of crop mixtures (Niqueux 1959). For each crop, this compared the yield in mixtures against the yield grown as a sole crop and then summed the total yield. The same method was used by Norman (1974) and it became the land equivalent ratio (LER).[3] Work based in India at the International Crops Research Institute for the Semi-Arid Tropics then led to formal elaboration of the LER concept by Willey (1979a, 1979b). Since the 1980s, advocacy of intercropping practices has become the standard fare of such organizations as the Information Centre for Low–External-Input and Sustainable Agriculture (ILEIA) and the International Institute for Environment and Development.

The LER measures the combined yield of mixed crops against individual sole-crop yields under comparable ecological conditions. Most measurements therefore have been made on experimental plots rather than in farmers' fields. The total yield obtained in intercropping (also called polyculture) is almost always higher because low crops planted between high crops make more effective use of the planting space. Sometimes the intercrop yield is higher by more than 70 percent. There are problems with the simple LER, discussed at length by Mead and Willey (1980) and Mead (1986), who experimented with a number of ingenious modifications. These problems arise mainly from disregarding the objective of the farmer, whether this be to maximize total productivity or to gain maximum yield of one crop in the mixture. Most data cited by Innis (1997) and other writers are single-year comparisons that do not measure interannual variability. Where there are data over successive years, they suggest greater yield stability with intercropping than with sole cropping.

Mixed cropping has many ecologically significant advantages. As discussed by Norman (1974) and Gliessman (1986, 1998), they include better use of light, water, and nutrients, reducing competition by mixing together crops that make their maximum demand on the environment at different times. By denying pests and pathogens a large uniform body of the nutrients they need, intercropping offers a measure of protection from pests and disease. It gives more complete protection of the soil from erosion by rain-splash. By ensuring a more continuous cover, it restricts the growth of weeds. In some mixtures, weeds also are controlled by allelopathy, the selective toxic effect on other plants of chemicals released by the growing crop plant. By combining crops that mature at different times, it is possible to make maximum use of the total growing season. Plants with different root architecture tap different levels in the soil and draw on different parts of the nutrient pool. Where nitrogen-fixing legumes form part of the mixture, there are benefits to all plants that use soil enriched in nitrogen by the plant residues and by the larger bacterial population attracted by the legumes. The maize–beans–squash combination common throughout the central American region often is cited as a clear example of this beneficial effect. In measured instances, the intercropped maize yield alone may be higher than the monocultural yield, within a total LER that approaches 2.0 (Gliessman 1998).

Very importantly, intercropping offers much greater security of production because crops best suited to different weather conditions and grown together will give a return from the field even if one or more crops fail. To many writers, this security is the most valuable attribute of intercropping. It may be planned and done in the same way every year, but very often this is not so. Richards (1989:40) showed how intercropping patterns can be developed by farmers as the season wears on so that "the crop mix—the layout of different crops in the field—is not a design but a result, a completed performance."

Intercropping is by no means all of managed agrobiodiversity. Agroforestry is another major form, with trees cultivated for their fruit or nuts, for medicine, for their wood, and to improve the soil. It has been mentioned at many points in this book, especially in chapter 8. A distinction must be made between the often complex systems developed by small farmers and the more simply designed alley cropping, with nitrogen-fixing hedgerow species, that has been extensively advocated in modern times. It has a large literature (e.g., Nair 1989). The emphasis in alley cropping is more on management of soils and their fertility than on productivity, which may be reduced because of competition (Sanchez 1995). Often, therefore, this form of agroforestry has been rejected by farmers.

PREFERENCE FOR MONOCROPPING

Although intercropping probably goes back to the earliest agriculture (chapter 4), so does monoculture. Modern farmers commonly explain intercropping in terms of the need to conserve space. Intercropping is harder work than monocropping. Unless crops are grown in rows, they leave little space in the field for the farmer and his or her tools. Most of Wilken's (1987) Mexican and Central American informants would have preferred separate monocropping had sufficient land been available. Taking issue with most of the modern literature, but specifically with Conway (1997), Wood (1998) argued that monocropping mimics the low species diversity of early successional growth after disturbance and has the same characteristic of selectively high primary productivity. In addition, he pointed to the stability of many natural stands with low diversity. Although some of his examples are in ecologically constrained environments, not all are, and they include the natural stands of wild wheat in the Near East and wild sorghum in sub-Saharan Africa, which provided the material for domestication. In his distinctive view, there is no intrinsic reason why this productive method should be ecologically vulnerable. These arguments are further developed by Wood and Lenné (1999b) and Lenné and Wood (1999). Farmers use intercropping for a variety of reasons unconnected with ecosystem sustainability. Where diversity is needed, it can be provided temporally through rotations, as in so many systems. Farmers' preference for monoculture is in no way illogical in agroecosystem terms.

Specialization of farming enterprises having few (or only one or two) food- or income-generating crops or products has greatly increased with the application of modern agrotechnology, but it is not new, nor is it confined to capital-intensive enterprises. Although they usually contain diversified elements, many small farms put most of their inputs into a limited range of activities and on the larger part of their land. Monocropping, or segregation of crops in different parts of the field, characterizes even some shifting cultivation systems, especially in South America, and it is a particular feature of almost all wet-rice lands and of dry field systems in which one particular crop is preferred. Examples include sweet potatoes in the agriculture of montane New Guinea (chapter 1) and maize over large areas in both the Old and New Worlds. In its core areas, the Green Revolution was responsible for widespread replacement of intercropping by monocropping, and the latter has increased everywhere as more and more farmers have undertaken production for the market. In central Mexico, for example, most fields now grow maize only, and the older mixtures of maize, beans, and other crops are confined to smaller fields near the home. Specialization maximizes production of the most preferred or profitable crop. It also simplifies decision making and could readily arise from the codification of field types discussed in chapter 3, given economic and technological conditions that reward simplification of the farm enterprise. Yet diversity survives, and in Asia it resurged as the Green Revolution passed its apogee (chapter 12).

■ Biophysical Diversity and Its Management: Alternatives to Herbicides

Research on total management of biophysical diversity has not advanced as far as research on intercropping. The question of soil and water conservation, especially on sloping land, has a very large literature, some of it touched on in chapter 9, but the total management of land together with its biota has much less. This has exposed one area of weakness, already introduced in chapter 13. It arises from the very recent reduction in the amount and depth of tillage in the agriculture of the North. A lot of the modern literature concerns minimum-tillage practices, and it is significant that an early but very detailed American statement made the point that conservation tillage became more possible for farmers as the real price of herbicides declined (Unger and McCalla 1980).

Tillage has been a part of the history of agriculture, and with it came the evolution of tools for land tillage, whether from the ard to the moldboard plow or through the great elaboration of hoe design. Breaking the ground, overturning the topsoil, and removing surface compaction, tillage has been thought of as a means of soil rejuvenation, and in many systems it has been practiced as the main means with which to get the ground into best condition for the crop. An important function has been clearing weeds. Eliminating grass is essential where a deep root mat exists or where the tough rhizomes of some savanna grasses must be broken up before a crop can be planted. Yet modern agricultural science, and some indigenous farmers' science as well, has increasingly viewed tillage, especially deep tillage, as doing more harm than good. This is partly because it encourages erosion but also because of oxidation of organic matter, the destruction of mycorrhiza, and the low density of soil mesofauna, especially worms, that can survive.[4] Conservation tillage practices continue to be advocated strongly (Gliessman 1998). Unfortunately, minimum tillage, the preferred option, leaves the problem of weed management without its most common traditional solution. Meanwhile, herbicides and herbicide-resistant crops offer a "quick, perhaps ephemeral, 'chemical fix' for the problems, meeting short-term, but not the longer-term, needs of agriculture" (Gressel 1998:297).

Alternatives to herbicide use do exist. They include tightly spaced rotations, grazing management, cover crops, high-density planting of leafy crops, and the use of shade and allelopathy to keep down weeds. All have been used with success. So also has shallow plowing, which undercuts the weeds without turning the soil. In Australia, for example, this labor-saving method has been widely adopted because it also conserves soil moisture and helps improve soil structure to permit greater infiltration. But most of the alternative systems add to the demand for labor, whereas herbicides reduce it; none of the alternatives have been developed and demonstrated on farmers' land sufficiently to gain widespread adoption. Although herbicides have been slower to gain adoption than insecticides among farmers of the South, labor saving gives them increasing popularity.

A group of Fijian farmers whom I knew in the 1970s provide an example. They were former plantation workers, from smaller islands, who leased land from a chief on the large island of Taveuni. Newly settled on fertile land that had not been cultivated since the nineteenth century, they set out to grow taro and kava *(Piper methysticum)* for market sale. When I first knew them, not long after they had cleared their blocks from almost 90 years of regrowth, they lived in shacks, using income to buy herbicide and spraying equipment before investing in permanent houses. They said they did this to approximately double the productivity of their labor (Brookfield et al. 1978). Ten years later, these innovative people had built themselves a village of solid houses, supplied with water by the reopening and diversion of an ancient irrigation channel. They continued to use herbicides.

■ Diversity in Farm Management

MODERN IDEAS ABOUT SUSTAINABLE LAND MANAGEMENT

It is useful in this context to consider the remedies proposed to deal with the unsustainable and often profitless outcomes of modernist agriculture in the North. The recommendations do not include any return to the labor-intensive methods of the past, but they draw on methods widely used in agrodiverse farming in the South. They involve reducing use of chemical inputs that can harm the environment, human health, and wildlife. Beyond that they involve a mixture of modern innovations, principally minimum tillage and soil and water conservation practices, with principles derived from agrodiversity as it has been described and analyzed in this book.

Recommendations found in a range of sources already cited focus on the use of biological processes and improvements in management. They are all designed to make farming less vulnerable to pests and diseases, pollution, and erosion and to stabilize farm incomes. They are strategies to improve resilience. Principles drawn from the sort of agrodiversity described throughout this book include the following:

- Shifting from nutrient management based on fertilizer application to increased use of natural processes such as biological nitrogen fixation, crop rotation and nutrient recycling, and the use of manure and compost.
- Making more appropriate matches between crop choice and cropping patterns on one hand and the ecological conditions of the farm site on the other, rather than seeking to override biophysical diversity with uniform management to suit a narrow set of crops and livestock.
- Making precise the application of water and other external inputs in both space and time to avoid waste and the problems that arise from oversupply. Sometimes this is called precision farming.

- Managing pests by monitoring and encouraging natural enemies, including their necessary breeding conditions in conserved nonfarm spaces, rather than seeking to control the pests (i.e., IPM).
- Diversifying the farm structure and management through use of rotations, agroforestry, and, where appropriate, intercropping.
- Total farm planning, carried to the level of cooperative management over an area such as a catchment.
- Profitable and efficient production with emphasis on improved farm management and use of information and the conservation of soil, water, energy, and biological resources.

To these are already being added the culture and application of soil microorganisms capable of not only forming organic matter but also improving nitrogen fixation, water use and phosphate uptake, in addition to other benefits (Okon, Bloemberg, and Lugtenberg 1998).

All this applies just as much to farmers in the South as in the North. It relies on farmers' understanding of the resources they manage. Good local knowledge is the foundation of success, just as poor knowledge is the foundation of failure, and this is as true in the South as it is in the North. Farmers' understanding of local ecological processes changes as their production patterns change, but the processes themselves do not, so that old lessons sometimes have to be relearned in new ways. We have seen considerable differences in the rate of learning in this book. Striking examples of very rapid learning in Africa were presented in chapter 11. Learning calls for constant experiment, and new discovery of ecological conditions and dynamics arises with it. In the literature on small farmers' practices, experimentation was first remarked on by de Schlippe (1956), but in recent years it has come to be recognized as an integral property of diverse systems of farming in all places and at all times.

RESILIENCE AND SUSTAINABILITY

In all parts of the agrotechnological domain, the resilience of agrodiverse practices underlies their long-term sustainability. It similarly underlies the manner in which diversity can be reproduced after long periods in which uniform management has been practiced. Diversity enhances resilience in the face of climatic shocks, disease and pest outbreaks, and variability in the returns from sale of a range of crops. Modern use of the principles of agrodiversity in the North has been shown to be profitable, once a few years of adjustment have been overcome. New crops and technologies can be used quickly, leading to quite rapid adaptation (e.g., Committee on the Role of Alternative Farming Methods 1989). Old and unprofitable crops and practices can be discarded without doing damage to the basic structure of the system. This does not always happen, and if the productive capacity of the land is worn down without effective adaptations being made, resilience is weakened (chapter 9). However, resilience can be restored, and even destroyed soils can be rebuilt by farmers' hard work. In central Mexico in May 1999, I saw a field that had been rebuilt by its holder from a gullied landscape, entirely by carrying topsoil from one place to another, making silt traps at appropriate points in the field, and creating contour barriers to check new erosion.

■ The Organizational Domain: An Area of Weakness

SUCCESSFUL AND UNSUCCESSFUL FARMERS

Good management of the farm—its land, crops, and resources of labor and working livestock—is central to tapping the resilient properties of agrodiversity. Not all farmers achieve high-quality management, and there is always heterogeneity in an agricultural landscape that may appear, at first sight, to be essentially homogenous in its practices. One of the central objectives of the PLEC project, mentioned several times in this book, is to locate the expert farmers and assist them in showing other farmers how their practices can be improved and resources better managed.

This is fine where there is only slow change in the organizational structure of agriculture. This is not so everywhere. In the North, most farms practicing alternative agriculture are small. In the South all but a very few farms, as distinct from plantations, are small. Yet farms are not enduring elements of organizational diversity. They may be enlarged or reduced in size, combined into larger farms, or subdivided. Farms may be combined into collectives, as in China in the 1950s, then reestablished as individual enterprises 30 years later, though on different land. Farmers can do well or can fall on bad times. Those with least resources can go under during periods of severe adversity, periods from which the regional system as a whole may seem to recover remarkably well. Small farms are particularly vulnerable. If a whole group of farms fails, or if the conditions within which the farms of a region operate are undermined, there may be widespread loss of all resilient elements. Under the worst circumstances, whole areas and all the agrodiversity and agrobiodiversity that they contain may be abandoned for the foreseeable future.

THE VULNERABILITY OF SMALL FARMERS AS INDIVIDUALS

As individuals, small farmers have few defenses against larger shocks to the social and economic viability of the systems within which they operate. Their collaboration in resource management seldom extends to a higher level than that of the village or group of villages. Political decisions usually are imposed on them, and they have limited political voice. Powerful organizations of farmers, able to bargain effectively for their members, have been formed only in the developed countries of the North. We saw the effective influence of the American Corn Growers' Association on the decisions of a major multinational earlier in this chapter. Small farmers in developing countries cannot organize in this way and often have only the defenses of noncooperation or silent resistance (Scott 1985).

The option of direct resistance has sometimes been adopted, as in the burning of Monsanto trial fields by Indian farmers in 1998–99. Historically there have been many armed revolts by peasant farmers in many parts of the world. Unfortunately, their record of success is dismal. Most revolts have been crushed easily, and they have succeeded only when they have acquired powerful allies among townspeople or, in modern times, national political movements. Small farmers are ill-equipped to react collectively to major onslaughts on their way of life. This is why so many observers have recently seen them as so vulnerable to the well-organized and strongly backed onslaught of the seed–chemical juggernaut.

■ The Conditions for Success of Diversity

THE IMPORTANCE OF HOW WE VIEW FARMERS

It is very important for the future that prevailing views about farmers be revised; businesses, governments, and even many of their own defenders must cease to see them as mere passive recipients of what is to come. There is a myth in countries of the North that farmers, and especially the farmers of the South, are destroying the environment and are inefficient and often irrationally stupid. In this view, the great majority of farmers are conservative, in the pejorative sense that they are unwilling to accept good advice, and prefer to stick with has been done in the past. This myth plays very much into the hands of commerce and its allies and has already been responsible for a lot of harm during the Green Revolution period.

Along with the large majority of those whose work I cite in this book, I see a fair proportion of farmers as being innovative and adaptive. Their conservatism is not denied, but it can involve holding on to what may still be useful at a later time, such as the qualities of the natural landscape or memories of old ways of doing things, preserved only here and there to be rediscovered when useful. This has been one rather remarkable finding of the PLEC project in west Africa, where the scientists have sought out these old ways and, once they had praised them, found enthusiasm among farmers for readopting them in the interest of restoring the quality and productivity of the land. Supporters of the rights of small farmers must properly understand the resilience and adaptability of their farming practices. Those who respect their capacity for innovation are more likely to find ways to assist them in new crises that are not of their own making.

BACK TO SCIENCE

Although its early promise has not yet been achieved, biotechnology is a truly exciting new venture that ultimately can offer enormous benefits if it is handled with ecological and social sensitivity. Farmers' experimentation must become allied to formal sector experimentation. Links must be forged between the evolving science of agrodiversity and innovative biotechnology. All aspects of biotechnology, at all levels, must be demystified so that what is good for the farmer and what is practical and convenient can be identified. Developments have advanced too fast for anything like this to take place yet. Unfortunately, power always goes along with possession of whatever essential ingredient of progress is in shortest supply. High-technology science became that ingredient during the 1990s, when it also came to be owned largely by a small number of very big corporations, together with a few governments.

The problem has been compounded by the prevailing ethos of the two decades during which biotechnology has evolved. When profit is the bottom line, evaluation on the basis of the good of people as a whole is unlikely to be easy. When control over a highly reductionist science became power, it was hard for ecologically and socially sensitive approaches to make headway. Some recent trends reported in this book are more hopeful, and a good deal of faith can be put in the resilience of agrodiversity itself and the ability of those who practice diverse farming to sustain the securities that it provides. Fortunately, there is also some hope from the emergence of another body of allies of small farmers and their diverse practices. That takes us to the future, and the epilogue.

Notes

1. Many of the developments selected for discussion in this chapter are multiply reported, on the Internet and increasingly in the print media. The latter include not only specialized journals but also sections of the popular press, and even *The Economist*. Basic Web sources are as given in chapter 13. Specific references are cited only where appropriate. Also important are articles and news in issues of the revitalized ILEIA newsletter, *LEISA*.

2. Meeting in late 1998, the CGIAR adopted a resolution emphatically opposing GURTs, and one or two governments declared a total ban. The international organizations, including the important Subsidiary Body on Scientific, Technical and Technological Advice of the Convention on Biological Diversity, still remained divided over GURTS in mid-1999 because of a small number of governments that unreservedly supported the powerful biotechnology industry of Northern countries.

3. The land equivalent ratio (LER) follows:

$$\text{LER} = \sum \left(\frac{M_a}{S_a} + \frac{M_b}{S_b} + \cdots \frac{M_n}{S_n} \right),$$

where M is the yield of each crop from a–n per unit area in mixtures, and S is the yield in sole crop. When LER > 1.0 there is a gain in total yield by intercropping, sometimes called overyielding.

4. In southern Ghana, the old practice was to slash the herbage and then plant crops in holes or broadcast among the mulch. Invasion of grasses has led to widespread use of the hoe, especially by migrant tenant farmers coming from grassland areas. Over years, this has been seen to diminish the seedbank of woody plants in the soil and to harm soil structure, possibly by reducing organic matter.

Looking at the Future

■ Allies of Agrodiversity

WHO STANDS UP FOR SMALL FARMERS?

Much of the information presented about a fast-changing situation in chapters 13 and 14 is ephemeral, and interpretation will become outdated quickly. This concluding attempt at crystal-ball gazing is even less likely to endure. Yet it is necessary to try. Many observers are very pessimistic about the future of diversity in agriculture, but there is now powerful scientific support for diversity, coming both from ecological scientists and from the biotechnologists who are more abundantly endowed than others with wide vision of genetic diversity and a social conscience. There is powerful new support of a political nature from some nongovernment organizations and from within the United Nations system itself. Given the organizational weakness of small farmers as a body, this support is invaluable and, if sustained, may be critical for the future.

The support arises from the still-rising global concern for the conservation of biodiversity, especially from the more recent surge of concern for conservation and enrichment of agrobiodiversity. Some of this latter goes back to the time of the Brundtland report (World Commission 1987) and the 1992 Convention on Biological Diversity, but its major emergence is more recent, coming together in critical meetings that took place in 1996. The first of these was the Leipzig conference on Conservation and Sustainable Use of Plant Genetic Resources for Food and Agriculture (FAO 1996). The Global Plan of Action developed at Leipzig had a lot to say about the *in situ* conservation of crop plants on farms. In sympathy with writers cited in chapters 13 and 14, it noted that the range of germplasm available on farms in the regions of diversity has not nearly been tapped, nor has the range held *ex situ* in genebanks. The plan called for new interest in old-fashioned mass selection and simple plant breeding on farms to achieve higher and more reliable yields without significant biodiversity loss (FAO 1996). The aim should be diversity-rich agricultural production. Also, emphasis was placed on the role of farmers in actively conserving the agrobiodiversity they and their ancestors have created.

The second initiative followed this, and was the third meeting of the Conference of Parties to the Convention on Biological Diversity. Specifically discussing the conservation and sustainable use of agricultural biological diversity (i.e., agrobiodiversity), the meeting laid stress on the contribution of traditional farming communities and their agricultural practices to the conservation and enhancement of biodiversity. Among the practices that should be encouraged were organic farming, integrated pest management, biological control, no-till agriculture, multicropping, intercropping, crop rotation, agroforestry, and promoting the use of underused crops. Again, stress was placed on small farmers and local communities (Conference of Parties to the CBD 1996). The resolution also sought "the mobilization of farming communities, including indigenous and local communities for the development, maintenance and use of their knowledge and practices in the conservation and sustainable use of biological diversity in the agricultural sector with special reference to gender roles" (Conference of Parties 1996:Annex II, III/11, para 17c). This was part of the new environment in which independently generated efforts to develop projects working with farmers, which had hitherto been received grudgingly, began to gain international support and funding. In terms of projects that focus on agrodiversity as guardian of agrobiodiversity, the People, Land Management and Environmental Change project was the first international effort to receive Global Environmental Facility support, in April 1997.

The new spirit of cooperation with small farmers has been taken up by other organizations, initiating projects in the areas of soil biological fertility and improved management of semiarid areas. At the beginning of 2000, the Global Environmental Facility itself was considering the addition of a focal area on agricultural biodiversity in addition to its existing three on biodiversity in general, global climatic change, and international waters. A significant increase in support for action and research may result. All this is in contrast to the emphasis on a narrow range of engineered genotypes still being pressed by those who put their faith in modernist farming.

Importantly, an aroused awareness of the importance of the small farmer has moved into new approaches to rural development. Biodiversity conservation is now seen as linked with agricultural development, and United Nations agencies recognize that they are mandated to help governments meet their obligations under the Convention on Biological Diversity. It should follow that institutional and financial constraints to the performance of rural communities must be eased. Writing in a World Bank document, Pagiola et al. (1997) remarked on the very wide incidence of national policies that stifle the ability of farmers to innovate productively and in a conservationist manner. They called for a major effort to ease these sti-fling policies. Hitherto, there has not been much international appreciation of the fact that successive waves of development thinking, whatever their important dif-ferences in other ways, have not been so different in their outcomes for the South's rural people. It will take time for such new approaches to have a full impact at na-tional level within the developing countries. Nevertheless, new approaches that are more sensitive to the small farmer, as major user of the environment, are coming to be embraced within the core of the United Nations system. This is important be-cause of the financial leverage that the World Bank and other agencies wield. Small farmers now have powerful new allies.

FARMERS AND OTHER SCIENTISTS

The 1996 Global Plan of Action also called for the collaboration of scientists with farmers to place scientific work in context. It called for scientists to be in-volved specifically in the "monitoring, evaluation and improvement of on-farm ef-forts" (FAO 1996:Annex 2, para 45). Thrupp, Hecht, and Browder (1997) de-scribed a number of initiatives in which scientists have worked with farmers, going back to the 1980s, but the sum of effort is still small. Scientists who work with farmers, instead of merely telling them what they should do, are among the most important allies of agrodiversity. Such scientists, many of whom are cited in this book, have access to the channels of international communication, whether or not they are among the select few invited to many international conferences. Like me, most developing country scientists who work with farmers are outside that latter circle, but they often have access to those who help make decisions in their own countries.

Molecular biologists and ecological scientists need to communicate better if a sensible mixed-technology future agriculture is to be achieved. A major book on the new biotechnology of tissue culture and transgenics, edited by Altman (1998), calls for a leading role for biotechnology in supporting and enlarging the scope of traditional plant and animal breeding as well as a great many other areas. One can search it at length without finding more than marginal reference to risky conse-quences for whole agroecosystems. Public acceptance, rather than the possibility of unforeseen consequences, is the big problem. By contrast, a more cautionary study of biosafety risks from transgenics urged that "agricultural biotechnology should be evaluated in the context of efforts to make agriculture more sustainable and less en-vironmentally damaging" (Rissler and Mellon 1996:21). They further indicate that although biotechnology could play a positive role in the transition to more sus-tainable practices, not all introductions are likely to be beneficial.

The evolving ecological science of agriculture, in hand with socially aware biotechnologists, could be a major force for the future. Ecological approaches lack financial resources and corporate backing, and they receive far less government support than do the high-technology elements of biological science. Yet by underpinning the recent development of alternative agricultures in the North, they have gained substantial minority support among farmers, the public, a wide range of scientists, and even some government agencies.[1] This sort of science will not go away, and it is now far stronger than it was in the 1960–85 period of the Green Revolution. Biological husbandry already has a strong body of scientific support.

Biotechnology has scarcely begun and will evolve rapidly in the coming decades. Public resistance to introduction of transgenic crops, however powerful at present, probably will do no more than delay their advance. The intermediate technology approaches discussed in chapter 14 may become of greater importance because they can be more readily diffused and are not crowded in by patent legislation. Agrodiverse farmers should be able to profit from a more benign generation of transgenic introductions, but they will profit much more from crop-breeding activities that demand lower levels of technology. Biotechnologists at national level have an important role to play. It cannot yet be said that national biotechnology will be any more sensitive to the real needs of small farmers, or to agrodiversity and biodiversity, than are the scientists who serve the corporations. But developing country scientists are much closer to the farmers than are those who work in corporation laboratories.

One point is important. The fact that biotechnology is already a dominant force, and will become more dominant, does not mean that dependence on inorganic chemicals necessarily will remain entrenched. There is now a large body of evidence from ecological science and agronomy about the importance of maintaining and rebuilding organic matter in the soil. The seed–chemical multinationals may have begun as chemical companies, but they may not remain so. Nor may they be able to convince any large body of scientists or thinking farmers that a continuation of heavy chemicalization is sustainable. The seed–chemical juggernaut has its own serious weakness here. Almost everything that it has advocated until now involves further use and expansion of agrotechnologies that have been discredited widely. Fortunately, some biotechnology research is already moving in other directions.

■ In Conclusion

DIVERSITY AND SUSTAINABILITY

Before reaching a final view, I turn back to an old paper of my own, written for a conference in Oslo in 1990 and published in the place decreed by the organizers (Brookfield 1991). Few readers of this book will have seen it. I was seeking an agenda for research in sustainable development. After rejecting elitist steady-state (no growth) notions of sustainability—in any case based on the equilibrium ecology fallacy—I argued that sustainability of the land resources currently under use demands not only that reproduction of qualities by natural renewal not be drawn down but also that there be substantial investment in improving land capability. This depends on human artifice allied to natural processes, and this has to happen in a context of uncertainty in both natural and economic conditions.

I also followed Lowrance, Hendrix, and Odum (1986) in distinguishing agronomic sustainability at field level from microeconomic sustainability at farm level. They went on to characterize ecological sustainability at landscape level, where the objective is to maintain life-support capacity over long time scales, and macroeconomic sustainability, which determines the conditions of production. This fourfold approach usefully shifts interest recursively between management of specific resources and the local and larger organizational systems within which management takes place. I used this to focus attention on the sort of political economy that must be involved in advances toward sustainable management of the natural environment. All this led me further to propose that human inventiveness and adaptability, which have been of major importance in the past, will be of even greater significance in a future that will embrace significant climatic change.

The latter consideration may be critical in evaluating the future utility of agrodiversity. Possessing a range of strategies, already known and in place, is essential for achieving the adaptability that global environmental change will demand. Agrodiversity contains this range. It is based on a large body of resilient agrotechnologies, and its adaptable practitioners have been able to profit from a great number of germplasm introductions, new methods, and commercialization. Although with losses, it has survived right through the twentieth century. Now it has to cope with global climatic change and the further inroads of globalization. What is important is to ensure that these strategies, as well as the genetic diversity they manage, are conserved so that they can be used and further modified.

FUTURE NEEDS

Agrodiversity has to change, as it has in the past, in order to survive. This is where the Cuban model, discussed in chapter 13, perhaps warrants closer and more dispassionate study than it has received. Note should also be taken of Swaminathan's eclectic approach, willing to try all sorts of new innovation with precautionary care but also willing to preserve what is valuable in the old systems. Too much of agrodiversity is labor-intensive, and temporary and permanent migration from farms to the cities have created labor shortages that threaten the use of many conservationist and productive practices. It cannot be said with certainty that these trends will be offset by continuing population growth. Agrodiversity will need to become a complex of practices that demand less labor, thus necessarily adopting some elements that now characterize modernist farming. Small farmers in the South and North need to become more alike, learning from one another. This will not be easy to achieve.

Technically intensive monoculture is the greatest enemy of diversity, and it continues to spread along with the market economy. Notwithstanding some encouraging initiatives toward restoring old practices among farmers who enjoy the support of scientists, the practice of intercropping may well become more rare in future. Fortunately, diversity in the whole agroecosystem can be sustained in other ways, discussed in chapter 14. Small farmers can achieve a lot by themselves, but only their allies can ensure that they will have success. In another book, specifically on land degradation, Blaikie and I concluded more than a decade ago that "while good management of the land can be achieved ultimately only at the interface where a spade or plough is dug into the soil, the whole of society and economy are involved in determining what will and what will not work" (Blaikie and Brookfield 1987:250). With the sole exception that digging a spade or plow into the soil may no longer be seen as the best treatment, this remains just as true today. Agrodiversity may be helped to survive by its own resilient properties, but it would survive much more securely with stronger public backing.

At the end of a new book written in praise of agrodiversity, I have to agree with the doubters that its future is more unclear than at any previous time. It is impossible to know the future, but an understanding of the resilience of an important set of agricultural systems can help us interpret what is going on now. Much of the change that is taking place in several regions of the world is best interpreted neither as the abandonment of ancient conservationist practices nor as discovery or adoption of a new set of ways. It is simply rapid change in response to new conditions, and thus in line with everything that has happened in the last many hundred years. Much more change will have to take place in the future, and farmers need all the help, scientific and political, that they can get. To come back finally to a question posed in chapter 3, I conclude that the adaptive dynamism of agrodiversity is its most essential property for survival, let alone for restoration of what has been lost. Fortunately, this adaptability is now widely appreciated, and I hope this book will advance that appreciation further.

Note

1. Small research programs into diverse and resilient agricultural systems are funded in some developed countries. One such is Australia, where a competition for funds to work in a resilient agricultural system subprogram, begun in 1997, was announced in 1999 (Rural Industries Research and Development Corporation 1999).

Aarnink, W., S. Bunning, L. Collette, and P. Mulvany. 1999. *Sustaining agricultural biodiversity and agro-ecosystem functions.* International Technical Workshop organized jointly by FAO and Secretariat of the Convention on Biological Diversity with the support of the Government of The Netherlands, 2–4 December 1998. Rome: Food and Agriculture Organization of the United Nations.

Adams, W. M. 1989. Definition and development in African indigenous irrigation. *Azania* 24: 21–27.

Adams, W. M. and M. J. Mortimore. 1997. Agricultural intensification and flexibility in the Nigerian Sahel. *Geographical Journal* 163(2): 150–60.

Adams, W. M., T. Potkanski, and J. E. G. Sutton. 1994. Indigenous farmer-managed irrigation in Sonjo, Tanzania east. *Geographical Journal* 160(1): 17–32.

Alcorn, J. B. 1981. Haustec noncrop resource management. *Human Ecology* 9(4): 395–417.

Alcorn, J. B. 1984a. Development policy, forests, and peasant farms: Reflections on Haustec-managed forests' contributions to commercial production and resource conservation. *Economic Botany* 38(4): 389–406.

Alcorn, J. B. 1984b. *Haustec Mayan ethnobotany.* Austin: University of Texas.

Alcorn, J. B. 1989. Process as resource: The traditional agricultural ideology of Bora and Haustec resource management and its implications for research. In D. A. Posey and W. Balée, eds., *Resource management in Amazonia: Indigenous folk strategies,* 63–77. Advances in Economic Botany Vol. 7. Bronx: New York Botanical Garden.

Alcorn, J. B. 1990. Indigenous agroforestry strategies meeting farmers' needs. In A. B. Anderson, ed., *Alternatives to deforestation in Amazonia: Steps toward sustainable use of the Amazon rainforest,* 141–51. New York: Columbia University Press.

Alexander, I. 1989. Mycorrhizas in tropical forests. In J. Proctor, ed., *Mineral nutrients in tropical forest and savanna ecosystems,* 169–88. Oxford, U.K.: Blackwell Scientific.

Alexander, M. J. 1996. The effectiveness of small-scale irrigated agriculture in the reclamation of mine land soils on the Jos Plateau of Nigeria. *Land Degradation and Development* 7(1): 77–84.

Allan, W. 1949. *Studies in African land usage in northern Rhodesia.* Rhodes–Livingstone Papers No. 15. Cape Town: Oxford University Press for the Rhodes–Livingstone Institute.

Allan, W. 1965. *The African husbandman.* Edinburgh: Oliver & Boyd.

Allan, W., M. Gluckman, D. U. Peters, and C. G. Trapnell. 1948. *Land holding and land usage among the plateau Tonga of Mazabuka district.* Rhodes–Livingstone Institute Papers No. 14. London: Rhodes–Livingstone Institute.

Allen, B., R. M. Bourke, and R. L. Hide. 1995a. Agricultural Systems in Papua New Guinea Project: Approaches and methods. *PLEC News and Views* 5: 16–25.

Allen, B., R. M. Bourke, and R. L. Hide. 1995b. The sustainability of Papua New Guinea agricultural systems: The conceptual background. *Global Environmental Change: Human and Policy Dimensions* 5(4): 297–312.

Allen, B. and R. Crittenden. 1987. Degradation and a pre-capitalist political economy: The case of the New Guinea highlands. In P. Blaikie and H. Brookfield, eds., *Land degradation and society,* 145–56. London: Methuen.

Allen, H. 1974. The Bagundji of the Darling Basin: Cereal gatherers in an uncertain environment. *World Archaeology* 5: 309–22.

Almekinders, C. J. M., L. O. Fresco, and P. C. Struik. 1995. The need to study and manage variation in agro-ecosystems. *Netherlands Journal of Agricultural Science* 43(2): 127–42.

Altieri, M. A. 1995 (2d ed.). *Agroecology: The science of sustainable agriculture.* Boulder, Colo.: Westview.

Altieri, M. A. n.d. The CGIAR and biotechnology: Can the renewal keep the promise of a research agenda for the rural poor? http://nature.berkeley.edu/~agroeco3/cgiar/cgiar-bio.html [November 25, 1999].

Altman, A., ed. 1998. *Agricultural biotechnology.* New York: Marcel Dekker.

Amanor, K. S. 1994. *The new frontier: Farmer responses to land degradation.* London: Zed.

Amblard, S. 1996. Agricultural evidence and its implication on the Dhars Tichitt and Oualata, southeastern Mauritania. In G. Pwiti and R. Soper, eds., *Aspects of African archaeology,* 421–27. Harare: University of Zimbabwe Publications.

American Society of Agronomy. 1978. *Diversity of soils in the tropics.* ASA Special Publication No. 34. Madison, Wis.: American Society of Agronomy; Soil Science Society of America.

Ames, O. 1939. *Economic annuals and human cultures.* Cambridge, Mass.: Botanical Museum of Harvard University.

Andam, C. J. 1995. The rices of the Banaue terraces in Ifugao. *Philippine Technology Journal* 20(4): 22–27.

Anderson, A. B. 1990. Extraction and forest management by rural inhabitants in the Amazon estuary. In A. B. Anderson, ed., *Alternatives to deforestation in Amazonia: Steps toward sustainable use of the Amazon rainforest,* 65–85. New York: Columbia University Press.

Anderson, A. B., P. H. May, and M. J. Balick. 1991. *The subsidy from nature: Palm forests, peasantry and development on an Amazonian frontier.* New York: Columbia University Press.

Anderson, A. B. and D. A. Posey. 1989. Management of a tropical scrub savanna by the Gorotire Kayapó of Brazil. In D. A. Posey and W. Balée, eds., *Resource management in Amazonia: Indigenous folk strategies,* 159–73. Advances in Economic Botany Vol. 7. Bronx: New York Botanical Garden.

Anderson, D. 1984. Depression, dust bowl, demography, and drought: The colonial state and soil conservation in east Africa during the 1930s. *African Affairs* 83(332): 321–43.

Anderson, E. 1952. *Plants, man and life.* Berkeley: University of California Press.

Anderson, J. R., ed. 1994. *Agricultural technology: Policy issues for the international community.* Wallingford, UK: CAB International in association with the World Bank.

Anderson, P. C. 1998. History of harvesting and threshing techniques for cereals in the prehistoric Near East. In A. B. Damania, J. Valkoun, G. Willcox, and C. O. Qualset, eds., *The origins of agriculture and crop domestication,* 145–59. Aleppo, Syria: ICARDA.

Anderson, S. D., F. L. T. Marques, and M. F. M. Noguiera. 1999. The usefulness of traditional technology for rural development: The case of tidal energy near the mouth of the Amazon. In C. Padoch, J. M. Ayres, M. Pinedo-Vasquez, and A. Henderson, eds., *Várzea: Diversity, development, and conservation of Amazonia's whitewater floodplains,* 329–44. Advances in Economic Botany 13. Bronx: New York Botanical Garden.

Andriesse, J. P. and T. T. Koopmans. 1984/85. A monitoring study on nutrient cycles in soils for shifting cultivation under various climatic conditions in tropical Asia. I. The influence of simulated burning on form and availability of plant nutrients. *Agriculture, Ecosystems and Environment* 12: 1–16.

Andriesse, J. P. and R. M. Schelhaas. 1987a. A monitoring study on nutrient cycles in soils used for shifting cultivation under various climatic conditions in tropical Asia. II. Nutrient stores in biomass and soil: Results of baseline studies. *Agriculture, Ecosystems and Environment* 19: 285–310.

Andriesse, J. P. and R. M. Schelhaas. 1987b. A monitoring study on nutrient cycles in soils used for shifting cultivation under various climatic conditions in tropical Asia. III. The effects of land clearing through burning on fertility level. *Agriculture, Ecosystems and Environment* 19: 311–32.

Asch, D. L. and N. E. Asch. 1985. Prehistoric plant cultivation in west-central Illinois. In R. I. Ford, ed., *Prehistoric food production in North America,* 149–203. Ann Arbor: Museum of Anthropology, University of Michigan.

Balée, W. 1989. The culture of Amazonian forests. In D. W. Posey and W. Balée, eds., *Resource management in Amazonia,* 1–21. Advances in Economic Botany Vol. 7. Bronx: New York Botanical Garden.

Balée, W. 1992. People of the fallow: A historical ecology of foraging in lowland South America. In K. H. Redford and C. Padoch, eds., *Conservation of neotropical forests: Working from traditional resource use,* 35–57. New York: Columbia University Press.

Balée, W. L. 1994. *Footprints of the forest: Ka'apor ethnobotany.* New York: Columbia University Press.

Balfour, E. B. 1943. *The living soil: Evidence of the importance to human health of soil vitality, with special reference to national planning.* London: Faber & Faber.

Balfour, E. B. 1948 (rev. ed.). *The living soil: Evidence of the importance to human health of soil vitality, with special reference to national planning.* London: Faber & Faber.

Balfour, E. 1977. Towards a sustainable agriculture: The living soil. An address given by the late Lady Eve Balfour to an IFOAM conference in Switzerland in 1977. Canberra Organic Growers Society, 1995. http://www.netspeed.com.au/cogs/cogbal.htm [July 28, 1999].

Bandara, M. 1977. Hydrological consequences of agrarian change. In B. H. Farmer, ed., *Green revolution?: Technology and change in rice-growing areas of Tamil Nadu and Sri Lanka,* 323–39. London: Macmillan.

Baron, R. C., J. B. McCormick, and O. A. Zubeir. 1983. Ebola virus disease in southern Sudan: Hospital dissemination and intrafamilial spread. *Bulletin of the World Health Organization* 61: 997–1003.

Barry, A. K., S. Fofana, A. Diallo, and I. Boiro. 1996. *Systèmes de production et changements de l'environnement dans le Sous-Bassin de Kollangui-Pita.* Conakry, Guinea: WAPLEC-Guinée.

Bayliss-Smith, T. P. and S. Wanmali, eds. 1984. *Understanding green revolutions: Agrarian change and development planning in south Asia.* Cambridge, U.K.: Cambridge University Press.

Beardsley, R. K., J. W. Hall, and R. E. Ward. 1959. *Village Japan.* Chicago: University of Chicago Press.

Beckerman, S. 1983. Barí swidden gardens: Crop segregation patterns. *Human Ecology* 11(1): 85–102.

Beckerman, S. 1987. Swidden in Amazonia and the Amazon rim. In B. L. Turner II and S. B. Brush, eds., *Comparative farming systems,* 156–87. New York: Guilford.

Bellagio Apomixis Conference. 1998. *Bellagio Apomixis Declaration. Designing a research strategy for achieving asexual seed production in cereals,* Rockefeller Foundation's Bellagio Conference and Study Center, Italy, 27 April–1 May, 1998.

Belsky, A. J. and R. G. Amundson. 1992. Effects of trees on understorey vegetation and soils at forest–savanna boundaries in east Africa. In P. A. Furley, J. Proctor, and J. A. Ratter, eds., *Nature and dynamics of forest–savanna boundaries,* 353–66. London: Chapman & Hall.

Benneh, G. 1970. The Huza strip farming system of the Krobo of Ghana. *Geographia Polonica* 19: 188–206.

Bicknell, P. 1972. Zande savagery. In A. Singer and B. V. Street, eds., *Zande themes: Essays presented to Sir Edward Evans-Pritchard,* 41–63. Oxford, U.K.: Blackwell Scientific.

Bicknell, R. A. and K. B. Bicknell. 1999. Who will benefit from apomixis? *Biotechnology and Development Monitor* 37: 17–20.

Birch, H. F. 1960. Nitrification in soils after different periods of dryness. *Plant and Soil* 12(1): 81–96.

Birch, R. G. 1997. Plant transformation: Problems and strategies for practical application. *Annual Review of Plant Physiology and Plant Molecular Biology* 48: 297–326.

Blackburn, R. 1972. Cuba and the super-powers. In E. de Kadt, ed., *Patterns of foreign influence in the Caribbean,* 121–47. London: Oxford University Press for the Royal Institute of International Affairs.

Blaikie, P. and H. Brookfield. 1987. *Land degradation and society.* London: Methuen.

Blumler, M. A. 1998. Biogeography of land-use impacts in the Near East. In K. S. Zimmerer and K. R. Young, eds., *Nature's geography: New lessons for conservation in developing countries,* 215–36. Madison: University of Wisconsin Press.

Boardman, J. and M. Bell. 1992. Past and present soil erosion: Linking archaeology and geomorphology. In M. Bell and J. Boardman, eds., *Past and present soil erosion: Archaeological and geographical perspectives,* 1–8. Oxford, U.K.: Oxbow.

Boehler, J.-M. 1983. Economie agraire et société rurale dans la plaine d'Alsace aux XVIIe et XVIIIe siècles: L'amorce des mutations. In J.-M. Boehler, D. Lercy, and J. Vogt, eds., *Histoire de l'Alsace rurale,* 177–226. Strasbourg: Librairie Istra.

Bonnemaison, J. 1986. *L'arbre et la pirogue, les fondements d'une identité territoire, histoire et société dans L'Archipel de Vanuatu (Mélanésie).* Collection Travaux et Documents No 201. Paris: Éditions de l'Orstrom.

Boorstein, E. 1968. *The economic transformation of Cuba.* New York: Modern Reader.

Boserup, E. 1965. *The conditions of agricultural growth: The economics of agrarian change under population pressure.* Chicago: Aldine.

Boserup, E. 1970. Present and potential food production in developing countries. In W. Zelinsky, L. A. Kosinski, and R. M. Prothero, eds., *Geography and a crowding world: A symposium on pop-*

ulation pressure upon physical and social resources in the developing lands, 100–14. New York: Oxford University Press.

Boserup, E. 1981. *Population and technology.* Oxford, U.K.: Blackwell Scientific.

Botkin, D. B. 1990. *Discordant harmonies: A new ecology for the twenty-first century.* New York: Oxford University Press.

Brookfield, H. C. 1964. The ecology of highland settlement: Some suggestions. *American Anthropologist* 66(4): 20–38.

Brookfield, H. C. 1972a. *Colonialism, development and independence. The case of the Melanesian Islands in the South Pacific.* Cambridge, U.K.: Cambridge University Press.

Brookfield, H. 1972b. Intensification and disintensification in Pacific agriculture: A theoretical approach. *Pacific Viewpoint* 13: 30–48.

Brookfield, H. C. 1973. Full circle in Chimbu: A study of trends and cycles. In H. C. Brookfield, ed., *The Pacific in transition. Geographical perspectives on adaptation and change,* 127–60. London: Edward Arnold.

Brookfield, H. 1975. *Interdependent development.* London: Methuen.

Brookfield, H. 1984. Intensification revisited. *Pacific Viewpoint* 25: 15–44.

Brookfield, H. C. 1986. Intensification intensified [Review of I. S. Farrington, ed., *Prehistoric intensive agriculture in the Tropics,* BAR International Series 232]. *Archaeology in Oceania* 21(3): 177–81.

Brookfield, H. 1991. Environmental sustainability with development: What prospects for a research agenda? *European Journal of Development Research* 3(1): 42–66.

Brookfield, H. 1993. What PLEC is about. *PLEC News and Views* 1: 2–5.

Brookfield, H. 1996a. Meanings of "agrodiversity." *PLEC News and Views* 7: 13–15.

Brookfield, H. 1996b. Untying the Chimbu circle: An essay in and on hindsight. In H. Levine and A. Ploeg, eds., *Work in progress: Essays in honour of Paula Brown Glick,* 63–84. Frankfurt-am-Main: Peter Lang.

Brookfield, H. 1997. *Working with plants, and for them: Indigenous fallow management in perspective.* Unpublished paper delivered at a regional conference on Indigenous Strategies for Intensification of Shifting Cultivation in Southeast Asia, held in Bogor, Indonesia, June 23–27, 1997.

Brookfield, H. 1998. *Intensification, and alternative approaches to agricultural change.* Unpublished paper delivered at the Resource Management in Asia–Pacific "Intensification Workshop," held in Canberra, Australia, 4–6 November 1998. Canberra: Asia Pacific Press (forthcoming).

Brookfield, H., Abdul Samad Hadi, and Zaharah Mahmud. 1991. *The city in the village: The in-situ urbanization of villages, villagers and their land around Kuala Lumpur, Malaysia.* Singapore: Oxford University Press.

Brookfield, H. C. and P. Brown. 1963. *Struggle for land: Agriculture and group territories among the Chimbu of the New Guinea highlands.* Melbourne: Oxford University Press.

Brookfield, H. C., B. Denis, R. D. Bedford, P. S. Nankivell, and J. B. Hardaker. 1978. *Taveuni: Land, population and production.* UNESCO and UNFPA: Population and Environment Project in the Eastern Islands of Fiji. Island Reports 3. Canberra: Development Studies Centre, Australian National University for UNESCO.

Brookfield, H. C. and D. Hart. 1971. *Melanesia: A geographical interpretation of an island world.* London: Methuen.

Brookfield, H. and G. S. Humphreys. 1992. *Mission to the Pacificland Network of IBSRAM: General report on Vanuatu, Fiji and Western Samoa.* Bangkok: IBSRAM.

Brookfield, H. and G. S. Humphreys. 1998. An ancient land management practice in Raga, North Pentecost, Vanuatu. In D. Guillaud, M. Seysset, and A. Walter, eds., *Le voyage inachevé . . . a Joël Bonnemaison.* 305–10. Paris: ORSTROM.

Brookfield, H. and C. Padoch. 1994. Appreciating agrodiversity: A look at the dynamism and diversity of indigenous farming practices. *Environment* 36(5): 6–11, 37–45.

Brookfield, H., L. Potter, and Y. Byron. 1995. *In place of the forest: Environmental and socio-economic transformation in Borneo and the eastern Malay Peninsula.* Tokyo: United Nations University Press.

Brookfield, H. and M. Stocking. 1998. *Agrodiversity in demonstration site areas: Guidelines for description, analysis and the making of a database.* Tokyo: United Nations University.

Brookfield, H. and M. Stocking. 1999. Agrodiversity: Definition, description and design. *Global Environmental Change: Human and Policy Dimensions* 9: 77–80.

Brookfield, H., M. Stocking, and M. Brookfield. 1999. Guidelines on agrodiversity in demonstration site areas. *PLEC News and Views* (Special Issue) 13: 17–31.

Brown, P. 1972. *The Chimbu: A study of change in the New Guinea highlands.* Cambridge, Mass.: Schenkman.

Brown, P. 1995. *Beyond a mountain valley: The Simbu of Papua New Guinea.* Honolulu: University of Hawaii Press.

Brown, P. 1999. Coffee: The mechanism of transition to a money economy. In L. Plotnicov and R. Scaglion, eds., *Consequences of cultivar diffusion,* 119–29. Pittsburgh: Department of Anthropology, University of Pittsburgh.

Brown, P. and H. C. Brookfield. 1959. Chimbu land and society. *Oceania* 30: 1–75.

Brown, P., H. C. Brookfield, and R. Grau. 1990. Land tenure and transfer in Chimbu, Papua New Guinea: 1958–1984. A study in continuity and change, accommodation and opportunism. *Human Ecology* 18(1): 21–49.

Brundrett, M. 1991. Mycorrhizas in natural ecosystems. *Advances in Ecological Research* 21: 171–313.

Brunner, K. 1995. Continuity and discontinuity of Roman agricultural knowledge in the early Middle Ages. In D. Sweeney, ed., *Agriculture in the Middle Ages: Technology, practice, and representation,* 21–40. Philadelphia: University of Pennsylvania Press.

Brush, S. B. 1977. *Mountain, field, and family: The economy and human ecology of an Andean valley.* Philadelphia: University of Pennsylvania Press.

Brush, S. B. 1992. Ethnoecology, biodiversity, and modernization in Andean potato agriculture. *Journal of Ethnobiology* 12(2): 161–85.

Brush, S. B. 1995. In situ conservation of landraces in centers of crop diversity. *Crop Science* 35(2): 346–54.

Brush, S. B. 1999. *Genes in the field: Conserving plant diversity on farms.* Boca Raton, Fla.: Lewis.

Bunders, J. F. G., B. Haverkort, and W. Hiemstra, eds. 1996. *Biotechnology: Building on farmers' knowledge.* London: Macmillan.

Buol, S. W. and W. Couto. 1978. Fertility management interpretations and soils surveys of the tropics. In American Society of Agronomy, ed., *Diversity of soils in the tropics,* 65–75. Madison, Wis.: American Society of Agronomy and Soil Science Society of America.

Burnham, C. P. 1989. Pedological processes and nutrient supply from parent material in tropical soils. In J. Proctor, ed., *Mineral nutrients in tropical forest and savanna ecosystems,* 27–41. Oxford, U.K.: Blackwell Scientific.

Butzer, K. W. 1976. *Early hydraulic civilization in Egypt: A study in cultural ecology*. Chicago: University of Chicago Press.

Butzer, K. W. 1989. Cultural ecology. In G. C. Gaile and C. J. Willmott, eds., *Geography in America*, 192–208. Columbus, Ohio: Merrill.

Butzer, K. W. 1992. The Americas before and after 1492: An introduction to current geographical research. *Annals of the Association of American Geographers* 82(3): 345–68.

Butzer, K. W. 1996. Ecology in the long view: Settlement histories, agrosystemic strategies, and ecological performance. *Journal of Field Archaeology* 23(2): 141–50.

Byerlee, D. 1996. Modern varieties, productivity, and sustainability: Recent experience and emerging challenges. *World Development* 24(4): 697–718.

Caine, N. and P. K. Mool. 1982. Landslides in the Kolpu Khola drainage, Middle Mountains Nepal. *Mountain Research and Development* 2(2): 157–73.

Cairns, M. 1997. *Modification of fallow vegetation to increase swidden productivity: Understanding farmer strategies in Southeast Asia.* Unpublished paper delivered at a regional conference on Indigenous Strategies for Intensification of Shifting Cultivation in Southeast Asia, held in Bogor, Indonesia, June 23–27, 1997.

Cairns, M., S. Keitzar, and A. Yaden. 1997. *Shifting forests in North-eastern India: Management of* Alnus nepalensis *as an improved fallow species in Nagaland.* Unpublished paper delivered at a regional conference on Indigenous Strategies for Intensification of Shifting Cultivation in Southeast Asia, held in Bogor, Indonesia, June 23–27, 1997.

Carneiro, R. L. 1960. Slash and burn agriculture: A closer look at its implications for settlement patterns. In A. F. C. Wallace, ed., *Men and cultures: Selected papers,* 229–34. Philadelphia: University of Pennsylvania Press.

Carr, C. and R. H. Meyers. 1973. The agricultural transformation of Taiwan: The case of Ponlai rice, 1922–42. In R. T. Shand, ed., *Technical change in Asian agriculture,* 28–50. Canberra: Australian National University Press.

Carson, B. 1985. *Erosion and sedimentation processes in the Nepalese Himalayas.* ICIMOD Occasional Paper No. 1. Kathmandu: International Centre for Integrated Mountain Development.

Carson, R. 1962. *Silent spring.* Boston: Houghton Mifflin.

Cederroth, S. 1995. *Survival and profit in rural Java: The case of an east Javanese village.* Nordic Institute of Asian Studies Monograph Series No. 63. Richmond, Surrey: Curzon.

Centeno, M. A. and M. Font, eds. 1997. *Toward a new Cuba?: Legacies of a revolution.* Boulder, Colo.: L. Rienner.

Chambers, J. D. and G. E. Mingay. 1966. *The agricultural revolution 1750–1880.* London: B.T. Batsford.

Chambers, R. 1977. Men and water: The organisation and operation of irrigation. In B. H. Farmer, ed., *Green Revolution?: Technology and change in rice-growing areas of Tamil Nadu and Sri Lanka,* 340–63. London: Macmillan.

Chambers, R., A. Pacey, and L. A. Thrupp. 1989. *Farmer first: Farmer innovation and agricultural research.* London: Intermediate Technology Publications.

Chang, T. T. 1976. The origin, evolution, cultivation, dissemination, and diversification of Asian and African rices. *Euphytica* 25: 425–41.

Chang, T. T. 1989. Domestication and the spread of the cultivated rices. In D. R. Harris and G. C. Hillman, eds., *Foraging and farming: The evolution of plant exploitation,* 408–17. London: Unwin Hyman.

Chang, T. T. 1993. Sustaining and expanding the "Green Revolution" in rice. In H. Brookfield and

Y. Byron, eds., *South-east Asia's environmental future: The search for sustainability,* 201–10. Tokyo: United Nations University Press and Kuala Lumpur: Oxford University Press.

Chang, T. T., C. R. Adair, and T. H. Johnston. 1984. The conservation and use of rice genetic resources. *Advances in Agronomy* 35: 37–91.

Chang, T. T., G. C. Loresto, and O. Tagumpay. 1972. Agronomic and growth characteristics of upland and lowland rices. In International Rice Research Institute, ed., *Rice breeding,* 645–61. Los Baños, Philippines: IRRI.

Chang, T. T. and D. A. Vaughan. 1991. Conservation and potentials of rice genetic resources. In Y. P. S. Bajaj, ed., *Biotechnology in agriculture and forestry,* Vol. 14, *Rice,* 532–52. Berlin: Springer-Verlag.

Childe, V. G. 1936. *Man makes himself.* London: Watts.

Chin, S. C. 1985. Agriculture and resource utilization in a lowland rainforest Kenyah community. *Sarawak Museum Journal* 34(NS 56), Special Monograph No. 4.

Chinnappa, N. 1977. Adoption of the new technology in North Arcot District. In B. H. Farmer, ed., *Green Revolution?: Technology and change in rice-growing areas of Tamil Nadu and Sri Lanka,* 92–123. London: Macmillan.

Christanty, L. 1990. Home gardens in tropical Asia, with special reference to Indonesia. In K. Landauer and M. Brazil, eds., *Tropical home gardens,* 9–20. Tokyo: United Nations University Press.

Clarke, W. C. 1971. *Place and people: An ecology of a New Guinean community.* Canberra: Australian National University Press.

Clements, F. E. 1916. *Plant succession: An analysis of the development of vegetation.* Washington, D.C.: Carnegie Institute of Washington.

Clout, H. D. 1977. Agricultural changes in the eighteenth and nineteenth centuries. In H. D. Clout, ed., *Themes in the historical geography of France,* 407–46. London: Academic Press.

Cocking, E. C. 1986. The tissue culture revolution. In L. A. Withers and P. G. Alderson, eds., *Plant tissue culture and its agricultural applications,* 3–20. London: Butterworths.

Coghlan, A. 1999. Relative values: Crossing rice with its wild cousins works wonders. *New Scientist* 164(2214): 13.

Cohen, D. J. 1998. The origins of domesticated cereals and the Pleistocene–Holocene transition in east Asia. *Review of Archaeology* 19(2): 22–29.

Colfer, C. J. P. and R. G. Dudley. 1993. *Shifting cultivators of Indonesia: Marauders or managers of the forest?* Community Forestry Study Series No. 6. Rome: Food and Agriculture Organization of the United Nations.

Colfer, C. J. P., N. Peluso, and C. S. Chin. 1997. *Beyond slash and burn: Building on indigenous management of Borneo's tropical rain forests.* Advances in Economic Botany Vol. 11. Bronx: New York Botanical Garden.

Colfer, C. J. P. and H. Soedjito. 1996. Food, forests, and fields in a Bornean rain forest: Towards appropriate agroforestry development. In C. Padoch and N. L. Peluso, eds., *Borneo in transition: People, forests, conservation and development,* 162–86. Kuala Lumpur: Oxford University Press.

Committee on the Role of Alternative Farming Methods in Modern Production Agriculture, Board on Agriculture, National Research Council. 1989. *Alternative agriculture.* Washington, D.C.: National Academy Press.

Conference of the Parties to the Convention on Biological Diversity. 1996. *Report of the third meeting of the Conference of the Parties to the Convention on Biological Diversity, Buenos Aires, Ar-*

gentina, 3–14 November 1996. Secretariat of the Convention on Biological Diversity. http://www.biodiv.org/cop3/docs.html [November 1999].

Conklin, H. C. 1949. Preliminary report on the field work on the islands of Mindoro and Palawan, Philippines. *American Anthropologist* 51: 268–73.

Conklin, H. C. 1954. An ethnoecological approach to shifting agriculture. *Transactions of the New York Academy of Sciences* Series 2, 17: 133–42.

Conklin, H. C. 1957. *Hanunóo agriculture in the Philippines*. FAO Forestry Development Paper No. 12. Rome: Food and Agriculture Organization of the United Nations.

Conklin, H. C. 1961. The study of shifting cultivation. *Current Anthropology* 2(1): 27–61.

Conklin, H. C. 1980. *Ethnographic atlas of Ifugao*. New Haven, Conn.: Yale University Press.

Conway, G. R. 1985. Agroecosystem analysis. *Agricultural Administration* 20: 31–35.

Conway, G. R. 1997. *The Doubly Green Revolution: Food for all in the 21st century*. London: Penguin.

Correa, C. M. 1999. *Access to plant genetic resources and intellectual property rights*. Background Study Paper No. 8. Rome: Commission on Genetic Resources for Food and Agriculture, FAO.

Cowgill, G. L. 1975. On causes and consequences of ancient and modern population changes. *American Anthropologist* 77: 505–25.

Cox, T. S. and D. Wood. 1999. The nature and role of crop biodiversity. In D. Wood and J. M. Lenné, eds., *Agrobiodiversity: Characterization, utilization and management*, 35–57. New York: CABI.

Cramb, R. A. 1985. The importance of secondary crops in Iban hill-rice farming. *Sarawak Museum Journal* 34(NS 55): 37–45.

Cressey, G. B. 1958. Qanats, karez and foggaras. *Geographical Review* 48: 27–44.

Cromwell, E. 1999. *Agriculture, biodiversity and livelihoods: Issues and entry points*. London: Overseas Development Institute. http://nt1.ids.ac.uk/eldis/agbio.htm [November 18, 1999].

Cronon, W. 1983. *Changes in the land. Indians, colonists and the ecology of New England*. New York: Hill & Wang.

Cronon, W. 1996. The trouble with wilderness: Or, getting back to the wrong nature. In W. Cronon, ed., *Uncommon ground: Rethinking the human place in nature*, 69–90. New York: W.W. Norton.

Crouch, M. L. 1998. How the terminator terminates: An explanation for the non-scientist of a remarkable patent for killing second generation seeds of crop plants. http://www.bio.indiana.edu/people/terminator.html [April 24, 1999].

Dalal-Clayton, D. B. and D. A. Robinson. 1993. The experimental use of correspondence analysis to assess the relationship between soils and geomorphology in eastern Zambia. *Catena* 20(1/2): 141–60.

Dalrymple, D. G. 1976. *Development and spread of high-yielding varieties of wheat and rice in the less developed nations*. Washington, D.C.: Foreign Development Division, Economic Research Service, U.S. Department of Agriculture in cooperation with U.S. Agency for International Development.

Dalrymple, D. G. 1986. *Development and spread of high-yielding rice varieties in developing countries*. Washington, D.C.: Bureau for Science and Technology, Agency for International Development.

Darwin, C. 1881. *The formation of vegetable mould through the action of worms with observations on their parts*. London: John Murray.

Datoo, B. A. 1978. Toward a reformulation of Boserup's theory of agricultural change. *Economic Geography* 54: 134–44.

de Foresta, H. and G. Michon. 1993. Creation and management of rural agroforests in Indonesia: Potential applications in Africa. In C. M. Hladick, A. Hladick, O. F. Linares, H. Pagezy, A. Semple, and M. Hadley, eds., *Tropical forests, people and food: Biocultural interactions and applications to development*, 709–24. Man and the Biosphere Series Vol. 13. Paris: UNESCO and Carnforth, UK: Parthenon.

de Jong, W. 1996. Swidden–fallow agroforestry in Amazonia: Diversity at close distance. *Agroforestry Systems* 34: 277–90.

de Schlippe, P. 1956. *Shifting cultivation in Africa: The Zande system of agriculture*. London: Routledge & Kegan Paul.

Deere, C. D., N. Perez, and E. Gonzales. 1994. The view from below: Cuban agriculture in the "Special Period in Peacetime." *Journal of Peasant Studies* 21(2): 194–234.

Denevan, W. M. 1992. The pristine myth: The landscape of the Americas in 1492. *Annals of the Association of American Geographers* 82(3): 369–85.

Denevan, W. M. and C. Padoch. 1987. *Swidden–fallow agroforestry in the Peruvian Amazon*. Advances in Economic Botany Vol. 5. Bronx: New York Botanical Garden.

Denevan, W. M. and J. M. Treacy. 1987. Young managed fallows at Brillo Nuevo. In W. M. Denevan and C. Padoch, eds., *Swidden–fallow agroforestry in the Peruvian Amazon*, 8–46. Advances in Economic Botany Vol. 5. Bronx: New York Botanical Garden.

Denevan, W. M., J. M. Treacy, J. B. Alcorn, C. Padoch, J. Denslow, and S. Flores Paitán. 1984. Indigenous agroforestry in the Peruvian Amazon: Bora Indian management of swidden fallows. *Interciencia* 9: 346–57.

Denslow, J. S. 1988. The tropical rain forest setting. In J. S. Denslow and C. Padoch, eds., *People of the tropical rain forest*, 25–36. Berkeley: University of California Press in association with Smithsonian Institution Traveling Exhibition Service.

Dickie, A. 1991. Systems of agricultural production in southern Sudan. In G. M. Craig, ed., *The agriculture of the Sudan*, 280–307. Oxford, U.K.: Oxford University Press.

Diemont, W. H., A. C. Smiet, and Nurdin. 1991. Re-thinking erosion on Java. *Netherlands Journal of Agricultural Science* 39: 213–24.

Donahue, R. L., R. W. Miller, and J. C. Shickluna. 1983 (5th ed.). *Soils: An introduction to soils and plant growth*. Englewood Cliffs, N.J.: Prentice Hall.

Donkin, R. A. 1979. *Agricultural terracing in the aboriginal New World*. Viking Fund Publications in Anthropology 56. Tucson: University of Arizona Press.

Doolittle, W. E. 1984. Agricultural change as an incremental process. *Annals of the Association of American Geographers* 74(1): 124–37.

Dove, M. R. 1980. The swamp rice swiddens of the Kantu' of West Kalimantan. In J. I. Furtado, ed., *Tropical ecology and development*, 953–56. Kuala Lumpur: International Society for Tropical Ecology.

Dove, M. R. 1985. *Swidden agriculture in Indonesia: The subsistence strategies of the Kalimantan Kantu'*. New York: Mouton.

Dove, M. R. 1996. Rice-eating rubber and people-eating governments: Peasant versus state critiques of rubber development in colonial Borneo. *Ethnohistory* 43(1): 33–63.

Dumaresq, D. and R. Greene. 1997. *From farmer to consumer: The future for organic agriculture in Australia*. RIRDC Research Paper Series. Barton, Australia: Rural Industries Research and Development Corporation.

Dupuis, J. 1960. *Madras et le nord du Coromandel: Étude des conditions de la vie Indienne dans un cadre géographique.* Paris: Librairie d'Amérique et d'Orient Adrien-Maisonneuve.

Eden, M. J., W. Braz, L. Herrera, and C. McEwan. 1984. *Terra preta* soils and their archaeological context in the Caquetá basin of southeast Colombia. *American Antiquity* 49: 124–40.

Eder, J. F. 1982. No water in the terraces: Agricultural stagnation and social change at Banaue, Ifugao. *Philippine Quarterly of Culture and Society* 10: 101–16.

Evans-Pritchard, E. E. 1937. *Witchcraft, oracles and magic among the Azande.* Oxford, U.K.: Clarendon.

Evans-Pritchard, E. E. 1951. *Essays in social anthropology.* London: Cohen & West.

Evans-Pritchard, E. E. 1958. An historical introduction to a study of Zande society. *African Studies* 17(1): 1–15.

Evans-Pritchard, E. E. 1960. A contribution to the study of the Zande culture. *Africa* 30(4): 309–24.

Evans-Pritchard, E. E. 1963. A further contribution to the study of the Zande culture. *Africa* 33(3): 183–97.

Evans-Pritchard, E. E. 1965. *The position of women in primitive societies and other essays in social anthropology.* London: Faber & Faber.

Evans-Pritchard, E. E. 1971. *The Azande: History and political institutions.* Oxford, U.K.: Clarendon.

Evans-Pritchard, E. E., ed. 1974. *Man and woman among the Azande.* New York: Free Press.

Fairhead, J. 1993. Representing knowledge: The "new farmer" in research fashions. In J. Pottier, ed., *Practising development: Social science perspectives,* 187–204. London: Routledge.

Fairhead, J. and M. Leach. 1996. *Misreading the African landscape: Society and ecology in a forest–savanna mosaic.* Cambridge, U.K.: Cambridge University Press.

FAO. 1996. *Report on the state of the world's plant genetic resources for food and agriculture prepared for the International Technical Conference on Plant Genetic Resources, Leipzig, Germany, 17–23 June 1996.* Rome: Food and Agriculture Organization of the United Nations.

FAO. 1998. *The state of the world's plant genetic resources for food and agriculture.* Rome: Food and Agriculture Organization of the United Nations.

Farmer, B. H., ed. 1977. *Green Revolution?: Technology and change in rice-growing areas of Tamil Nadu and Sri Lanka.* London: Macmillan.

Feil, D. 1987. *The Evolution of highland Papua New Guinea societies.* Cambridge, U.K.: Cambridge University Press.

Fernandez-Cornejo, J., C. Greene, R. Penn, and D. Newton. 1998. Organic vegetable production in the U.S.: Certified growers and their practices. *Alternative Agriculture* 13(2): 69–78.

Field, M. J. 1943–44. The agricultural system of the Manya-Krobo of the Gold Coast. *Africa* 14: 54–65.

Finkel, E. 1999. Australian Center develops tools for developing world. *Science* 285(5433): 1481–83.

Flowers, N. M., D. R. Gross, M. L. Ritter, and D. W. Werner. 1982. Variation in swidden practices in four central Brazilian Indian societies. *Human Ecology* 10(2): 203–17.

Foth, H. D. 1984 (7th ed.). *Fundamentals of soil science.* New York: Wiley.

Fowler, C., E. Lachkovics, P. R. Mooney, and H. Shand. 1988. The laws of life: Another development and the new biotechnologies. *Development Dialogue* 1–2.

Fowler, C. and P. Mooney. 1990. *Shattering: Food, politics, and the loss of genetic diversity.* Tucson: University of Arizona Press.

Fox, J. J. 1991. Managing the ecology of rice production in Indonesia. In J. Hardjono, ed., *Indonesia: Resources, ecology, and environment,* 61–84. Singapore: Oxford University Press.

Fox, J. J. 1993a. Ecological policies for sustaining high production in rice: Observations on rice intensification in Indonesia. In H. Brookfield and Y. Byron, eds., *South-east Asia's environmental future: The search for sustainability,* 211–24. Tokyo: United Nations University Press and Kuala Lumpur: Oxford University Press.

Fox, J. J. 1993b. The rice baskets of east Java: The ecology and social context of sawah production. In H. W. Dick, J. J. Fox, and J. A. C. Mackie, eds., *Balanced development: East Java and the new order,* 120–57. Singapore: Oxford University Press.

Francis, C. A., ed. 1986. *Multiple cropping systems.* New York: Macmillan.

Francks, P. 1983. *Technology and agricultural development: A case-study of technical and economic change in pre-war Japanese agriculture.* New Haven, Conn.: Yale University.

Freeman, J. D. 1955. *Iban agriculture: A report on the shifting cultivation of hill rice by the Iban of Sarawak.* London: Her Majesty's Stationery Office.

Freeman, J. D. 1970. *Report on the Iban.* London: Althone.

Freeman, J. D. 1992. *The Iban of Borneo.* Kuala Lumpur: S. Abdul Majeed.

Fritter, A. H. 1985. Functioning of vesicular–arbuscular mycorrhizas under field conditions. *New Phytologist* 99: 257–65.

Frost, P. 1996. The ecology of *miombo* woodlands. In B. Campbell, ed., *The Miombo in transition: Woodlands and welfare in Africa,* 11–57. Bogor, Indonesia: Center for International Forestry Research.

Fujisaka, S. 1997. Research: Help or hindrance to good farmers in high risk systems? *Agricultural Systems* 54(2): 137–52.

Fujisaka, S., L. Hurtado, and R. Uribe. 1996. A working classification of slash-and-burn agricultural systems. *Agroforestry Systems* 34(2): 151–69.

Fukui, H. 1979. An agroenvironmental consideration. In M. Kuchiba, Y. Tsubouchi, and N. Maeda, eds., *Three Malay Villages: A sociology of paddy growers in west Malaysia,* 309–26. Monographs of the Center for Southeast Asian Studies Kyoto University. Honolulu: University Press of Hawaii.

Fukuoka, M. 1978. *The one straw revolution. An introduction to natural farming.* Emmaus, Pa.: Rodale.

Fukuoka, M. 1987. *The road back to nature: Regaining the paradise lost.* Tokyo: Japan Publications.

Galinat, W. C. 1985. Domestication and diffusion of maize. In R. I. Ford, ed., *Prehistoric food production in North America,* 245–78. Ann Arbor: Museum of Anthropology, University of Michigan.

Gardner, J. C., K. Jamtgaard, and F. Kirschenmann. 1995. What is sustainable agriculture? In E. A. R. Bird, G. L. Bultena, and J. C. Gardner, eds., *Planting the future: Developing an agriculture that sustains land and community,* 45–65. Ames: Iowa State University Press.

Gasser, C. S. and R. T. Fraley. 1989. Genetically engineering for crop improvement. *Science* 244: 1293–99.

Geddes, W. R. 1954. *The Land Dayaks of Sarawak: A report on a social economic survey of the Land Dayaks of Sarawak presented to the Colonial Social Science Research Council.* London: Her Majesty's Stationery Office for the Colonial Office.

Geertz, C. 1963. *Agricultural involution: The process of ecological change in Indonesia.* Berkeley: University of California Press.

Ghassemi, F., A. J. Jakeman, and H. A. Nix. 1995. *Salinisation of land and water resources: Human causes, extent, management, and case studies.* Sydney: NSW University Press.

Gibson, T. 1986. *Sacrifice and sharing in the Philippines: Religion and society among the Buid of Mindoro*. London: Althone.

Glazovsky, N. F. 1995. The Aral Sea basin. In J. X. Kasperson, R. E. Kasperson, and B. L. Turner II, eds., *Regions at risk: Comparison of threatened environments,* 92–139. Tokyo: United Nations University Press.

Gliessman, S. R. 1986. Plant interactions in multiple cropping systems. In C. A. Francis, ed., *Multiple cropping systems,* 82–95. New York: Macmillan.

Gliessman, S. R. 1998. *Agroecology: Ecological processes in sustainable agriculture*. Chelsea, Mich.: Ann Arbor Press.

Goldman, A. and J. Smith. 1995. Agricultural transformations in India and northern Nigeria: Exploring the nature of green revolutions. *World Development* 23(2): 243–63.

Golson, J. 1977. No room at the top: Agricultural intensification in the New Guinea highlands. In J. Allen, J. Golson, and R. Jones, eds., *Sunda and Sahul: Prehistoric studies in southeast Asia, Melanesia and Australia,* 601–38. London: Academic Press.

Golson, J. 1982. The Ipomoean revolution revisited: Society and the sweet potato in the upper Wahgi valley. In A. Strathern, ed., *Inequality in New Guinea highlands society,* 109–36. Cambridge, U.K.: Cambridge University Press.

Golson, J. 1991. Bulmer phase II: Early agriculture in the New Guinea highlands. In A. Pawley, ed., *Man and a half. Essays in Pacific anthropology and ethnobiology in honour of Ralph Bulmer,* 484–91. Auckland, N.Z.: The Polynesian Society.

Golson, J., R. J. Lampert, J. M. Wheeler, and W. R. Ambrose. 1967. A note on carbon dates for horticulture in the New Guinea highlands. *Journal of Polynesian Society* 76: 369–71.

Gómez-Ibáñez, D. A. 1975. *The western Pyrenees: Differential evolution of the French and Spanish borderland*. Oxford, U.K.: Clarendon.

Greenland, D. J. 1975. Bringing the Green Revolution to the shifting cultivator. *Science* 190(4217): 841–44.

Greig, J. 1989. *Archaeobotany*. Handbooks for Archaeologists No. 4. Strasbourg: European Science Foundation.

Gressel, J. 1998. Biotechnology of weed control. In A. Altman, ed., *Agricultural biotechnology,* 295–325. New York: Marcel Dekker.

Grigg, D. B. 1974. *The agricultural systems of the world: An evolutionary approach*. Cambridge, U.K.: Cambridge University Press.

Grigg, D. B. 1979. Ester Boserup's theory of agrarian change: A critical review. *Progress in Human Geography* 3: 64–83.

Grigg, D. B. 1984. The agricultural revolution on Western Europe. In T. P. Bayliss-Smith and S. Wanmali, eds., *Understanding green revolutions: Agrarian change and development planning in south Asia,* 1–17. Cambridge, U.K.: Cambridge University Press.

Grove, A. T. 1952. Land use and soil conservation in the Jos plateau. *Geological Survey of Nigeria Bulletin* 22: 23–28.

Grove, A. T. and J. E. G. Sutton. 1989. Agricultural terracing south of the Sahara. *Azania* 24: 113–22.

Grove, R. 1997. *Ecology, climate and empire: Colonialism and global environmental history 1400–1940*. Cambridge, U.K.: White Horse Press.

Guo Huijun, Dao Zhiling, and H. Brookfield. 1996. Agrodiversity and biodiversity on the ground and among the people: Methodology from Yunnan. *PLEC News and Views* 6: 14–22.

Guo Huijun and C. Padoch. 1995. Patterns and management of agroforestry systems in Yunnan:

An approach to upland rural development. *Global Environmental Change: Human and Policy Dimensions* 5(4): 273–79.

Guo Huijun, Xia Yongmei, and C. Padoch. 1997. Alnus nepalensis–*based agroforestry systems in Yunnan, southwest China.* Unpublished paper delivered at a regional conference on Indigenous Strategies for Intensification of Shifting Cultivation in Southeast Asia, held in Bogor, Indonesia, June 23–27, 1997.

Guthman, J. 1998. Regulating meaning, appropriating nature: The codification of California organic agriculture. *Antipode* 30(2): 135–54.

Gyasi, E., G. T. Agyepong, E. Ardayfio-Schandorf, L. Enu-Kwesi, L. Nabila, and E. Owusu-Bennoah. 1995. Production pressure and environmental change in the forest–savanna zone of southern Ghana. *Global Environmental Change: Human and Policy Dimensions* 5(4): 355–66.

Gyasi, E. A. and J. I. Uitto, eds. 1997. *Environment, biodiversity, and agricultural change in west Africa.* Tokyo: United Nations University Press.

Haberle, S. 1994. Anthropogenic indicators in pollen diagrams: Problems and prospects for late Quaternary palynology in New Guinea. In J. G. Hather, ed., *Tropical archaeobotany: Applications and new developments,* 172–201. London: Routledge.

Hagen, E. E. 1962. *On the theory of social change: How economic growth begins.* Homewood, Ill.: Dorsey.

Halstead, P. 1992. Agriculture in the Bronze Age Aegean: Towards a model of palatial economy. In B. Wells, ed., *Agriculture in ancient Greece. Proceedings of the Seventh International Symposium at the Swedish Institute at Athens, 16–17 May 1990,* 105–17. Stockholm: Svenska Institutet i Athen.

Hames, R. 1983. Monoculture, polyculture, and polyvariety in tropical swidden cultivation. *Human Ecology* 11(1): 13–34.

Hardjono, J. 1983. Rural development in Indonesia: The "top-down" approach. In D. A. M. Lea and D. P. Chaudri, eds., *Rural development and the state: Contradictions and dilemmas in developing countries,* 38–65. London: Methuen.

Hardjono, J. 1987. *Land, labour and livelihood in a west Java village.* Yogyakata, Indonesia: Gadjah Mada University Press.

Harlan, H. V. 1957. *One man's life with barley.* New York: Exposition Press.

Harlan, J. R. 1975. *Crops and Man.* Madison, Wis.: American Society of Agronomy.

Harlan, J. R. 1992 (2d ed.). *Crops and man.* Madison, Wis.: American Society of Agronomy.

Harlan, J. R. 1995. *The living fields: Our agricultural heritage.* Cambridge, U.K.: Cambridge University Press.

Harris, D. R. 1971. The ecology of swidden cultivation in the Upper Orinoco rain forest, Venezuela. *Geographical Review* 61(4): 475–95.

Harris, D. R. 1972. The origins of agriculture in the tropics. *American Scientist* 60: 180–93.

Harris, D. R. 1973. The prehistory of tropical agriculture: An ethnoecological model. In C. Renfrew, ed., *The explanation of culture change: Models in prehistory,* 391–417. London: Duckworth.

Harris, D. R. 1976. Traditional systems of plant food production and the origins of agriculture in west Africa. In J. R. Harlan, J. M. de Wet, and A. B. L. Stemler, eds., *Origins of African plant domestication,* 311–56. The Hague: Mouton.

Harris, D. R. 1989. An evolutionary continuum of people–plant interactions. In D. R. Harris and G. C. Hillman, eds., *Foraging and farming: The evolution of plant exploitation,* 11–26. London: Unwin Hyman.

Harris, D. R. 1996. The origins and spread of agriculture and pastoralism in Eurasia: An overview. In D. R. Harris, ed., *The origins and spread of agriculture and pastoralism in Eurasia,* 552–73. London: UCL Press.

Harris, D. R. 1998a. The origins of agriculture in southwest Asia. *Review of Archaeology* 19(2): 5–11.

Harris, D. R. 1998b. The spread of Neolithic agriculture from the Levant to western central Asia. In A. B. Damania, J. Valkoun, G. Willcox, and C. O. Qualset, eds., *The origins of agriculture and crop domestication,* 65–82. Aleppo, Syria: ICARDA.

Harris, D. R. and G. C. Hillman, eds. 1989. *Foraging and farming: The evolution of plant exploitation.* London: Unwin Hyman.

Harrison, P. D. and B. L. Turner II, eds. 1978. *Pre-Hispanic Maya agriculture.* Albuquerque: University of New Mexico Press.

Harriss, B. 1977. Rural electrification and the diffusion of electric water lifting technology in North Arcot District, India. In B. H. Farmer, ed., *Green Revolution?: Technology and change in rice-growing areas of Tamil Nadu and Sri Lanka,* 182–203. London: Macmillan.

Harriss, J. 1977a. Bias in perception of agrarian change in India. In B. H. Farmer, ed., *Green Revolution?: Technology and change in rice-growing areas of Tamil Nadu and Sri Lanka,* 30–36. London: Macmillan.

Harriss, J. 1977b. Implications of changes in agriculture for social relationships at the village level: The case of Randam. In B. H. Farmer, ed., *Green Revolution?: Technology and change in rice-growing areas of Tamil Nadu and Sri Lanka,* 225–45. London: Macmillan.

Haverkort, B., J. van der Kamp, and A. Waters-Bayer, eds. 1991. *Joining farmers' experiments.* Leusden, The Netherlands: ILEIA and London: Intermediate Technology Publications.

Hayami, Y. 1973. Development and diffusion of high yielding rice varieties in Japan, Korea and Taiwan, 1890–1940. In R. T. Shand, ed., *Technical change in Asian agriculture,* 9–27. Canberra: Australian National University Press.

Hazell, P. B. R. and C. Ramasamy, eds. 1991. *The Green Revolution reconsidered: The impact of high-yielding rice varieties in south India.* Baltimore: Johns Hopkins University Press.

Hecht, S. B., A. B. Anderson, and P. May. 1988. The subsidy from nature: Shifting cultivation, successional palm forests and rural development. *Human Organization* 47(1): 25–35.

Hecht, S. B. and D. A. Posey. 1989. Preliminary results on soil management techniques of the Kayapó Indians. In D. A. Posey and W. Balée, eds., *Resource management in Amazonia: Indigenous folk strategies,* 174–88. Advances in Economic Botany Vol. 7. Bronx: New York Botanical Garden.

Hernández Bermejo, J. E. and E. García Sánchez. 1998. Economic botany and ethnobotany in al-Andalus (Iberian Peninsula: Tenth–fifteenth centuries), an unknown heritage of mankind. *Economic Botany* 52(1): 15–26.

Hidayati, D. 1994. *Striving to reach "Heaven's Gate": Javanese adaptions of swamp and upland environments in Kalimantan.* Ph.D. thesis, Australian National University, Canberra.

Hill, P. 1956. *The Gold Coast cocoa farmer: A preliminary survey.* London: Oxford University Press.

Hill, P. 1963. *The migrant cocoa-farmers of southern Ghana.* London: Cambridge University Press.

Hill, R. D. 1977. *Rice in Malaya: A study in historical geography.* Kuala Lumpur: Oxford University Press.

Hillman, G. C. and M. S. Davies. 1990. Measured domestication rates in wild wheats and barley under primitive cultivation, and their archaeological implications. *Journal of World Prehistory* 4(2): 157–222.

Hillman, G. C. and M. S. Davies. 1992. Domestication rates in wild wheats and barley under primitive cultivation: Preliminary results and archaeological implications of field measurements of selection coefficient. In P. C. Anderson, ed., *Préhistoire de l'agriculture: Nouvelle approches expérimentales et ethnographiques.* Paris: Centre National de la Recherche Scientifique.

Hiraoka, M. 1995. Land use changes in the Amazon estuary. *Global Environmental Change: Human and Policy Dimensions* 5(4): 323–36.

Ho, R. 1964. The environment. In Wang Gungwu, ed., *Malaysia: A survey,* 25–43. New York: Praeger.

Hoadley, M. C. 1994. *Towards a feudal mode of production: West Java, 1680–1800.* Singapore: Social Issues in Southeast Asia, Institute of Southeast Asian Studies.

Högberg, P. 1989. Root symbioses of trees in savanna. In J. Proctor, ed., *Mineral nutrients in tropical forest and savanna ecosystems,* 121–36. Oxford, U.K.: Blackwell Scientific.

Högberg, P. and G. D. Piearce. 1986. Mycorrhizas in Zambian trees in relation to host taxonomy, vegetation type and successional patterns. *Journal of Ecology* 74: 774–85.

Hole, F. 1998. The spread of agriculture to the eastern arc of the Fertile Crescent: Food for the herders. In A. B. Damania, J. Valkoun, G. Willcox, and C. O. Qualset, eds., *The origins of agriculture and crop domestication,* 83–92. Aleppo, Syria: ICARDA.

Holling, D. G. 1978. *Adaptive environmental assessment and management.* Chichester, U.K.: Wiley.

Hong, E. 1987. *Natives of Sarawak: Survival in Borneo's vanishing forests.* Pulau Pinang, Malaysia: Institut Masyarakat.

Hopkins, D. P. 1948. *Chemicals, humus and the soil.* Brooklyn, N.Y.: Chemical Publishing Co.

Hourani, A. H. 1991. *A history of the Arab peoples.* London: Faber & Faber.

Howard, A. 1940. *An agricultural testament.* London: Oxford University Press.

Hudson, N. W. 1992. *Land husbandry.* London: Batsford.

Humphreys, G. S. 1996. *Cultivation history from an open archaeological site in Vanuatu.* Poster paper for ASSSI and NZSSS National Soils Conference, July 1996.

Humphreys, G. S. and H. C. Brookfield. 1991. The use of unstable steeplands in the mountains of Papua New Guinea. *Mountain Research and Development* 11(4): 295–318.

Hunter, J. M. and D. F. Hayward. 1965. Drainage patterns of the Densu Basin, Ghana. *Bulletin of the Ghana Geographical Association* 10(1): 66–82.

Hüttermann, A. and A. Majcherczyk. 1998. Bioremediation: Pesticides and other agricultural chemicals. In A. Altman, ed., *Agricultural biotechnology,* 367–86. New York: Marcel Dekker.

Igarashi, T. 1985. Some notes on the subsistence in a Sundanese village. In S. Suzuki, O. Soemarwoto, and T. Igarashi, eds., *Human ecological survey in rural west Java in 1978 to 1982: A project report,* 9–77. Tokyo: Nissan Science Foundation.

Imamura, K. 1996. Jomon and Yayoi: The transition to agriculture in Japanese history. In D. R. Harris, ed., *The origins and spread of agriculture and pastoralism in Eurasia,* 552–73. London: UCL Press.

Innis, D. Q. 1997. *Intercropping and the scientific basis of traditional agriculture.* London: Intermediate Technology Publications.

International Rice Research Institute (IRRI). 1995. *World rice statistics.* Manila: IRRI.

Isely, D. 1982. Leguminosae and *Homo sapiens. Economic Botany* 36(1): 46–70.

Ives, J. D. and P. Ives. 1987. The Himalayan Ganges problem. Proceedings of a conference, Mohonk Mountain House, New Paltz, New York, USA. *Mountain Research and Development* 7(3): 181–344.

Ives, J. D. and B. Messerli. 1989. *The Himalayan dilemma: Reconciling development and conservation.* New York: Routledge.

Izikowitz, K. G. 1951. Lamet: Hill peasants in French Indochina. *Etnologiska Studier* Vol. 17.

James, C. 1998. Global review of commercialized transgenic crops. ISAAA Briefs No. 8. Ithaca, N.Y.: International Service for the Acquisition of Agri-Biotech Applications.

Jana, S. 1999. Some recent issues on the conservation of crop genetic resources in developing countries. *Genome* 42(4): 562–69.

Janos, D. P. 1980. Mycorrhizae influence tropical succession. *Biotropica* 12(Supplement): 56–64.

Janos, D. P. 1996. Mycorrhizas, succession and the rehabilitation of deforested lands in the humid tropics. In J. C. Frankland, N. Magan, and G. M. Dadd, eds., *Fungi and environmental change: Symposium of the British Mycological Society held at Cranfield University, March 1994,* 129–62. Cambridge, U.K.: Cambridge University Press for the British Mycological Society.

Jarvis, D. I. and T. Hodgkin, eds. 1998. Strengthening the scientific basis of in situ conservation of agricultural biodiversity on-farm. Options for data collecting and analysis. *Proceedings of a Workshop to Develop Tools and Procedures for In Situ Conservation On-Farm, 25–29 August 1997, Rome, Italy.* Rome: International Plant Genetic Resources Institute.

Jefferson, R. A. 1994. Apomixis: A social revolution for agriculture? *Biotechnology and Development Monitor* 19: 14–16.

Jensen, E. 1974. *The Iban and their religion.* Oxford, U.K.: Clarendon.

Jirström, M. 1996. *In the wake of the Green Revolution. Environmental and socio-economic consequences of intensive rice agriculture: The problems of weeds in Muda, Malaysia.* Meddelanden fran Lunds Universitets Geografiska Institutioner Avhandlingar. Lund, Sweden: Lund University Press.

Johnson, A. 1983. Machiguenga gardens. In R. B. Hames and W. T. Vickers, eds., *Adaptive responses of native Amazonians,* 29–63. New York: Academic Press.

Johnson, K., E. A. Olson, and S. Manandhar. 1982. Environmental knowledge and response to natural hazards in mountainous Nepal. *Mountain Research and Development* 2(2): 175–88.

Jones, J. A. 1989. Soil genesis, morphology, and classification. *Soil Science Society of America* 53: 1748–58.

Jones, J. A. 1990. Termites, soil fertility and carbon cycling in dry tropical Africa: A hypothesis. *Journal of Tropical Ecology* 6: 291–305.

Jordan, C. F. 1985. *Nutrient cycling in tropical forest ecosystems.* Chichester, U.K.: Wiley.

Jordan, C. F. 1986. Shifting cultivation, Case Study No. 1: Slash and burn agriculture near San Carlos de Rio Negro, Venezuela. In C. F. Jordan, ed., *Amazonian rainforests: Ecosystem disturbance and recovery. Case studies of ecosystem dynamics under a spectrum of land-use intensities,* 9–23. New York: Springer-Verlag.

Jordan, C. F., ed. 1987. *Amazonian rain forests. Case studies of ecosystem dynamics under a spectrum of land-use intensities.* Ecological Studies, Vol. 60. New York: Springer-Verlag.

Juma, C. 1989. *The gene hunters. Biotechnology and the scramble for seeds.* London: Zed.

Kajale, M. D. 1994. Archaeobotanical investigations on a multicultural site at Adam, Maharashtra, with special reference to the development of tropical agriculture in parts of India. In J. G. Hather, ed., *Tropical archaeobotany: Applications and new developments,* 34–50. London: Routledge.

Kaltoft, P. 1999. Values about nature in organic farming practice and knowledge. *Sociologia Ruralis* 39(1): 39–53.

Kamal Salih. 1969. *The growth-pole model of economic development: Preliminary testing of two hypotheses derived from the model.* B.A. honors thesis, Monash University, Melbourne.

Karyono. 1990. Home gardens in Java: Their structure and function. In K. Landauer and M. Brazil, eds., *Tropical home gardens,* 138–46. Tokyo: United Nations University Press.

Kea, R. A. 1982. *Settlements, trade and polities in the seventeenth-century Gold Coast*. Baltimore: John Hopkins University Press.

Kellman, M. 1989. Mineral nutrient dynamics during savanna–forest transformation in Central America. In J. Proctor, ed., *Mineral nutrients in tropical forest and savanna ecosystems*, 137–51. Oxford, U.K.: Blackwell Scientific.

Kerridge, E. 1967. *The agricultural revolution*. London: Allen & Unwin.

Kienholz, H., H. Hafner, G. Schneider, and R. Tamrakar. 1983. Mountain hazards mapping in Nepal's middle mountains: Maps of land use and geomorphic damages (Kathmandu–Kakani area). *Mountain Research and Development* 3: 195–220.

Kiome, R. M. and M. Stocking. 1995. Rationality of farmer perception of soil erosion: The effectiveness of soil conservation in semi-arid Kenya. *Global Environmental Change: Human and Policy Dimensions* 5(4): 281–95.

Kirch, P. V. 1985. Intensive agriculture in prehistoric Hawaii: The wet and the dry. In I. S. Farrington, ed., *Prehistoric intensive agriculture in the tropics*, Vol. II, 435–54. BAR International Series 232. Oxford, U.K.: British Archaeological Reports.

Klock, J. 1995a. Indigenous woodlot management and ethnobotany in Ifugao, Philippines. *International Tree Crops Journal* 8(2/3): 95–106.

Klock, J. S. 1995b. Agricultural and forest policies of the American colonial regime in Ifugao Territory, Luzon, Philippines, 1901–1945. *Philippine Quarterly of Culture and Society* 23: 3–19.

Koepf, H. H. 1989. *The biodynamic farm*. Hudson, N.Y.: Anthroposophic Press.

Kooistra, L. I. 1996. *Borderland farming: Possibilities and limitations of farming in the Roman period and early Middle Ages between the Rhine and Meuse*. Assen, The Netherlands: Van Gorcum.

Kunstadter, P. 1978. Subsistence agricultural economies of Lua and Karen hill farmers, Mae Sariang District, northwestern Thailand. In P. Kunstadter, E. C. Chapman, and S. Sabhasri, eds., *Farmers in the forest: Economic development and marginal agriculture in northern Thailand*, 74–133. Honolulu: University Press of Hawaii.

Kurimoto, E. 1996. People of the river: Subsistence economy of the Anywaa (Anuak) of western Ethiopia. In *Essays in northeast African studies*, 29–57. Senri Ethnological Studies No. 43. Osaka, Japan: National Museum of Ethnology.

Lahjie, A. 1996. Traditional land use and Kenyah Dayak farming systems in East Kalimantan. In C. Padoch and N. L. Peluso, eds., *Borneo in transition: People, forests, conservation and development*, 150–61. Kuala Lumpur: Oxford University Press.

Lal, R. 1993. Soil erosion and conservation in west Africa. In D. Pimentel, ed., *World soil erosion and conservation*, 7–25. Cambridge, U.K.: Cambridge University Press.

Lal, R. 1995. *Sustainable management of soil resources in the humid tropics*. Tokyo: United Nations University Press.

Lamers, J. P. A. and P. R. Feil. 1995. Farmers' knowledge and management of spatial soil and crop growth variability in Niger, west Africa. *Netherlands Journal of Agricultural Science* 43(4): 375–89.

Lappé, M. and B. Bailey. 1999. *Against the grain: The genetic transformation of global agriculture*. London: Earthscan.

Larcher, W. 1995 (3d ed.). *Physiological plant ecology*. Berlin: Springer-Verlag.

Lawrence, D. C., D. Astiani, M. Syazhaman-Karwur, and I. Fiorentino. 1997. *Does tree diversity affect soil fertility? A critical hypothesis and initial findings in the alternative fallow management systems of West Kalimantan*. Unpublished paper delivered at a regional conference on Indigenous Strategies for Intensification of Shifting Cultivation in Southeast Asia, held in Bogor, Indonesia, June 23–27, 1997.

Lawton, R. M. 1978. A study of the dynamic ecology of Zambian vegetation. *Journal of Ecology* 66: 175–98.

Lawton, R. M. 1982. Natural resources of miombo woodland and recent changes in agricultural and land-use practices. *Forest Ecology and Management* 4: 287–97.

Le Riverend, J. 1967. *Economic history of Cuba*. Havana: Book Institute.

Leach, M. and R. Mearns, eds. 1996. *The lie of the land: Challenging received wisdom on the African environment*. London: International African Institute.

Leaf, M. J. 1972. *Information and behaviour in a Sikh village: Social organisation reconsidered*. Berkeley: University of California Press.

Leaf, M. J. 1983. The Green Revolution and cultural change in a Punjab village, 1965–1978. *Economic Development and Cultural Change* 31: 227–70.

Leaf, M. J. 1984. *Song of hope: The Green Revolution in a Punjab village*. New Brunswick, N.J.: Rutgers University Press.

Leaf, M. J. 1987. Intensification in peasant farming: Punjab in the Green Revolution. In B. L. Turner II and S. B. Brush, eds., *Comparative farming systems*, 248–75. New York: Guilford.

Lenné, J. M. and D. Wood. 1999. Optimizing biodiversity for productive agriculture. In D. Wood and J. M. Lenné, eds., *Agrobiodiversity: Characterization, utilization and management*, 447–470. New York: CABI.

Lewis, M. W. 1992. *Wagering the land: Ritual, capital, and environmental degradation in the Cordillera of northern Luzon, 1900–1986*. Berkeley: University of California Press.

Lian, F. J. 1987. *Farmers' perceptions and economic change: The case of Kenyah farmers of the Fourth Division, Sarawak*. Ph.D. thesis, Australian National University, Canberra.

Life Sciences Knowledge Center. 1999. Home page. http://www.biotechknowledge.com/ [August 2, 1999].

Lim Teck Ghee. 1977. *Peasants and their agricultural economy in colonial Malaya 1874–1941*. Kuala Lumpur: Oxford University Press.

Lineberger, R. D. n.d. The many dimensions of plant tissue culture research. http://aggie-horticulture.tamu.edu/tisscult/pltissue/pltissue.html [November 23, 1999].

Lipton, M. 1968. The theory of the optimising peasant. *Journal of Development Studies* 4(3): 327–51.

Lipton, M. and R. Longhurst. 1989. *New seeds and poor people*. London: Unwin Hyman.

Lirb, H. J. 1993. Partners in agriculture: The pooling of resources in rural Sociates in Roman Italy. In H. Sancisi-Weerdenburg, R. J. van der Spek, H. C. Teitler, and H. T. Wallinga, eds., *De agricultura: In memoriam Pieter Willem de Neeve (1945–1990)*, 263–95. Amsterdam: Gieben.

Liu, D. and H. Sun, eds. 1981. *Proceedings of symposium on Qinghai-Xizang (Tibet) Plateau*. Beijing: Science Press.

Lopez, V. B. 1976. *The Mangyans of Mindoro: An ethnohistory*. Quezon City: University of the Philippines Press.

Lopez-Gonzaga, V. 1983. *Peasants in the hills: A study of the dynamics of social change among the Buhid swidden cultivators in the Philippines*. Quezon City: University of the Philippines Press.

Louette, D. 1999. Traditional management of seed and genetic diversity: What is a landrace? In S. B. Brush, ed., *Genes in the field: Conserving plant diversity on farms*, 109–142. Boca Raton, Fla.: Lewis.

Lowitt, R. 1985. Agricultural policy and soil conservation: Comment. *Agricultural History* 59: 320–25.

Lowrance, R., P. F. Hendrix, and E. P. Odum. 1986. A hierarchical approach to sustainable agriculture. *American Journal of Alternative Agriculture* 1(4): 169–73.

Lu, Y. and M. Stocking. 1998. *Sustainable land use: Developing and testing a methodology for soil erosion and conservation assessment.* Presented at International Workshop, PLEC, Mbarara, Uganda, 29 March–4 April, 1998.

M. S. Swaminathan Research Foundation. 1999. Home page. http://www.mssrf.org/ [July 1999].

Magdoff, F. 1992. *Building soils for better crops: Organic matter management.* Lincoln: University of Nebraska Press.

Mainguet, M. and R. Letolle. 1997. The ecological crisis of the Aral Sea basin in the frame of a new time scale: The anthropo-geological scale. *Naturwissenschaften* 84(8): 331–39.

Manning, C. and S. Jayasuriya. 1997. Survey of recent developments. *Bulletin of Indonesian Economic Studies* 32(2): 3–43.

Marshall, F. 1998. Early food production in Africa. *Review of Archaeology* 19(2): 47–58.

Matsudo, H. 1996. Riverbank cultivation in the Lower Omo Valley: The intensive farming system of the Kara, southwestern Ethiopia. In *Essays in northeast African studies,* 1–28. Senri Ethnological Studies No. 43. Osaka, Japan: National Museum of Ethnology.

A. Maulud bin Mohd Yusof. 1976. *Cultural adaptations in an urbanizing Malay community.* Ph.D. thesis, Rice University, Houston, Tex.

Mayer, E. 1985. Production zones. In S. Masuda, I. Shimada, and C. Morris, eds., *Andean ecology and civilization,* 45–84. Tokyo: University of Tokyo Press.

Maxted, N., B. Ford-Lloyd, and J. G. Hawkes, eds. 1997. *Plant genetic conservation: The in situ approach.* London: Chapman & Hall.

McBride, K. A. and R. E. Dewar. 1987. Agriculture and cultural evolution: Causes and effects in the Lower Connecticut River Valley. In W. F. Keegan, ed., *Emergent horticultural economies of the eastern woodlands,* 305–28. Carbondale: Center for Archaeological Investigations, Southern Illinois University.

McCann, J. C. 1995. *People of the plow: An agricultural history of Ethiopia, 1800–1990.* Madison: University of Wisconsin Press.

McGrath, D. G. 1987. The role of biomass in shifting cultivation. *Human Ecology* 15(2): 221–42.

McTainsh, G. H. 1980. Harmattan dust deposition in northern Nigeria. *Nature* 286: 587–88.

McTainsh, G. H. 1984. The nature and origin of the Aeolian mantles of central northern Nigeria. *Geoderma* 33: 13–37.

McTainsh, G. H. 1987. Desert loess in northern Nigeria. *Zeitschrift für Geomorphologie* 31(2): 145–65.

Mead, R. 1986. Statistical methods for multiple cropping. In C. A. Francis, ed., *Multiple cropping systems,* 317–50. New York: Macmillan.

Mead, R. and R. W. Willey. 1980. The concept of "land equivalent ratio" and advantages in yields from intercropping. *Experimental Agriculture* 16(3): 217–28.

Meadow, R. H. 1998. Pre- and proto-historic agricultural and pastoral transformations in northwest south Asia. *Review of Archaeology* 19(2): 12–21.

Meggitt, M. J. 1965. *The lineage system of the Mae-Enga of New Guinea.* Edinburgh: Oliver & Boyd.

Menzies, N. 1988. Three hundred years of taungya: A sustainable system of forestry in south China. *Human Ecology* 16(4): 361–76.

Michon, G. and F. Mary. 1990. Transforming traditional home gardens and related systems in west Java (Bogor) and west Sumatra (Maninjau). In K. Landauer and M. Brazil, eds., *Tropical home gardens,* 169–85. Tokyo: United Nations University Press.

Michon, G. and F. Mary. 1994. Conversion of traditional village gardens and new economic strategies of rural households in the area of Bogor, Indonesia. *Agroforestry Systems* 25: 31–58.

Milne, G. 1936. Normal erosion as a factor in soil profile development. *Nature* 138: 548–49.

Milne, G. 1947. A soil reconnaissance journey through parts of Tanganyika Territory, December 1935 to February 1936. *Journal of Ecology* 35: 192–265.

Mitton, R. 1983. *The lost world of Irian Jaya.* Melbourne: Oxford University Press.

Miyamoto, M. 1988. *The Hanunóo-Mangyan: Society, religion and law among a mountain people of Mindoro Island, Philippines.* Senri Ethnological Studies No. 22. Osaka, Japan: National Museum of Ethnology.

Miyawaki, Y. 1996. Cultivation strategy and historical change of sorghum varieties in the Hoor of southwestern Ethiopia. In *Essays in northeast African studies,* 77–120. Senri Ethnological Studies No. 43. Osaka, Japan: National Museum of Ethnology.

Modjeska, N. 1982. Production and inequality: Perspectives from central New Guinea. In A. Strathern, ed., *Inequality in New Guinea highlands societies,* 50–108. Cambridge, U.K.: Cambridge University Press.

Moon, O. 1989. *From paddy field to ski slope: The revitalisation of tradition in Japanese village life.* Manchester: Manchester University Press.

Mooney, P. R. 1983. The law of the seed: Another development and plant genetic resources. *Development Dialogue* Vols. 1–2.

Mooney, P. R. 1996. The parts of life: Agricultural biodiversity, indigenous knowledge, and the role of the Third System. *Development Dialogue* Special Issue.

Moore, H. L. and M. Vaughan. 1994. *Cutting down trees. Gender, nutrition, and cultural change in the northern province of Zambia 1890–1990.* Portsmouth, N.H.: Heinemann.

Moore, R. H. 1990. *Japanese agriculture: Patterns of rural development.* Boulder, Colo.: Westview.

Moran, E. F. 1995. Rich and poor ecosystems of Amazonia: An approach to management. In T. Nishizawa and J. I. Uitto, eds., *The fragile tropics of Latin America: Sustainable management of changing environments,* 45–67. Tokyo: United Nations University Press.

Moran, E. 1996. Nurturing the forest: Strategies of native Amazonians. In R. F. Ellen and K. Fukui, eds., *Redefining nature: Ecology, culture, and domestication,* 531–55. Washington, D.C.: Berg.

Morgan, W. T. W., ed. 1979. *The Jos plateau: A survey of environment and land use, being a selective abridgement of a report to the government of Nigeria from the Land Resources Development Centre of the U.K. Overseas Development Administration.* Occasional Publications (New Series) No. 14. Durham, U.K.: Department of Geography, University of Durham.

Mortimore, M. 1998. *Roots in the African dust: Sustaining the sub-Saharan drylands.* Cambridge, U.K.: Cambridge University Press.

Moss, R. P. 1969. The ecological background to land-use studies in tropical Africa, with special reference to the West. In M. F. Thomas and G. W. Whittington, eds., *Environment and land use in Africa,* 193–238. London: Methuen.

Mubyarto. 1993. Discussant comments [following J. J. Fox, Ecological policies for sustaining high production in rice: Observations on rice intensification in Indonesia]. In H. Brookfield and Y. Byron, eds., *South-east Asia's environmental future: The search for sustainability,* 223. Tokyo: United Nations University Press and Kuala Lumpur: Oxford University Press.

Nagatsuka, T. 1989. *The soil: A portrait of rural life in Meiji Japan.* London: Routledge.

Nair, P. K. R. 1989. *Agroforestry systems in the tropics.* Dordrecht, The Netherlands: Kluwer Academic.

National Research Council. 1993. *Sustainable agriculture and the environment in the humid tropics.* Washington, D.C.: National Academy Press.

National Research Council. 1996. *Lost crops of Africa,* Vol. I, *Grains.* Washington, D.C.: National Academy Press.

Netting, R. McC. 1965. Household organization and intensive agriculture: The Kofyar case. *Africa* 35: 422–29.

Netting, R. McC. 1968. *Hill farmers of Nigeria: Cultural ecology of the Kofyar of the Jos plateau.* Seattle: University of Washington Press.

Netting, R. McC. 1993. *Smallholders, householders: Farm families and the ecology of intensive, sustainable agriculture.* Stanford, Calif.: Stanford University Press.

Netting, R. McC., G. D. Stone, and M. P. Stone. 1993. Agricultural expansion, intensification, and market participation among the Kofyar, Jos Plateau, Nigeria. In B. L. Turner II, G. Hyden, and R. W. Kates, eds., *Population growth and agricultural change in Africa,* 206–49. Gainesville: University Press of Florida.

Netting, R. McC. and M. P. Stone. 1996. Agro-diversity on a farming frontier: Kofyar smallholders on the Benue plains of central Nigeria. *Africa* 66(1): 52–70.

Ngo, T. H. G. M. 1996. A new perspective on property rights: Examples from the Kenyah of Kalimantan. In C. Padoch and N. L. Peluso, eds., *Borneo in transition: People, forests, conservation and development,* 137–49. Kuala Lumpur: Oxford University Press.

Nibbering, J. W. S. M. 1991a. Crisis and resilience in upland land use in Java. In J. Hardjono, ed., *Indonesia: Resources, ecology, and environment,* 104–32. Singapore: Oxford University Press.

Nibbering, J. W. S. M. 1991b. *Hoeing in the hills: Stress and resilience in an upland farming system in Java.* Ph.D. thesis, Australian National University, Canberra.

Nibbering, J. W. 1997. Upland cultivation and soil conservation in limestone regions on Java's south coast. In P. Boomgaard, F. Colombijn, and D. Henley, eds., *Paper landscapes: Explorations in the environmental history of Indonesia,* 153–83. Leiden, The Netherlands: KITLV Press.

Nibbering, J. W. and A. Schrevel. 1982. *The role of additional activities in rural Java: A case study of two villages in Malang regency.* Utrecht, The Netherlands: University of Utrecht.

Nicolaisen, J. and I. Nicolaisen. 1997. *The pastoral Tuareg: Ecology, culture and society.* London: Thames & Hudson.

Niemeijer, D. 1996. The dynamics of African agricultural history: Is it time for a new development paradigm? *Development and Change* 27(1): 87–110.

Niggli, U. 1998. Organic agriculture. JCVP Schweiz. http://www.youngvision.ch/orgagricult.htm [June 30, 1999].

Niqueux, M. 1959. Choix de variétés d'arachides au Tchad. III. Essais de culture associée d'arachides et de sorghos au Tchad. *L'Agronomie Tropicale* 14: 501–2.

Nishida, M. 1983. The emergence of food production in Neolithic Japan. *Journal of Anthropological Archaeology* 2: 305–22.

Norman, D. W. 1974. Rationalizing mixed cropping under indigenous conditions: The example of northern Nigeria. *Journal of Development Studies* 11: 3–21.

Norman, D. and M. Collinson. 1985. *Farming systems research in theory and practice.* ACIAR Proceedings. Canberra: Australian Centre for International Agricultural Research.

Nye, P. H. and D. J. Greenland. 1960. *The soil under shifting cultivation.* Technical Communication No. 51, Commonwealth Bureau of Soils, Harpenden. Farnham Royal, Bucks, U.K.: Commonwealth Agricultural Bureaux.

Odum, E. P. 1954. *Fundamentals of ecology.* Philadelphia: Saunders.

O'Hara, S. L., F. A. Street-Perrot, and T. P. Bert. 1993. Accelerated soil erosion around a Mexican highland lake caused by pre-Hispanic agriculture. *Nature* 362: 48–51.

Ohnuki-Tierney, E. 1993. *Rice as self: Japanese identities through time.* Princeton, N.J.: Princeton University Press.

Okigbo, B. N. and D. J. Greenland. 1976. Intercropping systems in tropical Africa. In R. I. Papendinck, P. A. Sanchez, and G. B. Triplett, eds., *Multiple cropping,* 63–101. Madison, Wis.: American Society of Agronomy, Crop Science Society of America, Soil Science Society of America.

Okon, Y., G. V. Bloemberg, and B. J. J. Lugtenberg. 1998. Biotechnology of biofertilization and phytostimulation. In A. Altman, ed., *Agricultural biotechnology,* 327–49. New York: Marcel Dekker.

Oldeman, L. R., R. T. A. Hakkeling, and W. G. Sombroek. 1990 (2d rev. ed.). *World map of the status of human-induced soil degradation: An explanatory note.* Wageningen, The Netherlands: International Soil Reference and Information Centre.

Orejuela, J. E. 1992. Traditional productive systems of the Awa (Cuaiquer) Indians of southwestern Columbia and neighbouring Ecuador. In K. Redford and C. Padoch, eds., *Conservation of neotropical forests: Working from traditional resource use,* 58–82. New York: Columbia University Press.

Östberg, W. 1995. *Land is coming up: The Burunge of central Tanzania and their environments.* Stockholm: Department of Social Anthropology, Stockholm University.

Overton, M. 1985. The diffusion of agricultural innovations in early modern England: Turnips and clover in Norfolk and Suffolk, 1580–1740. *Transactions of the Institute of British Geographers* NS 10: 205–21.

Overton, M. 1996. *Agricultural revolution in England: The transformation of the agrarian economy, 1500–1850.* Cambridge, U.K.: Cambridge University Press.

Owusu-Bennoah, E. 1997. Soils. In E. A. Gyasi and J. I. Uitto, eds., *Environment, biodiversity, and agricultural change in west Africa,* 58–63. Tokyo: United Nations University Press.

Padoch, C. 1982. *Migration and its alternatives among the Iban of Sarawak.* The Hague: Martinus Nijhoff.

Padoch, C. 1985. Labour efficiency and intensity of land use in rice production: An example from Kalimantan. *Human Ecology* 13(3): 271–89.

Padoch, C. 1986. Agricultural site selection among permanent field farmers: An example from East Kalimantan, Indonesia. *Journal of Ethnobiology* 6(2): 279–88.

Padoch, C. 1988a. Agriculture in interior Borneo. *Expedition* 30(1): 18–28.

Padoch, C. 1988b. People of the floodplain and forest. In J. S. Denslow and C. Padoch, eds., *People of the tropical rain forest,* 127–41. Berkeley: University of California Press in association with Smithsonian Institution Traveling Exhibition Service.

Padoch, C., J. Chota Inuma, W. de Jong, and J. Unruh. 1985. Amazonian agroforestry: A market oriented system in Peru. *Agroforestry Systems* 3: 47–58.

Padoch, C. and W. de Jong. 1987. Traditional agroforestry practices of native and Ribereño farmers in the lowland Peruvian Amazon. In H. L. Goltz, ed., *Agroforestry: Realities, possibilities and potentials,* 179–94. Dordrecht, The Netherlands: Martinus Nijhoff.

Padoch, C. and W. de Jong. 1989. Production and profit in agroforestry: An example from the Peruvian Amazon. In J. O. Browder, ed., *Fragile lands of Latin America: Strategies for sustainable development,* 102–14. Boulder, Colo.: Westview.

Padoch, C. and W. de Jong. 1990. Santa Rosa: The impact of the forest products trade on an Amazonian place and population. In G. T. Prance and M. J. Balick, eds., *New directions in the study of plants and people: Research contributions from the institute of economic botany,* 151–58. Advances in Economic Botany Vol. 8. Bronx: New York Botanical Garden.

Padoch, C. and W. de Jong. 1991. The house gardens of Santa Rosa: Diversity and variability in an Amazonian agricultural system. *Economic Botany* 45(2): 166–75.

Padoch, C. and W. de Jong. 1992. Diversity, variation and change in Ribereño agriculture. In K. Redford and C. Padoch, eds., *Conservation of neotropical forests: Working from traditional resource use,* 158–74. New York: Columbia University Press.

Padoch, C. and W. de Jong. 1995. Subsistence- and market-oriented agroforestry in the Peruvian Amazon. In T. Nishizawa and J. I. Uitto, eds., *The fragile tropics of Latin America: Sustainable management of changing environments,* 226–37. Tokyo: United Nations University Press.

Padoch, C., E. Harwell, and A. Susanto. 1998. Swidden, sawah and in-between: Agricultural transformation in Borneo. *Human Ecology* 26(1): 3–20.

Padoch, C. and C. Peters. 1993. Managed forest gardens in West Kalimantan, Indonesia. In C. S. Potter, J. I. Cohen, and D. Janczewski, eds., *Perspectives on biodiversity: Case studies of genetic resource conservation and development,* 167–76. Washington, D.C.: American Association for the Advancement of Science Press.

Padoch, C. and M. Pinedo-Vasquez. 1999. Farming above the flood in the várzea of Amapá: Some preliminary results of the Projeto Várzea. In C. Padoch, J. M. Ayres, M. Pinedo-Vasquez, and A. Henderson, eds., *Várzea: Diversity, development, and conservation of Amazonia's whitewater floodplains,* 345–54. Advances in Economic Botany 13. Bronx: New York Botanical Garden.

Pagiola, S. 1999. *The global environmental benefits of land degradation control on agricultural land.* World Bank Environment Paper No 16. Washington, D.C.: World Bank.

Pagiola, S., J. Kellenberg, L. Vidaeus, and J. Srivastava. 1997. *Mainstreaming biodiversity in agricultural development.* World Bank Environment Paper No. 15. Washington, D.C.: World Bank.

Pahlman, C. 1992. Soil erosion? That's not how we see the problem. In W. Hiemstra, C. Reijntjes, and E. van der Werf, eds., *Let farmers judge: Experiences in assessing sustainability of agriculture,* 43–61. London: Intermediate Technology Publications.

Palte, J. G. L. 1989. *Upland farming on Java, Indonesia: A socio-economic study of upland agriculture and subsistence under population pressure.* Nederlandse Geografische Studies 97. Amsterdam: Geografisch Instituut Rijksuniversiteit Utrecht.

Parker, E. 1992. Forest islands and Kayapó resource management in Amazonia: A reappraisal of the *Apêtê. American Anthropologist* 94(2): 406–28.

Parsons, J. J. 1985. Raised fields farmers as pre-Columbian landscape engineers: Looking north from the San Jorge (Columbia). In I. S. Farrington, ed., *Prehistoric intensive agriculture in the tropics,* Vol. I, 149–65. BAR International Series 232. Oxford, U.K.: British Archaeological Reports.

Paton, T. R. 1978. *The formation of soil material.* London: Allen & Unwin.

Paton, T. R., G. S. Humphreys, and P. B. Mitchell. 1995. *Soils: A new global view.* London: University College London Press.

Paul, E., M. Goto, and S. McCouch. 1999. RiceGenes: An information system for rice research. http://www.nal.usda.gov/pgdic/Probe/v4n3_4/ricegenes.html [November 30, 1999].

Peacock, E. 1998. Historical and applied perspectives on prehistoric land use in eastern North America. *Environment and History* 4(1): 1–29.

Pearsall, D. M. 1994. Investigating New World tropical agriculture: Contributions from phytolith analysis. In J. G. Hather, ed., *Tropical archaeobotany: Applications and new developments,* 115–38. London: Routledge.

Pearse, A. C. 1980. *Seeds of plenty, seeds of want: A critical analysis of the Green Revolution.* Oxford, U.K.: Clarendon.

Peluso, N. L. 1996. Fruit trees and family trees in an anthropogenic forest: Ethics of access, property zones, and environmental change in Indonesia. *Comparative Studies in Society and History* 38(3): 510–48.

Peluso, N. L. and C. Padoch. 1996. Changing resource rights in managed forests of West Kalimantan. In C. Padoch and N. L. Peluso, eds., *Borneo in transition: People, forests, conservation and development*, 121–36. Kuala Lumpur: Oxford University Press.

Pereira, W. 1993. *Tending the earth: Traditional, sustainable agriculture in India*. Bombay: Earthcare.

Peters, C. M. 1996. Illipe nuts (*Shorea* spp.) in West Kalimantan: Use, ecology and management potential of an important forest resource. In C. Padoch and N. L. Peluso, eds., *Borneo in transition: People, forests, conservation and development*, 230–44. Kuala Lumpur: Oxford University Press.

Peters, C. M. 1997. Observations on the sustainable exploitation of non-timber tropical forest products: An ecologist's perspective. In M. Ruiz Rérez and J. E. M. Arnold, eds., *Current issues on non-timber forest products research: Proceedings of the workshop "Research on NTFP,"* Hot Springs, Zimbabwe, 28 August–2 September 1995, 19–39. Bogor, Indonesia: CIFOR; ODA.

Peters, D. U. 1950. *Land usage in the Serenje district: A survey of land usage and the agricultural system of the Lala of the Serenje Plateau*. Rhodes–Livingstone Papers No. 19. Oxford, U.K.: Oxford University Press.

Pfeiffer, J. E. 1976. A note on the problem of basic causes. In J. R. Harlan, J. M. de Wet, and A. B. L. Stemler, eds., *Origins of African plant domestication*, 23–28. The Hague: Mouton.

Phillips-Howard, K. D. 1994. Agricultural intensification and the threat to soil fertility in Africa: Evidence from the Jos plateau, Nigeria. *Geographical Journal* 160(3): 252–65.

Pierik, R. L. M. 1987. *In vitro culture of higher plants*. Dordrecht, The Netherlands: Martinus Nijhoff.

Pimentel, D. 1993. *World soil erosion and conservation*. Cambridge, U.K.: Cambridge University Press.

Pinedo-Vasquez, M. 1999. Changes in soil formation and vegetation on silt bars and backslopes of levees following intensive production of rice and jute. In C. Padoch, J. M. Ayres, M. Pinedo-Vasquez, and A. Henderson, eds., *Várzea: Diversity, development, and conservation of Amazonia's whitewater floodplains*, 301–11. Advances in Economic Botany 13. Bronx: New York Botanical Garden.

Pinedo-Vasquez, M. and F. Rabelo. 1999. Sustainable management of an Amazonian forest for timber production: A myth or reality? *PLEC News and Views* 12(April): 20–28.

Pinedo-Vasquez, M., D. Zarin, and P. Jipp. 1995. Local management of forest resources in a rural community in north-east Peru. In T. Nishizawa and J. I. Uitto, eds., *The fragile tropics of Latin America: Sustainable management of changing environments*, 238–49. Tokyo: United Nations University Press.

Pingali, P. L., M. Hossain, and R. V. Gerpacio. 1997. *Asian rice bowls: The returning crisis?* New York: CAB International and IRRI.

Piperno, D. R. and D. R. Pearsall. 1993. Phytoliths in the reproductive structures of maize and teosinte: Implications for the study of maize evolution. *Journal of Archaeological Science* 20: 337–62.

Pleket, H. W. 1993. Agriculture in the Roman Empire in comparative perspective. In H. Sancisi-Weerdenburg, R. J. van der Spek, H. C. Teitler, and H. T. Wallinga, eds., *De agricultura: In memoriam Pieter Willem de Neeve (1945–1990)*, 317–42. Amsterdam: Gieben.

Plucknett, D. L. and N. J. H. Smith. 1986. Historical perspectives on multiple cropping. In C. A. Francis, ed., *Multiple cropping systems,* 20–39. New York: Macmillan.

Podolinsky, A. 1985. *Biodynamic agriculture introductory lectures,* Vol. 1. Sydney: Gavemer.

Polanyi, K. 1944. *The great transformation.* Boston: Beacon.

Porter, P. W. 1970. The concept of environmental potential as exemplified by tropical African research. In W. Zelinsky, L. A. Kosinski, and R. M. Prothero, eds., *Geography and a crowding world,* 187–217. New York: Oxford University Press.

Posey, D. A. 1983. Indigenous ecological knowledge and development of the Amazon. In E. F. Moran, ed., *The dilemma of Amazonian development,* 225–57. Boulder, Colo.: Westview.

Posey, D. A. 1984. A preliminary report on diversified management of tropical forest by the Kayapó Indians of the Brazilian Amazon. In G. T. Prance and J. A. Kallunki, eds., *Ethnobotany in the neotropics,* 112–26. Advances in Economic Botany Vol. 1. Bronx: New York Botanical Garden.

Posey, D. A. 1985a. Indigenous management of tropical forest ecosystems: The case of the Kayapó Indians of the Brazilian Amazon. *Agroforestry Systems* 3(2): 139–58.

Posey, D. A. 1985b. Native and indigenous guidelines for new Amazonian development strategies: Understanding biological diversity through ethnoecology. In J. Hemming, ed., *Change in the Amazon Basin: Symposium held at the 44th International Congress of Americanists, Manchester, 6–10 September 1982.* Vol. 1, 156–81. Manchester, U.K.: Manchester University Press.

Posey, D. 1991. Kayapo Indians: Experts in synergy. *ILEIA Newsletter* 4: 3–5.

Posey, D. A. 1992. Interpreting and applying the "reality" of indigenous concepts: What is necessary to learn from the natives? In K. Redford and C. Padoch, eds., *Conservation of neotropical forests: Working from traditional resource use,* 21–34. New York: Columbia University Press.

Posey, D. A. 1993. The importance of semi-domesticated species in post-contact Amazonia: Effects of the Kayapó Indians on the dispersal of flora and fauna. In C. M. Hladick, A. Hladick, O. F. Linares, H. Pagezy, A. Semple, and M. Hadley, eds., *Tropical forests, people and food: Biocultural interactions and applications to development,* 63–71. Man and the Biosphere Series Vol. 13. Paris: UNESCO and Carnforth, UK: Parthenon.

Postma, A. 1974. Development among the Mangyans of Mindoro: A privileged experience. *Philippine Quarterly of Culture and Society* 2: 21–37.

Potrykus, I., R. Bilang, J. Fütterer, C. Sautter, M. Schrott, and G. Spangenberg. 1998. Genetic engineering of crop plants. In A. Altman, ed., *Agricultural biotechnology,* 119–59. New York: Marcel Dekker.

Potter, L. and J. Lee. 1998. *Tree planting in Indonesia: Trends, impacts and directions.* Unpublished final report for CIFOR and AUSAID. Canberra: Australian Agency for International Development

Prausnitz, M. W. 1970. *From hunter to farmer and trader: Studies in the lithic industries of Israel and adjacent countries (from the Mesolithic to the Chalcolithic Age).* Jerusalem: Sivan Press.

Prescott-Allen, R. and C. Prescott-Allen. 1990. How many plants feed the world? *Conservation Biology* 4(4): 365–74.

Pretty, J. N. 1995. *Regenerating agriculture: Policies and practice for sustainability and self-reliance.* London: Earthscan.

Pretty, J. N. and P. Shah. 1997. Making soil and water conservation sustainable: From coercion and control to partnerships and participation. *Land Degradation and Development* 8(1): 39–58.

Pringle, R. 1970. *Rajahs and rebels: The Ibans of Sarawak under Brooke rule, 1841–1941.* London: Macmillan.

Qualset, C. O., A. B. Damania, A. C. A. Zanatta, and S. B. Brush. 1997. Locally based crop plant conservation. In N. Maxted, B. Ford-Lloyd, and J. G. Hawkes, eds., *Plant genetic conservation: The in situ approach,* 160–75. London: Chapman & Hall.

Querol, D. 1993. *Genetic resources: A practical guide to their conservation.* London: Zed and Penang, Malaysia: Third World Network.

Rackham, O. 1986. *The history of the countryside.* London: J.M. Dent.

Rackham, O. and J. A. Moody. 1992. Terraces. In B. Wells, ed., *Agriculture in ancient Greece. Proceedings of the Seventh International Symposium at the Swedish Institute at Athens, 16–17 May 1990,* 123–30. Stockholm: Svenska Institutet i Athen.

Ramakrishnan, P. S. 1992. *Shifting agriculture and sustainable development: An interdisciplinary study from north-eastern India.* Man and Biosphere Series Vol. 10. Paris: UNESCO and Carnforth, UK: Parthenon.

Ramasamy, C., P. B. R. Hazell, and P. K. Aiyasamy. 1991. North Arcot and the Green Revolution. In P. B. R. Hazell and C. Ramasamy, eds., *The Green Revolution reconsidered: The impact of high-yielding rice varieties in south India,* 11–28. Baltimore: Johns Hopkins University Press.

Rao, M. R. 1984. Prospects for sorghum and pearl millet in the cropping systems of northeast Brazil. Proceedings of Farming Systems Workshop, CIMMYT, ICRISAT, and INTSORMIL, Mexico, September 19, 1984. Mexico City: CIMMYT.

Reijntjes, C., B. Haverkort, and A. Waters-Bayer. 1992. *Farming for the future.* London: Macmillan and Leusden, The Netherlands: ILEIA.

Reining, C. C. 1966. *The Zande scheme: An anthropological case study of economic development in Africa.* Northwestern University African Studies No. 17. Evanston, Ill.: Northwestern University Press.

Remmers, G. G. A. and H. de Koeijer. 1992. The T'OLCHE', a Maya system of communally managed forest belts: The causes and consequences of its disappearance. *Agroforestry Systems* 18(2): 149–77.

Renfrew, C. 1972. *The emergence of civilisation: The Cyclades and the Aegean in the third millennium B.C.* London: Methuen.

Richards, A. I. 1939. *Land, labour and diet in northern Rhodesia. An economic study of the Bemba tribe.* London: Oxford University Press for International African Institute.

Richards, A. I. 1951. The Bemba of north eastern Rhodesia. In E. Colson and M. Gluckman, eds., *Seven tribes of British central Africa,* 164–93. London: Oxford University Press.

Richards, A. I. 1958. A changing pattern of agriculture in east Africa: The Bemba of northern Rhodesia. *Geographical Journal* 124: 302–31.

Richards, P. 1985. *Indigenous agricultural revolution: Ecology and food production in west Africa.* London: Hutchinson.

Richards, P. 1989. Agriculture as a performance. Is R and D directed at the wrong target? In R. Chambers, A. Pacey, and L. A. Thrupp, eds., *Farmer first: Farmer innovation and agricultural research,* 39–42. London: Intermediate Technology Publications.

Richter, D. D. and L. I. Babbar. 1991. Soil diversity in the tropics. *Advances in Ecological Research* 21: 315–89.

Rigg, J. 1989. The new rice technology and agrarian change: Guilt by association? *Progress in Human Geography* 13(3): 374–99.

Riley, T. J. 1987. Ridged-field agriculture and the Mississippian economic pattern. In W. F. Keegan, ed., *Emergent horticultural economies of the eastern woodlands,* 295–304. Carbondale: Center for Archaeological Investigations, Southern Illinois University.

Rissler, J. and M. Mellon. 1996. *The ecological risks of engineered crops.* Cambridge, Mass.: MIT Press.

Roberts, A. D. 1973. *A history of the Bemba: Political growth and change in north east Zambia before 1900.* London: Longman.

Roberts, A. 1976. *A history of Zambia.* London: Heinemann.

Robinson, W. and W. Schutjer. 1984. Agricultural development and demographic change: A generalization of the Boserup model. *Economic Development and Cultural Change* 32: 355–66.

Rocheleau, D. E., P. E. Steinberg, and P. A. Benjamin. 1995. Environment, development, crisis and crusade: Ukambani, Kenya, 1890–1990. *World Development* 23(6): 1037–51.

Rocheleau, D. E, K. Wachira, L. Malaret, and B. M. Wanjohi. 1989. Local knowledge for agroforestry and native plants. In R. Chambers, A. Pacey, and L. A. Thrupp, eds., *Farmer first: Farmer innovation and agricultural research,* 14–23. London: Intermediate Technology Publications.

Roder, W., O. Calvert, and Y. Dorji. 1992. Shifting cultivation systems practised in Bhutan. *Agroforestry Systems* 19(2): 149.

Roe, E. M. 1995. Except Africa: Postscript to a special section on development narratives. *World Development* 23(6): 1065–69.

Roosevelt, A. C. 1991. *Moundbuilders of the Amazon: Geophysical archaeology on Marajó, Brasil.* San Diego: Academic Press.

Roosevelt, A. C. 1999. Twelve thousand years of human–environment interaction in the Amazon floodplain. In C. Padoch, J. M. Ayres, M. Pinedo-Vasquez, and A. Henderson, eds., *Várzea: Diversity, development, and conservation of Amazonia's whitewater floodplains,* 371–92. Advances in Economic Botany 13. Bronx: New York Botanical Garden.

Rosset, P. M. and M. A. Altieri. 1997. Agroecology versus input substitution: A fundamental contradiction of sustainable agriculture. *Society and Natural Resources* 10(3): 283–95.

Rosset, P. M. and M. Benjamin. 1994. *The greening of the revolution: Cuba's national experiment with organic agriculture.* Melbourne: Ocean Press.

Rousseau, J. 1977. Kayan agriculture. *Sarawak Museum Journal* 25: 129–56.

Rural Advancement Foundation International (RAFI). 1998. Dead sea scroll? The USDA Terminator defence. http://www.rafi.org/translator/termtrans.html [June 5, 1999].

Rural Advancement Foundation International (RAFI). 1999a. Home page. http://www.rafi.org/.

Rural Advancement Foundation International (RAFI). 1999b. News releases: Biodiversity convention's terminator decision fails biodiversity and fails farmers. http://151.196.196.124/web/all news-one.shtml?dfl=allnews.db&tfl=allnews-one-frag.ptml&operation=display&ro1=recNo& rf1=31&rt1=31&usebrs=true [June 28, 1999].

Rural Advancement Foundation International (RAFI). 1999c. News releases: Terminator terminated? http://www.rafi.org/web/allnews-one.shtml?dfl=allnews.db&tfl=allnews-one-frag. ptml&operation=display&ro1=recNo&rf1=74&rt1=74&usebrs=true [October 4, 1999].

Rural Advancement Foundation International (RAFI). 1999d. News releases: Pharma-gedon. http://www.rafi.org/web/allnews-one.shtml?dfl=allnews.db&tfl=allnews-one-frag.ptml&opera tion=display&ro1=recNo&rf1=83&rt1=83&usebrs=true [December 22, 1999].

Rural Advancement Foundation International (RAFI). 2000. News releases: Terminator 2 years later. Suicide seeds on the fast track. http://www.rafi.org/web/allnews-one.shtml?dfl=allnews. db&tfl=allnews-one-frag.ptml&operation=display&ro1=recNo&rf1=97&rt1=97&usebrs= true [February 25, 2000].

Rural Industries Research and Development Corporation (RIRDC). 1999. Resilient agricultural systems. http://www.rirdc.gov.au/programs/ras.html [August 3, 1999].

Ruthenberg, H. 1980. *Farming systems in the tropics*. Cambridge, U.K.: Cambridge University Press.

Sachs, P. n.d. Humus: Still a mystery. Northeast Organic Farming Association, Rhode Island [article from the *Natural Farmer Magazine*]. http://users.ids.net/~nofari/tnf_hums.htm [July 6, 1999].

Saldarriaga, J. G. 1986. Recovery following shifting cultivation. Case study no. 2: A century of succession in the Upper Rio Negro. In C. F. Jordan, ed., *Amazonian rain forests: Ecosystem disturbance and recovery. Case studies of ecosystem dynamics under a spectrum of land-use intensities*, 24–33. New York: Springer Verlag.

Salisbury, F. B. and C. W. Ross. 1992 (4th ed.). *Plant physiology*. Belmont, Calif.: Wadsworth.

Salleh, A. 1998. Wearing out our genes? The case of transgenic cotton. In R. A. Hindmarsh, G. A. Lawrence, and J. Norton, eds., *Altered genes: Reconstructing nature*, 160–72. St. Leonards, Australia: Allen & Unwin.

Abdul Samad Hadi. 1981. *Population mobility in Negri Sembilan, peninsular Malaysia: Evidence from three villages in Kuala Pilah*. Ph.D. thesis, School of Social Sciences, Flinders University, Adelaide, Australia.

Sanchez, P. A. 1976. *Properties and management of soils in the tropics*. New York: Wiley Interscience.

Sanchez, P. A. 1995. Science in agroforestry. *Agroforestry Systems* 30: 5–55.

Sandor, J. A. and N. S. Eash. 1995. Ancient agricultural soils in the Andes of southern Peru. *Soil Science Society of America Journal* 59(1): 170–79.

Sandor, J. A., P. L. Gersper, and J. W. Hawley. 1990. Prehistoric agricultural terraces and soils in the Mimbres area, New Mexico. *World Archaeology* 22(1): 71–86.

Sasson, A. 1998. Biotechnology and food production: Relevance to developing countries. In A. Altman, ed., *Agricultural biotechnology*, 691–729. New York: Marcel Dekker.

Sather, C. 1990. Trees and tenure in Paku Iban society: The management of secondary forest resources in a long-established Iban community. *Borneo Review* 1: 16–40.

Sauer, C. O. 1952. *Agricultural origins and dispersals*. New York: American Geographical Society.

Sauer, J. D. 1993. *Historical geography of crop plants: A select roster*. Boca Raton, Fla.: CRC Press.

Sawit, M. H. and I. Manwan. 1991. The New Supra Insus rice intensification program: The case of the north coast of west Java and south Sulawesi. *Bulletin of Indonesian Economic Studies* 27(1): 81–103.

Schiefflin, E. L. 1975. Felling the trees on top of the crop: European contact and the subsistence ecology of the Great Papuan Plateau. *Oceania* 46(1): 25–39.

Schultz, J. 1976. *Land use in Zambia*. Munich: Weltforum Verlag.

Schultz, T. W. 1964. *Transforming traditional agriculture*. New Haven, Conn.: Yale University Press.

Schumpeter, J. A. 1934. *The theory of economic development*. Cambridge, Mass.: Harvard University Press.

Scientific and Technical Advisory Panel (STAP) of the Global Environment Facility (GEF). 1999. *Report of the STAP Expert Group Workshop on Land Degradation Linkages, Bologna, Italy, 14–16 June 1999*. Nairobi: STAP Secretariat, United Nations Environment Programme.

Scoones, I. 1992. Land degradation and livestock production in Zimbabwe's communal areas. *Land Degradation and Rehabilitation* 3: 99–113.

Scoones, I. 1994. *Living with uncertainty: New directions in pastoral development in Africa*. London: Intermediate Technology Publications.

Scoones, I., M. Melnyk, and J. N. Pretty. 1992. *The hidden harvest: Wild foods and agricultural systems*. London: IIED; SIDA; WWF.

Scoones, I. and C. Toulmin. 1998. Soil nutrient balances: What use for policy? *Agriculture, Ecosystems and Environment* 71: 255–268.

Scoones, I. and C. Toulmin. 1999. *Policies for soil fertility management in Africa.* London: Department for International Development.

Scott, G. A. J. 1986. Shifting cultivation where land is limited, Case study no. 3: Campa Indian agriculture in the Gran Pajonal of Peru. In C. F. Jordan, ed., *Amazonian rain forests: Ecosystem disturbance and recovery. Case studies of ecosystem dynamics under a spectrum of land-use intensities,* 34–45. New York: Springer-Verlag.

Scott, J. C. 1985. *Weapons of the weak: Everyday forms of peasant resistance.* New Haven, Conn.: Yale University Press.

Shaner, W. W., P. F. Philpp, and W. R. Schmel. 1981. *Farming systems research and development: Guidelines for developing countries.* Boulder, Colo.: Westview.

Sharma, R. 1997. *Alnus–cardamon agroforestry system: Potential for stabilizing upland shifting cultivation in the eastern Himalaya.* Unpublished paper delivered at a regional conference on Indigenous Strategies for Intensification of Shifting Cultivation in Southeast Asia, held in Bogor, Indonesia, June 23–27, 1997.

Sharma, R. and T. T. Poleman. 1993. *The new economics of India's Green Revolution: Income and employment diffusion in Uttar Pradesh.* Ithaca, N.Y.: Cornell University Press.

Sherratt, A. 1980. Water, soil and seasonality in early cereal production. *World Archaeology* 11: 313–30.

Sherratt, A. 1981. Plough and pastoralism: Aspects of the secondary products revolution. In I. Hodder, G. Isaac, and N. Hammond, eds., *Patterns of the past: Studies in honour of David Clarke,* 261–305. Cambridge, U.K.: Cambridge University Press.

Shiva, V. 1993. *Monocultures of the mind: Perspectives on biodiversity and biotechnology.* London: Zed.

Silberfein, M. 1989. *Rural changes in Machakos, Kenya: A historical geography perspective.* Lantham, Md.: University Press of America.

Simon, J. 1997. An organic coup in Cuba? *e-Amicus.* http://www.igc.apc.org/nrdc/nrdc/nrdc/eamicus/clip01/jscuba.html [June 16, 1999].

Singer, A. 1972. Ethnography and ecosystem. In A. Singer and B. V. Street, eds., *Zande themes: Essays presented to Sir Edward Evans-Pritchard,* 1–18. Oxford, U.K.: Blackwell Scientific.

Slater, G. 1918. *Some south Indian villages.* London: Oxford University Press.

Smaling, E. M. A., L. O. Fresco, and A. de Jager. 1996. Classifying, monitoring and improving soil nutrient stocks and flows in African agriculture. *Ambio* 25(8): 492–496.

Smith, B. D. 1987. The independent domestication of indigenous seed-bearing plants in eastern North America. In W. F. Keegan, ed., *Emergent horticultural economies of the eastern woodlands,* 3–47. Carbondale: Center for Archaeological Investigations, Southern Illinois University.

Smith, B. D. 1992. *Rivers of change: Essays on early agriculture in eastern North America.* Washington, D.C.: Smithsonian Institution Press.

Smith, C. T. 1967. *An historical geography of western Europe before 1800.* London: Longmans.

Smith, H. F. 1997. *Semi-commercial complex agroforestry: A case study from Kalimantan, Indonesia.* Ph.D. thesis, University of Adelaide, Australia.

Smith, H. F., P. J. O'Connor, S. E. Smith, and F. A. Smith. 1997. Vesicular–arbuscular mycorrhizas of durian and other species in forest gardens in West Kalimantan. *Forest soils and their management in south east Asia. Proceedings of the Third Conference of Tropical Soils: Soils of Tropical Forest Ecosystems. Balikpapan, Indonesia, 29 October–3 November 1995.*

Smith, N. J. H. 1980. Anthrosols and human carrying capacity in Amazonia. *Annals of the Association of American Geographers* 70: 553–66.

Smith, T. C. 1959. *The agrarian origins of modern Japan*. Stanford, Calif.: Stanford University Press.

Smith, T. C. 1988. *Native sources of Japanese industrialization, 1750–1920*. Berkeley: University of California Press.

Smith, T. C., R. Y. Eng, and R. T. Lundy. 1977. *Nakahara: Family farming and population in a Japanese village, 1717–1830*. Stanford, Calif.: Stanford University Press.

Socolow, R. H. 1999. Nitrogen management and the future of food: Lessons from the management of energy and carbon. *Proceedings of the National Academy of Sciences of the United States of America* 96(11): 6001–8.

Soemarwoto, O. and I. Soemarwoto. 1984. The Javanese rural ecosystem. In A. T. Rambo and P. E. Sajise, eds., *An introduction to human ecology research systems on agricultural systems in southeast Asia,* 254–87. Los Baños: University of Philippines.

Soil Association. 1999a. Briefing paper: The principal aims of organic agriculture and processing. http://www.soilassociation.org/SA/SAWebDoc.nsf/bdedfca988b2db3c85256207004f45a9/8e76ae2a4ffd1ad380256682003bd6a9!OpenDocument [June 29, 1999].

Soil Association. 1999b. History. http://www.soilassociation.org/SA/SAWeb.nsf/(SLDocsOpen)/History!OpenDocument [June 29, 1999].

Spate, O. H. K. and T. A. Learmonth. 1967. *India and Pakistan*. London: Methuen.

Spencer, J. E. 1966. *Shifting cultivation in southeastern Asia*. Berkeley: University of California Press.

Spencer, J. E. and G. A. Hale. 1961. The origin, nature and distribution of agricultural terracing. *Pacific Viewpoint* 2(1): 1–10.

Spillane, C. 1999. *Recent developments in biotechnology as they relate to plant genetic resources for food and agriculture*. Background Study Paper No. 9. Rome: Commission on Genetic Resources for Food and Agriculture, FAO.

Spooner, B., ed. 1972. *Population growth: Anthropological perspectives*. Cambridge, Mass.: MIT Press.

Steensberg, A. 1993. *Fire-clearance husbandry: Traditional techniques throughout the world*. Publication of the Royal Danish Academy of Science and Letters, Commission for Research on the History of Agricultural Implements and Field Structures No. 9. Herning, Denmark: Poul Kristensen.

Stemler, A. B. L. 1980. Origins of plant domestication in the Sahara and the Nile Valley. In M. Williams and H. Faure, eds., *The Sahara and the Nile,* 503–26. Rotterdam: A.A. Balkema.

Stocking, M. 1987. Measuring land degradation. In P. Blaikie and H. Brookfield, eds., *Land degradation and society,* 49–63. London: Methuen.

Stocking, M. 1995. Soil erosion and land degradation. In T. O'Riordan, ed., *Environmental science for environmental management*. Harlow, UK: Longman Scientific and Technical.

Stocking, M. A. 1996. Soil erosion: Breaking new ground. In M. Leach and R. Mearns, eds., *The lie of the land: Challenging received wisdom on the African environment,* 140–54. London: International African Institute.

Stone, G. D. 1996. *Settlement ecology: The social organization of Kofyar agriculture*. Tucson: University of Arizona Press.

Stone, G. D. 1997. Predatory sedentism: Intimidation intensification in the Nigerian savanna. *Human Ecology* 25(2): 223–42.

Stone, G. D. 1998. Keeping the home fires burning: The changed nature of householding in the Kofyar homeland. *Human Ecology* 26(2): 239–65.

Stone, G. D., M. P. Johnson-Stone, and R. McC. Netting. 1984. Household variability and inequality in Kofyar subsistence and cash-cropping economies. *Journal of Anthropological Research* 40: 90–108.

Stone, G. D., R. McC. Netting, and M. P. Stone. 1990. Seasonality, labour scheduling, and agricultural intensification in the Nigerian savanna. *American Anthropologist* 92(1,2): 7–23.

Stone, M. P., G. D. Stone, and R. McC. Netting. 1995. The sexual division of labor in Kofyar agriculture. *American Ethnologist* 22: 165–86.

Strathern, A. 1966. Despots and dictators in the New Guinea Highlands. *Man* 1: 356–67.

Stoorvogel, J. and E. M. A. Smaling. 1990. *Assessment of soil nutrient depletion in sub-Saharan Africa*. Wageningen, The Netherlands: Winand Staring Centre.

Strømgaard, P. 1984. Field studies of land use under Chitemene shifting cultivation, Zambia. *Geografisk Tidsskrift* 84: 78–85.

Strømgaard, P. 1985a. Biomass estimation equations for miombo woodland, Zambia. *Agroforestry Systems* 3: 3–13.

Strømgaard, P. 1985b. The infield–outfield system of shifting cultivation among the Bemba of south central Africa. *Tools and Tillage* 5(2): 67–84.

Strømgaard, P. 1988a. The grassland mound system of the Aisi-Mambwe of Zambia. *Tools and Tillage* 6(1): 33–46.

Strømgaard, P. 1988b. Soil and vegetation changes under shifting cultivation in the *miombo* of east Africa. *Geografiska Annaler* 70 B(3): 363–74.

Strømgaard, P. 1989. Adaptive strategies in the breakdown of shifting cultivation: The case of Mambwe, Lamba, and Lala of northern Zambia. *Human Ecology* 17(4): 427–44.

Strømgaard, P. 1990. Effects of mound-cultivation on concentration of nutrients in a Zambian miombo woodland soil. *Agriculture, Ecosystems and Environment* 32: 295–313.

Strømgaard, P. 1992. Immediate and long-term effects of fire and ash-fertilization on a Zambian miombo woodland soil. *Agriculture, Ecosystems and Environment* 41: 19–37.

Sugiyama, Y. 1987. Maintaining a life of subsistence in the Bemba village of northeastern Zambia. *African Study Monographs Supplementary Issue* 6: 15–32.

Sugiyama, Y. 1992. The development of maize cultivation and changes in the village life of the Bemba of northern Zambia. In *Africa 4*, 173–201. Senri Ethnological Studies No. 31. Osaka, Japan: National Museum of Ethnology.

Sutton, J. E. G. 1984. Irrigation and soil-conservation in African agricultural history, with a reconsideration of the Inyanga (Zimbabwe) and Engaruka irrigation works (Tanzania). *Journal of African History* 25: 25–41.

Sutton, J. E. G. 1989. Towards a history of cultivating the fields. *Azania* 24: 98–112.

Sutton, K. 1977. Reclamation of wasteland during the eighteenth and nineteenth centuries. In H. D. Clout, ed., *Themes in the historical geography of France*, 247–300. London: Academic Press.

Swift, M. J., B. M. Campbell, J. C. Hatton, and K. B. Wilson. 1989. Nitrogen cycling in farming systems derived from savanna: Perspectives and challenges. In M. Clarholm and L. Bergstrom, eds., *Ecology of arable land*, 63–76. Dordrecht, The Netherlands: Kluwer Academic.

Syers, J. K. and E. T. Craswell. 1995. Role of soil organic matter in sustainable agricultural systems. In R. D. B. Lefroy, G. J. Blair, and E. T. Craswell, eds., *Soil Organic Matter Management for Sustainable Agriculture. A workshop held in Ubon, Thailand, 24–26 August, 1994*, 7–14. ACIAR Proceedings No. 56. Canberra: Australian Centre for International Agricultural Research.

Tabor, S. R. 1992. Agriculture in transition. In A. Booth, ed., *The oil boom and after: Indonesian economic policy and performance in the Soeharto era,* 161–203. Singapore: Oxford University Press.

Tanksley, S. D. and S. R. McCouch. 1997. Seed banks and molecular maps: Unlocking genetic potential from the wild. *Science* 277(5329): 1063–66.

Tapp, N. and N. Menzies. 1997. *Forestry management strategies among Hmong and other upland cultivators of the southwest China borderlands.* Unpublished paper delivered at a regional conference on Indigenous Strategies for Intensification of Shifting Cultivation in Southeast Asia, held in Bogor, Indonesia, June 23–27, 1997.

Tengberg, A., M. A. Stocking, and M. Da Veiga. 1997. The impact of erosion on the productivity of a Ferralsol and a Cambisol in Santa Catarina, southern Brazil. *Soil Use and Management* 13: 90–96.

Thick, M. 1985. Market gardening in England and Wales. In J. Thirsk, ed., *The agrarian history of England and Wales, Vol. V. 1640–1750: II. Agrarian change,* 503–32. Cambridge, U.K.: Cambridge University Press.

Thirsk, J. 1984. Introduction. In J. Thirsk, ed., *The agrarian history of England and Wales, Vol. V. 1640–1750: I. Regional farming systems,* xix–xxxi. Cambridge, U.K.: Cambridge University Press.

Thirsk, J. 1997. *Alternative agriculture: A history from the Black Death to the present day.* Oxford, U.K.: Oxford University Press.

Thomas, P. J. and K. C. Ramakrishnan. 1940. *Some south Indian villages: A resurvey.* Madras: Oxford University Press.

Thrupp, L. A. 1998. *Cultivating diversity: Agrobiodiversity and food security.* Washington, D.C.: World Resources Institute.

Thrupp, L. A., S. Hecht, and H. Browder. 1997. *Linking biodiversity and agriculture: Challenges and opportunities for sustainable food security.* Washington, D.C.: World Resources Institute.

Thurston, D. H., M. Smith, G. Abawai, and S. Keari. 1994. *Tapado slash/mulch: How farmers use it and what researchers know about it.* Ithaca, N.Y.: Cornell International Institute for Food, Agriculture and Development.

Tiffen, M. 1996. Land and capital. In M. Leach and R. Mearns, eds., *The lie of the land: Challenging received wisdom on the African environment,* 168–85. London: International African Institute.

Tiffen, M., M. J. Mortimore, and F. Gichuki. 1994. *More people, less erosion: Environmental recovery in Kenya.* Chichester, U.K.: Wiley.

Tinker, P. B., J. S. I. Ingram, and S. Struwe. 1996. Effects of slash-and-burn agriculture and deforestation on climate change. *Agriculture, Ecosystems and Environment* 58(1): 13–22.

Tivy, J. 1990. *Agricultural ecology.* Harlow, UK: Longman Scientific and Technical.

Toulmin, C. 1992. *Cattle, women, and wells: Managing household survival in the Sahel.* Oxford, U.K.: Clarendon.

Trapnell, C. G. 1943. *The soils, vegetation and agriculture of north-eastern Rhodesia. Report of the ecological survey.* Lusaka, Zambia: Government Printer.

Trapnell, C. G. and J. N. Clothier. 1937. *The soils, vegetation and agricultural systems of north-western Rhodesia. Report of the ecological survey.* Lusaka, Zambia: Government Printer.

Trapnell, C. G., M. T. Friend, G. T. Chamberlain, and H. F. Birch. 1976. The effects of fire and termites on a Zambian woodland soil. *Journal of Ecology* 64: 577–88.

Treide, B. 1967. *Wildpflanzen in der ernährung der grundbevölkerungen Melanesiens.* Veröffentlichungen des Museums für Volkerunder zu Leipzig 16. Berlin: Akademie-Verlag.

Troll, C. 1968. The Cordilleras of the tropical Americas: Aspects of climatic, phytogeographic and agrarian ecology. In C. Troll, ed., *Geo-ecology of the mountainous regions of the tropical Americas,* 15–56. Colloquium Geographicum Band 9; Proceedings of the UNESCO Mexico Symposium, August 1–3, 1966. Bonn, Germany: Ferd Dummlers Verlag.

Turner, B. L. II and W. M. Denevan. 1985. Prehistoric manipulation of wetlands in the Americas: A raised field perspective. In I. S. Farrington, ed., *Prehistoric intensive agriculture in the tropics,* Vol. I, 11–30. BAR International Series 232. Oxford, U.K.: British Archaeological Reports.

Turner, B. L. II and W. E. Doolittle. 1978. The concept and measure of agricultural intensity. *Professional Geographer* 30: 297–301.

Turner, B. L. II, R. Q. Hanham, and A. V. Portararo. 1977. Population pressure and agricultural intensity. *Annals of the Association of American Geographers* 67: 384–96.

Turner, B. L., II, G. Hyden, and R. W. Kates, eds. 1993. *Population growth and agricultural change in Africa.* Gainesville: University Press of Florida.

Turner, M. D. 1998. The interaction of grazing history with rainfall and its influence on annual rangeland dynamics in the Sahel. In K. S. Zimmerer and K. R. Young, eds., *Nature's geography: New lessons for conservation in developing countries,* 237–61. Madison: University of Wisconsin Press.

Unger, P. W. and T. M. McCalla. 1980. Conservation tillage systems. *Advances in Agronomy* 33: 1–58.

United Nations Development Programme (UNDP). 1999. *Human development report.* New York: UNDP.

United States Department of Agriculture (USDA). 1998. *Why USDA's technology protection system (a.k.a. "Terminator") benefits agriculture: A discovery to spur new crop improvement.* Washington, D.C.: USDA.

United States Department of Agriculture (USDA). 1999. Biotechnology permits database. Animal and Plant Health Inspection Service, Plant Protection and Quarantine. http://www.aphis.usda.gov/biotech/statu.html [November 6, 1999].

United States Soil Conservation Service. 1981. *Soil taxonomy: A basic system of soil classification for making and interpreting soil surveys.* USDA Handbook No. 436. Kent, Ohio: Castle House Publications.

Unruh, J. D. 1990. Iterative increase of economic tree species in managed swidden fallows of the Amazon. *Agroforestry Systems* 11(2): 175–97.

USDA abandons three contentious issues in proposed standards. 1998. *Alternative Agriculture* 13(2): 95.

van Vliet-Lanoë, B., M. Helluin, J. Pellerin, and B. Valadas. 1992. Soil erosion in western Europe: From the last interglacial to the present. In M. Bell and J. Boardman, eds., *Past and present soil erosion: Archaeological and geographical perspectives,* 101–14. Oxford, U.K.: Oxbow Books.

Vavilov, N. I. 1926. Studies on the origin of cultivated plants. *Bulletin of Applied Botany and Plant Breeding (Leningrad)* 16(1): 1–243.

Vidal, J. 1999. The seeds of wrath. *Guardian Weekly* (Manchester), July 8–14: 16–17.

Visser, B. 1998. The CBDC programme: Community innovation systems and on-farm conservation. Presentation for the European Symposium on Plant Genetic Resources for Food and Agriculture, 30 June–3 July 1998, Braunschweig, Germany. http://www.cbdcprogram.org/BRAU_CBD.htm [December 9, 1999].

Visser, L. E. 1989. *My rice field is my child: Social and territorial aspects of swidden cultivation in Sahu, eastern Indonesia.* Dordrecht, The Netherlands: Foris.

Vita-Finzi, C. 1969. *The Mediterranean valleys: Geological changes in historical times*. Cambridge, U.K.: Cambridge University Press.

Voeks, R. A. 1998. Ethnobotanical knowledge and environmental risk: Foragers and farmers in northern Borneo. In K. S. Zimmerer and K. R. Young, eds., *Nature's geography: New lessons for conservation in developing countries,* 307–26. Madison: University of Wisconsin Press.

Vogt, J. 1953. Erosion des sols et techniques de culture en climat tempéré maritime en transition (France et Allemagne). *Revue de Géomorphologie Dynamique* 4: 157–83.

Vogt, J. 1983. Organisation des terroires et évolution des cultures. In J.-M. Boehler, D. Lercy, and J. Vogt, eds., *Histoire de l'Alsace rurale,* 227–36. Strasbourg: Librairie Istra.

Waddell, E. 1972. *The mound builders: Agricultural practices, environment and society in the central highlands of New Guinea*. Seattle: University of Washington Press.

Walker, D. and J. R. Flenley. 1979. Late Quaternary vegetational history of the Enga Province of upland Papua New Guinea. *Philosophical Transactions of the Royal Society of London* 286: 265–344.

Walker, D. and G. S. Hope. 1982. Late Quaternary vegetation history. In J. L. Gressit, ed., *Biogeography and ecology of New Guinea,* 263–85. The Hague: W. Junk.

Watson, A. M. 1974. The Arab agricultural revolution and its diffusion. *Journal of Economic History* 34: 8–35.

Watson, A. M. 1983. *Agricultural innovation in the early Islamic world: The diffusion of crops and farming techniques 700–1100*. Cambridge, U.K.: Cambridge University Press.

Watson, A. M. 1995. Arab and European agriculture in the Middle Ages: A case of restricted diffusion. In D. Sweeney, ed., *Agriculture in the Middle Ages: Technology, practice, and representation,* 62–75. Philadelphia: University of Pennsylvania Press.

Watson, J. P. 1964. A soil catena on granite in southern Rhodesia. *Journal of Soil Science* 15(2): 238–50.

Watters, R. F. 1971. *Shifting cultivation in Latin America*. FAO Forestry Development Paper 17. Rome: Food and Agriculture Organization of the United Nations.

Watts, M. 1983. *Silent violence: Food, famine and peasantry in northern Nigeria*. Berkeley: University of California Press.

Webster, R. 1965. The catena of soils on the northern Rhodesia plateau. *Journal of Soil Science* 16: 31–43.

Weeks, J. 1970. Uncertainty, risk, and wealth and income distribution in peasant agriculture. *Journal of Development Studies* 7(1): 28–49.

Wendorf, F., A. E. Close, R. Schild, K. Wasylikowa, R. A. Housley, J. R. Harlan, and H. Królik. 1992. Saharan exploitation of plants 8,000 years BP. *Nature* 359: 721–24.

Wendorf, F. and R. Schild. 1976. The use of ground grain during the late Paleolithic of the lower Nile valley, Egypt. In J. R. Harlan, J. M. de Wet, and A. B. L. Stemler, eds., *Origins of African plant domestication,* 269–88. The Hague: Mouton.

Wetterstrom, W. 1998. The origins of agriculture in Africa: With particular reference to sorghum and pearl millet. *Review of Archaeology* 19(2): 30–46.

Wheeler, D. L. 1995. The search for more productive rice. Special feature from the *Chronicle of Higher Education,* December 1, 1995. http://www.usis.usemb.se/sft/142/sf14209.htm [November 30, 1999].

Whitcombe, E. 1972. *Agrarian conditions in northern India,* Vol. 1, *The united provinces under British rule, 1860–1900*. Berkeley: University of California Press.

Whitcombe, E. 1973. The new agricultural strategy in Uttar Pradesh, India, 1968–70: Technical problems. In R. T. Shand, ed., *Technical change in Asian agriculture,* 183–201. Canberra: Australian National University Press.

White, B. 1983. "Agricultural involution" and its critics: Twenty years after. *Bulletin of Concerned Asian Scholars* 15(2): 18–31.

White, K. D. 1970. *Roman farming.* London: Thames & Hudson.

Whitmore, T. 1984. *Tropical rain forests of the Far East.* Oxford, U.K.: Oxford University Press.

Whitmore, T. C. 1989. Tropical forest nutrients, where do we stand? A tour de horizon. In J. Proctor, ed., *Mineral nutrients in tropical forest and savanna ecosystems,* 1–13. Oxford, U.K.: Blackwell Scientific.

Wiersum, K. F. 1982. Tree gardening and taungya on Java: Examples of agroforestry techniques in the tropics. *Agroforestry Systems* 1: 53–70.

Wild, A. 1989. Mineral nutrients: A soil scientist's view. In J. Proctor, ed., *Mineral nutrients in tropical forest and savanna ecosystems,* 441–56. Oxford, U.K.: Blackwell Scientific.

Wilken, G. C. 1987. *Good farmers: Traditional agricultural resource management in Mexico and Central America.* Berkeley: University of California Press.

Willcox, G. 1998. Archaeobotanical evidence for the beginnings of agriculture in southwest Asia. In A. B. Damania, J. Valkoun, G. Willcox, and C. O. Qualset, eds., *The origins of agriculture and crop domestication,* 25–38. Aleppo, Syria: ICARDA.

Willey, R. W. 1979a. Intercropping: Its importance and research needs. Part 1. Competition and yield advantages. *Field Crops Abstracts* 32(1): 1–10.

Willey, R. W. 1979b. Intercropping: Its importance and research needs. Part 2. Agronomy and research needs. *Field Crops Abstracts* 32(2): 73–85.

Williams, M. 1989. *Americans and their forests: A historical geography.* Cambridge, U.K.: Cambridge University Press.

Winarto, Y. T. 1995. State intervention and farmer creativity: Integrated pest management among rice farmers in Subang, west Java. *Agriculture and Human Values* 12(4): 47–57.

Winarto, Y. T. 1996a. Farmers' perspectives on integrated pest management. *Agricultural Research and Extension Newsletter* 34: 16–20.

Winarto, Y. T. 1996b. *Seeds of knowledge: The consequences of integrated pest management schooling on a rice farming community in west Java.* Ph.D. thesis, Australian National University, Canberra.

Wischmeier, W. H. 1976. Use and misuse of the universal soil loss equation. *Journal of Soil and Water Conservation* 31: 5–9.

Wiser, C. V. 1955. The foods of an Indian village of North India. *Annals of the Missouri Botanical Garden* 42(4): 301–412.

Wiser, C. V. 1978. *Four families of Karimpur.* Syracuse, N.Y.: Maxwell School of Citizenship and Public Affairs, Syracuse University.

Wiser, W. H. and C. V. Wiser. 1971. *Behind mud walls, 1930–1960 "with a sequel: The village in 1970 [by C. V. Wiser]".* Berkeley: University of California Press.

Withers, L. A. and F. Engelmann. 1998. In vitro conservation of plant genetic resources. In A. Altman, ed., *Agricultural biotechnology,* 57–88. New York: Marcel Dekker.

Wood, D. 1998. Ecological principles in agricultural policies: But which principles? *Food Policy* 23(5): 371–81.

Wood, D. and J. M. Lenné. 1997. The conservation of agrobiodiversity on-farm: Questioning the emerging paradigm. *Biodiversity and Conservation* 6(1): 109–29.

Wood, D. and J. M. Lenné. 1999a. Why agrodiversity? In D. Wood and J. M. Lenné, eds., *Agrobiodiversity: Characterization, utilization and management,* 1–13. New York: CABI.

Wood, D. and J. M. Lenné. 1999b. Agrodiversity and natural diversity: Some parallels. In D. Wood and J. M. Lenné, eds., *Agrobiodiversity: Characterization, utilization and management*, 425–445. New York: CABI.

World Commission [on Environment and Development]. 1987. *Our common future*. Oxford, U.K.: Oxford University Press.

Wyld, J. W. G. 1949. The Zande scheme. *Sudan Notes and Records* 30: 47–57.

Xiao, J. H., S. Grandillo, S. N. Ahn, S. R. McCouch, S. D. Tanksley, J. M. Li, and L. P. Yuan. 1996. Genes from wild rice improve yield. *Nature* 384(6606): 223–24.

Young, A. 1976. *Tropical soils and soil survey*. Cambridge, U.K.: Cambridge University Press.

Zarin, D. J., Guo Huijun, and L. Enu-Kwesi. 1999. Methods for the assessment of plant species diversity in complex agricultural landscapes: Guidelines for data collection and analysis from the PLEC Biodiversity Advisory Group. *PLEC News and Views* (Special Issue) 13: 3–16.

Zimmerer, K. S. 1994. Human geography and the "New Ecology": The prospect and promise of integration. *Annals of the Association of American Geographers* 84: 108–25.

Zimmerer, K. S. 1996. *Changing fortunes: Biodiversity and peasant livelihood in the Peruvian Andes*. Berkeley: University of California Press.

Zimmerer, K. S. 1998. Disturbances and diverse crops in the farm landscapes of highland South America. In K. S. Zimmerer and K. R. Young, eds., *Nature's geography: New lessons for conservation in developing countries*, 262–86. Madison: University of Wisconsin Press.

Zimmerer, K. S. 1999. Overlapping patchworks of mountain agriculture in Peru and Bolivia: Toward a regional–global landscape model. *Human Ecology* 27(1): 135–65.

Zimmerer, K. S. and K. R. Young. 1998. Introduction: The geographical nature of landscape change. In K. S. Zimmerer and K. R. Young, eds., *Nature's geography: New lessons for conservation in developing countries*, 3–34. Madison: University of Wisconsin Press.

Zohary, D. 1989. Domestication of the southwest Asian Neolithic crop assemblage of cereals, pulses, and flax: The evidence from the living plants. In D. R. Harris and G. C. Hillman, eds., *Foraging and farming: The evolution of plant exploitation*, 358–73. London: Unwin Hyman.

Zohary, D. and M. Hopf. 1993. *Domestication of plants in the Old World. The origin and spread of cultivated plants in west Asia, Europe, and the Nile Valley*. Oxford, U.K.: Clarendon.

Acacia mangium, 121

açai palm (*Euterpe oleracea*), 142–43, 155, 156

Accra (Ghana), 162

Actinomycetes, 99n3

Adams, W. M., 168, 169

adaptability, 21, 54, 162, 175n1, 248, 286; and external conditions, 38, 39, 198, 285; of farmers, 198, 236, 279; in organizational diversity, 53, 236; and resilience, 56; and shifting cultivation, 103, 134; and sustainability, 179

Africa, xi, xvi, 4; agricultural revolution in, xiv, 199; catenas in, 89, 123, 124; colonialism in, 23, 180; commercialization in, 214; crops in, 76, 92, 180, 208; development of agriculture in, 78n7, 138; domestication in, 62, 64, 132; droughts in, 158, 163, 204, 205, 213; farmer-driven transformation in, 198–213; field systems in, 72, 73; intercropping in, 196n1, 201–2, 272; irrigation systems in, 70, 170; literature on, xiii, 168–69, 201; local knowledge in, 58n3, 277, 279; population of, 154, 215; shallow valleys (*dambos*) in, 124, 132; soils in, 76, 82, 86–87, 89; south central, 181;
west, 76, 97–99, 153–55, 157, 161, 196n1, 218, 274. *See also particular countries*

Agno valley (Philippines), 20

agribusiness, 219, 254, 255–56, 279. *See also* multinational corporations

agricultural involution, 194

agricultural revolution, xiv, xvii, 135, 179–80, 183–87, 194, 196. *See also* Green Revolution

Agrobacterium, 250, 265

agrobiodiversity, 142, 241; adaptation in, 53; and agrodiversity, 50, 59; and agroecosystems, 271; in crops, 21, 40–41, 68, 189, 233–36, 255; and ecosystems, 41, 152; and farmers, 41, 235–36, 278; and farming spaces, 52; and genetic conservation, 249; and genetic erosion, 245, 246; and intercropping, 272–74; and large households, 205; and local knowledge, 57; and management diversity, 43; supporters of, 281–84; and transgenics, 282, 283, 284; and wild elements, 55

agrodiversity: as continuing strategy, 241; definitions of, 42–46, 80; elements of, 40–42, 140–41; history of, 59–79, 125,

agrodiversity (*Continued*)
186; origin of, 76–77; scientific basis of, 270–75; taxonomy of, 46–49, 50, 52; temporal aspects of, 42, 43; as term, xii

agroecosystems, 43, 46, 66, 175n1, 249, 271

agroforests, 51, 273–74, 282; in Amazonia, 152, 172, 173; in Ghana, 162–64; in Java, 151; in Peninsular Malaysia, 34–35, 38; management of, 55, 140, 156, 277; in Philippines, 17, 18–19; in Southeast Asia, 147, 149–50, 151; successional, 141–43, 156n3

Aisa Mambwe people, 131

Akamba people (Machakos, Kenya), 160, 201, 211–14, 217

Alcorn, Janis, 142

alfalfa (lucerne), 92

Allan, W., 131, 134

allelopathy, 273, 275

Allen, B., 22n4

Almekinders, Conny, 43, 44, 80

Alsace, 187–88

alternative agriculture, xvii, 95, 99n4, 262, 271, 278, 284; in Cuba, 260; in England, 254, 261n7; history of, 253–54, 261n8; in the North, 253–57; regulation of, 255–56; *vs.* modernist agriculture, 244, 253, 255–56

Altieri, M. A., 243, 256, 257, 271

Altman, A., 283

Amanor, K. S., 161, 163, 164, 165, 174

Amazon River, 69, 72

Amazonia, xii, 93, 172; fallow systems in, 144–49; managed forests in, 143, 152; population decline in, 141; research in, 126, 141–42; shifting cultivation in, 122n5; soils of, 86, 88, 97

Andean region, 22n7, 23–27, 45, 72, 172

Anderson, A. B., 142, 143, 152

Anderson, Edgar, 78n5

Anderson, J. R., 220

Anderson, P. C., 63, 65

Andriesse, J. P., 93, 126, 128

animals, wild, 55, 66, 129, 140, 144. *See also* livestock

anthropology, 27, 58n3

apêtê (forest islands; Amazonia), 153–54

apomixis (asexual reproduction), 267–68

Argentina, 250, 269, 270

arroyos (flat-floored gullies), 171, 182

Asia, xi, 76, 78n7, 92, 275; alternative agriculture in, 255; colonialism in, 180; genetic diversity in, 245; Green Revolution in, 236, 242; irrigation systems in, 70, 71; south, 199, 219, 230. *See also particular countries*

Australia, 10, 66–67, 245, 247, 275, 286n

Australian Research Council (ARC), xii

avocado, 94, 98

Azolla, 91

Azotobacter, 259

babassu palm (*Orbignya oleifera*), 156n2

Babbar, L. I., 86, 88

Bacillus thuringiensis, 252, 269

bacteria, 90, 95, 99n3, 124, 126

Bailey, B., 251, 270

Balée, W., 141, 153

Balfour, E. B., 260, 261n7

Bali, 22n6, 231

Bambara groundnut (*Vigna subterranea*), 29, 92, 202, 204, 207

Bambara people, 203, 217. *See also* Kala village (Mali)

bamboo, 94–95, 118, 146

bananas, 4, 29, 37, 98, 149, 207; in Amazonia, 122n5, 143, 173; in Mindoro, 118, 119

Bangweulu, Lake, 135

barley, 68, 92, 94, 269; in Andean region, 24, 26; domestication of, 61, 63; in Europe, 188; in Japan, 189, 193; in Karimpur, 224

Barry, A. K., 98

Bayliss-Smith, T. P., 222

Bayninan (Ifugao, Philippines), 16–20, 51

beans, 112, 129, 131, 188, 258, 274; broad (*Vicia faba*), 25, 26, 92; in protoagriculture, 66, 67

Beardsley, R. K., 190

Bemba (Zambia), 58n3, 123, 131, 132, 135, 136–37; *citemene* farming system in, 127, 129, 130

Benjamin, P. A., 212

Benneh, G., 176n4

Beranang (Peninsular Malaysia), 33, 34, 35, 36, 37

betel leaf, 112

beusani land use system, 50

Bidayuh (land Dayak) people, 106, 113–14, 115

Bie, Stein, 271

biodiversity, 53, 104, 142, 143, 212, 249; and agrobiodiversity, 41; in Borneo, 108, 112–13; and land management, 55, 113

biodynamic farming, 253–54, 256

biological husbandry, 253, 255, 259–60

biophysical diversity, 41, 43–44, 80, 183, 216, 235; and agrodiversity, 50, 59; management of, 275–76

biosafety, 268–70, 283

Biosafety Protocol, 270

biotechnology, 270, 271, 279–80; in Cuba, 259; and farmers, 283–84; non-commercial advances in, 265–68; patents on, 265–66, 267, 268; and plant breeding, 241, 249–53, 268; politics of, 262; terminator, 263–65, 269, 280n2

Birch, H. F., 125

black gram, 68, 92

Blaikie, P., 55, 158, 176n9, 286

Bolivia, 25

Borneo, xixn2, 105–21, 140, 141, 149; biodiversity in, 108, 112–13; cash crops in, 109, 112, 113; changes in, 120–21; land holding in, 109, 115–16, 121; local knowledge in, 76; migration in, 108, 115, 121, 174; population density in, 105, 106, 150; secondary crops in, 111–13; shifting cultivation in, 105–21, 146, 147, 174; soil diversity in, 88; *tembawang* in, 150–51

Boserup, Ester, 49, 60, 61, 208, 210, 215, 216; theory of agricultural change, 199–200

Brachystegia, 123

Brassica rapa, 269

Brassica spp., 68, 94, 269

Brazil, xi, 86, 142, 152, 262, 266. *See also* Amazonia

breadfruit, highland (*Ficus dammaropsis*), 9

Britain, xiii, 176n7, 183, 243; agricultural revolution in, 186–87, 188; alternative agriculture in, 254, 255, 261n7; colonies of, 29, 206; early field systems in, 73; opposition to transgenics in, 269

Brookfield, Harold C., 40, 47–48, 52, 55, 106, 158, 176n9

Brookfield, Muriel, xii

Browder, H., 283

Brown, Paula, 4, 6, 11, 12, 21n2

Brundtland report, 281

Brush, S. B., 25, 26

Buid people (Mindoro, Philippines), 116–20

Burkina Faso, 64

Burma, 148

Burunge people (Tanzania), 81, 82, 89, 95

burutí palm (*Mauritia flexuosa*), 155

butterflies, 252, 263

Butzer, K. W., 184

Byron, Y., 106

Caine, N., 276n8

Cairns, Malcolm, 147, 148

calcium, 126, 128, 131, 151

Canada, 250

cane grass (*Miscanthus*), 3, 7

cañihua (*Chenopodium pallidicaule*), 25

canola rape (*Brassica napus*), 250, 269

capital: creation of, 182–83, 216; and gender, 22n4; intensification of, 54; and investment, 22n4, 184, 188, 201, 208; and labor, 216; landesque, 55, 71, 171, 174, 175, 182, 214, 216; working, 174, 216

carbon, 125, 126, 128, 165, 244, 256

Caribbean, xi, 29

Carson, B., 166–67

Carson, R., 243

cash crops, xiv, 148, 169, 215, 218, 227; in Africa, 136, 162, 163, 204, 208; in Borneo, 109, 112, 113; coffee, 10, 15, 195; innovators in, 16; in Japan, 188; from market gardens, 36, 97, 183; and monoculture, 260n1; and organizational diversity, 45; and transgenics, 250–51; in Vanuatu, 75; of Zande, 29–30, 32

cassava. *See* manioc

Cassia siamea, 148

casuarina *Casuarina oligodon,* 3, 7, 10, 14

catenas, 29, 89, 90, 93. *See also* Africa

cation exchange capacity (CEC), 85, 95, 124, 151, 176n6

Cederroth, S., 234, 235

Center for the Application of Molecular Biology to International Agriculture (CAMBIA), 267

Centers for Production of Entomophages and Entomopathogens (Cuba), 259

Central African Republic, 28

Central America, 72, 79n10, 82, 92, 141, 245, 274

Centro Internacional de Majoramiento de Maiz y Trigo (CIMMYT), 219, 247, 267

cereals. *See* grain

Chad, 272

chakitaklla (foot-plow), 25, 85

Chambers, R., 161

Chang, T. T., 17, 234

chemical inputs, 194, 221, 241–42, 245, 260, 284; and pollution, 243, 276

chenopod (*Chenopodium berlandieri*), 67

chestnut, 65

chickpea, 63, 92

chilies, 29, 112

Chimbu (Papua New Guinea), 3–16, 32, 44, 49, 88, 166, 170; population of, 6, 22n3

Chimbu (Kuman) people, 5–6, 115, 217

Chin, C. S., 108, 146

Chin, S. C., 105, 106, 110, 111, 112, 115, 147

China, xi, xii, 92, 148, 170, 278; alternative agriculture in, 255; biotechnology in, 266; domestication in, 63, 65; rice in, xiii, 252; transgenics in, 270

chinampas, 72

Chinese fir (*Cunninghamia lanceolata*), 148

Christianity, 117, 122n4

Chromolaena odorata, 163, 176n6

cinnamon, 150

citemene farming system, 123, 124–38; large-circle, 127, 133; small-circle, 132, 133, 135

citrus, 94, 98, 258

Clarke, W. C., 156n1

Clements, F. E., 53

climate, 17, 59, 172, 273; and global warming, 104; micro-, 21, 41, 43, 71; and resilience, 277, 285; and soils, 84, 85

cloning, 249–50

Clothier, J. N., 124, 132

clover, 92, 188

cloves, 150, 151

cocoa, 122n7, 162, 163, 164

coconuts, 34

coco-yam. *See* taro

coffee, 150, 151, 152; as cash crop, 10, 15, 195; changes in distribution of, 13, 14; in Cuba, 258; in Ghana, 162; in Java, 195; in Peninsular Malaysia, 34; in Papua New Guinea, 9, 10–11; in Philippines, 18, 19

Coleus, indigenous. *See* Livingstone potato

Colfer, C.J.P., 105, 106, 108, 121, 146, 147

Colombia, 72

colonialism, 137, 138, 218–19; in Africa, 23, 28–29, 31–32, 180, 203, 206; in Andean region, 24, 25; in Asia, 180; British, 29, 206, 219; in Java, 195–96, 230; and peace, 6, 9, 16; in Philippines, 18, 20; and population growth, 158, 195; in Vanuatu, 75

Columella, 184

Combretum spp., 123

commercialization, 120, 149, 188, 242, 285; in Africa, 214; in Andean region, 25–27; in Peninsular Malaysia, 34, 38; and monocropping, 274; in Philippines, 20. *See also* cash crops

Commission on Plant Genetic Resources for Food and Agriculture (FAO), 246, 247

community, 24–25, 41–42

Community Biodiversity Development and Conservation Programme, 248

comparative method, 49–50

composting: in Africa, 165, 207, 211, 213; of grasses, 131, 133, 135, 137; in India, 225, 226; in Mediterranean region, 184

Congo (Zaire), 27, 127

Conklin, H. C., xv, 16, 17, 18, 20, 47, 49; on field types, 52; on Hanunóo, 32, 116–20,

156; on Ifugao, 51; on shifting
cultivation, 104; on taxonomy of
agrodiversity, 46

conservation, 45, 158, 208; of energy, 230,
277; of soils, 26, 31, 161, 166, 168–70,
175, 213, 251, 275–77; of water, 168,
213, 275, 276, 277. *See also* genetic
conservation

Consultative Group on International
Agricultural Research (CGIAR), 236n4,
246, 267

Convention on Biological Diversity (1992),
247, 270, 280n2; Conference of Parties
to, 281

Conway, G. R., 47, 274

cooperatives, regional, 11, 224–25, 228

Cordyline fruticosa, 7, 166

corn. *See* maize

Corn Growers' Association (U.S.A.), 264, 278

costs: of GURTS, 264; and minimum tillage,
253; of modernist agriculture, 243; of
production, 186, 241

cotton: domestication of, 63; in India, 225,
226, 227; in Japan, 190; monoculture of,
31, 260n2; transgenic, 250, 262, 269;
yields of, 242; and Zande, 30, 31, 32

Cowgill, G. L., 200

cowpea (*Vigna unguiculata*), 29, 92, 163, 165,
172, 208; interplanting of, 129, 202

Cramb, R. A., 111, 112

Crittenden, R., 22n4

Cromwell, E., 41, 247

Cronon, W., 66, 67

crop complexes, 34, 43, 51, 67, 164, 273; of
Zande, 30, 31

crop rotation, 227, 276, 282; in Africa, 135,
164, 207; in alternative agriculture, 254,
255; in Andean region, 25; in Cuba, 259;
and elimination of fallow, 186; in
Indonesia, 232; and management
diversity, 43, 277; in Mediterranean
region, 184

crop yields, 160, 186, 221, 234; and
agrobiodiversity, 282; in Amazonia, 126;
in Andean region, 25; and biophysical
diversity, 44; and biotechnology, 242,

264, 266; and commercial fertilizer, 191;
in Cuba, 259; declining, 162; and genetic
diversity, 244, 245; and herbicides, 269;
increase in, 193; and intercropping,
272–73; in Java, 169; and nutrients, 90;
in Philippines, 17–18; of rice, 37, 120,
194; stagnation of, 243–44; and
transgenics, 251, 265; of wheat, 219–20,
225; in Zambia, 130

cropping: alley, 164, 273–74; changes in, 227;
double, 191, 221, 229, 237n13; multiple,
191, 199, 221; relay, 184; standardized,
194; summer, 184; triple, 234, 237n13.
See also intercropping

crops: commercial, 120, 142, 149, 188, 242;
complementary, 143; diffusion of, xiv, 39,
54, 135; dry, 17, 33, 72–73, 176n10;
fiber, 30, 63, 68, 76, 98; fodder, 92, 185,
188, 191, 227; mixed, 9–10, 14, 31, 130,
138, 169, 224; secondary, 111–13; shade-
tolerant, 151, 162. *See also particular
species*

Cuba, 257–60, 285; and education, 258,
261n10

cucumbers, 129

cultivation: hoe, 78n7, 154, 164, 175n2, 203,
280n4; land rotation, 103–73, 157, 172,
188; and natural forest processes, 142–43;
permanent, 137, 146, 208; by tractor,
133, 226. *See also* plowing; tillage

Cuzco (Peru), 26

cyanobacteria, 91

Dalonguebougou. *See* Kala (Mali)

damar (resin), 149

dambos (shallow valleys), 124, 132

Dao Zhiling, 40

Darwin, Charles, 81, 89

Datoo, B. A., 199

Davies, M. S., 62

de Jong, W., 173

de Schlippe, Pierre, 27–32, 39nn2,3, 51, 52,
277

Dendrocalamus hamiltonii (bamboo), 94–95

Denevan, W. M., 73, 141, 144, 156nn3,4

Denmark, 255

Denslow, Julie, 86

desertification, 158

disease: and agrodiversity, 116, 212; biological control of, 258; in Borneo, 108; Ebola virus, 32; and intercropping, 273; and land management, 21; malaria, 6; and management diversity, 43, 276; and monocultural systems, xvii; and relocation, 29; and resilience, 277; resistance to, 245, 265; risk of, 221; sleeping sickness, 29; smallpox, 119

domestication: in Africa, 132; earliest sites of, 68–69; history of, 61–64; and land management, 70; of livestock, 62, 63, 64, 68; of plants, 5, 77n1, 132; of rice, 62, 63, 76

Donkin, R. A., 168

Doolittle, William, 171, 182, 200

Dove, M. R., 105, 106, 110, 111, 114, 115

drought, 18, 21, 230; in Africa, 158, 163, 204, 205, 213; resistance to, 264, 265

Dudley, R. G., 106

Dupuis, J., 237n11

durian (*Durio zibethinus*), 36, 37, 113, 146, 150

dusun area (Peninsular Malaysia), 34–35, 51

Dutch East India Company, 195

Eash, N. S., 168

Ebola virus, 32

ecology, 76, 99n2, 175n1; and agrodiversity, 270–71; of Amazonia, 141; and biotechnology, 283–84; cultural, xiv–xv, 55; equilibrium and nonequilibrium, 53, 248, 271, 284; and farming spaces, 9–10, 51, 52; of managed successional growth, 145–46; political, xvii, xviii

economy: and agricultural prices, 187, 188; cash, 12, 38, 121, 196, 202, 218; and farmers, 21, 241–43; and farming spaces, 52; global, 218; growth of, 181; and industry, 26, 30, 32; and innovation, 187; livestock, 11–12; market, 23, 24, 45, 188, 201, 285; and monocultural systems, xvii; new commercial, 210; political, 285, 286; and poverty, 193, 194; and rainfall, 205;

rural, 158, 189, 282. *See also* capital; income; investment

ecosystems: adaptation of, 175n1; and agrobiodiversity, 41, 152; and agrodiversity, 270–71; equilibrium of, 53; and habitat diversity, 245; and land degradation, 158; resilience of, 175; sustainability of, 274. *See also* agroecosystems

Eder, J. F., 20

Egypt, 62, 63, 70

electrification, rural, 229

elephant grass (*Pennisetum purpureum*), 28

employment: changes in, 229; nonagricultural, 38, 110, 190, 193–94, 228, 235; rural, 189–90, 228; temporary, 117

enclosure of fields, 186, 187

Eng, R. Y., 196n2

Enga (Papua New Guinea), 9, 217

environment: change in, 64, 66; and Cuban agriculture, 258; and farming practices, 118, 279; history of, 53; landrace adaptation to, 248; and modernist agriculture, 243; and transgenics, 268–70, 283. *See also* microenvironments

environmental movement, 255

Ethiopia, 78n6,7, 248

ethnographic present, xvi, 5, 20, 21n2, 46

Eurasia, xiii, 62

Europe: agricultural revolution in, xiv, 185, 186–87; alternative agriculture in, 254, 255; biological husbandry in, 257; crops in, 92, 180, 185, 187, 188; Eastern, 258; and genetic conservation, 245, 246, 247; management of woodland in, 151; Mediterranean, 78n9; modernist agriculture in, xiii; multinational corporations in, 242, 251; plowing in, 78n7, 175n2; soils in, 97; transgenics in, 262, 263, 269

Evans-Pritchard, Edward, 27, 28, 30, 39n2

experimentation, 126, 128, 130, 148, 213, 277; by farmers, xiv, 279

external forces, 42, 59, 211–12, 218–19; and adaptability, 38, 39, 198, 285; and settlement patterns, 29–30, 31, 32, 39. *See also* economy; politics; social conditions

Fairhead, James, 57, 153, 154

fallow systems: in Africa, 30, 98; in Amazonia, 144–49, 172; in Andean region, 24, 25, 26; in Borneo, 109–11, 121, 151; bush, 163; elimination of, 186; and field types, 51; green, 187; and land selection, 109–11; in Latin America, 116; and ley farming, 186, 187; and livestock, 25; management of, 43, 141, 144–49, 156n1, 164, 233; observations of, 47; in Papua New Guinea, 4, 7; in Philippines, 119; plowed, 187; rotational, 162, 166, 195; and shifting cultivation, 103, 104, 105, 187, 199; successional, 141, 144–49, 156n1; and swamp rice, 114, 154

family. *See* household

famine, 169, 185, 212

Farmer, B. H., 222

farmers: African, 98, 160, 162, 163, 173, 180, 204; and agrobiodiversity, 41, 45, 235–36; and alternative agriculture, 256–57; American Indian, 171; attitudes towards, 279; and change, 180, 205, 216, 217, 228, 229; cooperation between, 192, 197n4; cooperation with, 259, 278; and creation of landesque capital, 175; decision making of, 50–52, 188, 233–35; and ecology, 149, 233; economic hardships of, 241–43; English, 186; estuarine, 156; experiments of, xiv, 279; and farming systems, 132, 140; and fertilizer, 222; and genetic conservation, 245–48; and Green Revolution, 220, 233, 234, 279; household organization of, 201, 202; Indonesian, 174, 182, 231; ingenuity of, 157, 162, 163, 164, 165, 169–70; innovations of, xvii, 181, 185, 186, 196, 198–217; Iron Age, 138; Japanese, 192, 193, 194; Javanese, 233–35; knowledge of, 160–61; local knowledge of, 43, 57, 76, 87, 245, 277, 279, 282; management practices of, xviii, 149, 167–68, 183; and

markets, 145, 215; Mexican, 155, 182; and monoculture, 242; and multinational corporations, 252; organizational skills of, 200, 216–17; and pest control, 231–32, 233; of Roman period, 171, 183–84; and shifting cultivation, 103; small, xiii, 193, 194, 196, 200, 216, 218, 278–79, 281–86; and soils, 80–81, 87, 89, 99, 141; South American, 182; subsidies for, 241, 242, 257; supporters of, 281–84; technical skills of, 216

farming spaces: in Amazonia, 142, 173; in Andean region, 23–26; codification of, 50–52; and commercialization, 25–26; and field types, 51; and forests, 146; in India, 223–24; Peninsular Malaysia, 34–35; and land management, 39, 187; and local knowledge, 57; in Papua New Guinea, 9–10; and protoagriculture, 66; and Zande, 30

farming systems: *citemene*, 124–30; conservationist, 208; *fundikila*, 124–25, 130–32; history of, 77, 78n7, 138–39, 175, 183–87; household, 208; *ibala*, 130–32, 133, 137, 138; intensification of, 199–201; intensive, permanent, xiv, 207; and long-term change, 179, 182; mixed, 213; and modern science, 156, 183, 185, 216; multiple-use, 142; in North, 242–43; operative, 159; precision, 276; subsistence, xiv. *See also* alternative agriculture; modernist agriculture

farming systems research (FSR), 58n2, 132–39, 221

fencing, 12, 129, 133, 170, 171

fertilizers: and alternative agriculture, 256; alternatives to, 276; in Andean region, 25; and biophysical diversity, 43; black market in, 211; chemical, 222, 228, 231; commercial, 136, 191, 220–21; and crop yields, 136, 243; in Cuba, 258, 259, 260; on experimental farms, 130; and farmers, 222, 234, 242; in Ghana, 164; and GURTS, 263; inorganic, 169, 191, 210, 211, 213, 220, 225; in Japan, 190–91; manure as, 185, 186, 187, 190, 203, 209,

fertilizers (*Continued*)
211, 213, 214, 222; mineral, 186, 187; nitrogen, 222

Field, M. J., 176n4

fields: and agrodiversity, 156; archaeology of, 73–75; biophysical diversity in, 44; bush, 202, 204; commercial crop, 142; community, 73, 186; dry, 17, 18, 19, 72–73; enclosure of, 186, 187; and forests, 110, 156; management of, 143, 149; and manufacture of soils, 39; open, 9, 186, 187; raised, 72; and shifting cultivation, 108, 110, 142; terms for, xv; types of, 50, 51, 57; use of old, 143, 209; walled, 25; wet-rice, xvi, 17–18, 189; and wild plants, 76; and women's, 10; Zande, 30. *See also* terraced fields

Fiji, xixn1, 276

fire: and *citemene* farming system, 125, 127, 128, 130, 133; clearing with, xv, 119, 173; control of, 118; and domestication of plants, 76; dry season, 124, 125, 153, 163; effect on soil of, 105, 125–28, 141; and extraction of resin, 150; and *Imperata cylindrica* grassland, 174; and land management, 69; and protoagriculture, 64, 66; and savanna, 153, 154, 208; and shifting cultivation, 104, 105, 108, 123, 125; and swamp rice, 114

fish, 18, 36, 65, 66, 109, 129; and agrodiversity, 61, 76; and domestication of plants, 69

fixed-plot systems, 61, 75–76, 78n7

Flanders, 186, 188

Flenley, J. R., 5

flood plains, 172–73; and domestication of plants, 68, 69, 75–76, 77

floods: and land management, 21, 78n6, 155–56, 168, 171, 172–73, 182, 223; and rainfall, 171, 176n7

Food and Agriculture Organization (FAO; UN), 245, 246, 247, 269; and genetic conservation, 245, 246, 247, 248

foraging, 62, 63, 65, 66, 76–77, 109

forest islands, 144, 152, 153–54

forests: biodiversity of, 108; *chipya* savanna, 124; created, 152; deciduous, 162;

destruction of, 20, 162; fires in, 121; and germplasm, 151, 153, 155; in Japan, 190–91; managed, 140, 141–51, 152, 175; *miombo*, 123, 124, 133, 134, 135, 136, 137, 138; natural, 148–49, 150, 151, 152, 154, 162; in North America, 65; old, 108, 109–11, 119, 149; old secondary, 117, 143, 145, 146; and population, 141; resources in, 124, 191; reversion of agricultural land to, 37; secondary, 108, 109–11, 119, 148, 150, 153; semideciduous, 162, 163; shifting cultivation in, 103, 105, 123; successional, 141–51; timber from, 18, 37, 120–21; unmanaged, 55; village, 151; virgin, 141. *See also* agroforests

Fouta Djallon plateau (Guinea), 97–99, 160, 165

Fowler, C. E., 247

France, 29, 185, 187–88, 201, 243, 267

Francis, C. A., 272

Freeman, J. D., 105, 106, 110, 111

French Equatorial Africa, 28

Fresco, Louise, 43, 80

fruit, 18, 24, 34, 74, 151

Fujisaka, S., 104

Fukuoka, Masanobu, 254, 255

Fulani people, 98, 201–6, 211, 217

fundikila farming system, 123, 124–25, 130–32, 133, 134, 136, 137, 138

galerias filtrantes (Mexico), 78n8. *See also* qanats

Ganga canal system, 223

gardens: in Borneo, 112, 113; and dump heaps, 68–69; fencing of, 129; and field types, 51; forest, 149–50; fruit, 150; and horticulture, xv, 212; house, 47, 142, 149, 151, 156; market, 183; mixed, 9–10, 14, 47, 130; mounded maize, 132, 136; private, 205; second-year *citemene* (*icifwani*), 134; subsidiary, 137; village, 130, 131, 132–33, 138, 208; Zande, 30

Geddes, W. R., 106, 114

Geertz, Clifford, 194–95, 215

gender, 22n4, 41, 44, 130, 215, 282

genetic conservation, 246–49

genetic diversity, 244–49

genetic engineering, xvii, 220, 250, 256, 270; with nonplant genes, 265–66; in public domain, 265–67. *See also* transgenic crops

genetic erosion, 241, 244–46

Genetic Use Restriction Technologies (GURTS), 263–65, 280n2

Germany, 253

germplasm, 219, 221, 225, 285; disease-free, 250; diversity of, 281–82; and forests, 151, 153, 155; storage of, 246–49. *See also* seeds

Gerpacio, R. V., 220

Gersper, P. L., 171

Ghana, xi, 154, 280n4; Akuapem area of, 161, 162, 163, 164, 176n5; Krobo area of, 161, 163, 164, 174, 175n4, 176n5; land degradation in, 161–65

Gichuki, F., 160, 212, 214, 215

ginger, 38

Gliessman, S. R., 247, 257, 271, 273

Global Environmental Facility (GEF), xi, 282

Global Plan of Action (1996), 281, 283

global warming, 104

globalization, 180, 242, 285

Goldman, A., 215, 222

Golson, Jack, 5, 71

Gómez-Ibáñez, D. A., 185

goosefoot (*Chenopodium berlandieri*), 67

gourds, 30, 129, 207

government: and biological husbandry, 257; and biotechnology, 280, 283–84; colonial, 203; dependence on, 20; and drainage systems, 225; and farmers, 279; and innovation, 282–83; intervention by, 113, 180, 219, 224–26; and multinational corporations, 252; and science, 219; and seeds, 221; and transgenics, 263; and wet-rice production, 114

grain: coarse, 221; domesticated, 61–62, 65; farming based on, 187; and legumes, 92, 183; photoperiodic, 236n3; surpluses of, 242; *vs.* vegeculture, 76; wild, 65. *See also particular species*

grasses, 28, 118, 159, 172, 275, 280n4; clearing of, 127, 164; composting of, 131, 133, 135, 137; and domestication of crops, 61–62; *Imperata cylindrica*, 7, 37, 174, 176n6

grassland, 49, 120, 123, 137, 163

Greece, 78n9

green gram (*Vigna mungo*), 29, 68, 92

Green Revolution, xvii, 181, 194, 198, 201, 218–36, 241; and agrodiversity, 271, 275; and alternative agriculture, 284; in Asia, 236, 242; and farmers, 220, 233, 234, 279; in India, 222–30; in Java, 196, 230–31, 234; in Peninsular Malaysia, 37; literature of, 220; and monocropping, 274; in Philippines, 21; and rice, 218, 219, 220, 223, 228, 236

Greenland, D. J., 48, 60, 61, 86, 126, 158, 272

Greig, J., 74

Grigg, D. B., 52, 180, 199

groundnuts (*Arachis hypogaea*): in Africa, 29, 98, 131, 154, 208, 209; in Amazonia, 172; and eleusine millet, 31, 129; in India, 228, 230; in Papua New Guinea, 4, 9; in South America, 92; trade in, 135; yields of, 204. *See also* Bambara groundnut

Grove, A. T., 168

Guatemala, 47, 56, 182

Guinea Republic (West Africa), xi, 97–99, 153–55, 161, 164

Gunung Sewu (Java), 107, 169

Guo Huijun, 40

Gyasi, E. A., 161, 164

Haberle, S., 5

habitat diversity, 245

hacienda system, 24, 26

Hadi, Abdul Samad, 32, 37

haj (pilgrimage to Mecca), 37

Hakkeling, R. T. A., 158

Hale, G. A., 168

Hall, J. W., 190

Hanham, R. Q., 200

Hanunóo people (Mindoro, Philippines), 116–20, 140, 156n1

Hardjono, Joan, 234–35

Harlan, H. V., 64

Harlan, J. R., 78n5, 244

Harmattan dust, 82

Harris, D. R., 49, 50, 62, 63, 67, 69

Harriss, J., 237n10

Harwell, E., 114

Haverkort, B., 272

Hawley, J. W., 171

Hayward, D. F., 176n4

Hecht, S. B., 283

hemp, Deccan, 68

Hendrix, P. F., 285

herbicides, 108, 253; alternatives to, 275–76; and GURTS, 263; and transgenics, 250, 251, 265, 268–69

hibiscus, 202

Hill, P., 176n4

Hillman, G. C., 62

Himalayan region, 176n9

Hiraoka, M., 142

hoe culture, 78n7, 154, 164, 175n2, 203, 280n4

Holling, D. G., 271

Hong, E., 106

Hong Kong, 97

Hope, G. S., 5

Hopf, M., 67, 92

Hopkins, D. P., 260

horse gram, 68

horticulture. *See* gardens

Hossain, M., 220

Hourani, A. H., 185

household: as farming unit, 108, 194, 209, 225; large, 205–6, 209, 214; and subdivision of land, 192

Howard, A., 260

Humphreys, G. S., 7, 14, 74, 75, 82, 96, 159

Hunter, J. M., 176n4

hunting, 28, 109; and agrodiversity, 60, 76; and protoagriculture, 63, 64, 65, 66

Hurtado, L., 104

Hüttermann, A., 95

huza land purchase system (Ghana), 175n4, 176n5

hyacinth bean, 68, 92

hybridization, 78n5, 219, 246, 252, 266, 267

Hyden, G., 200

hyptis (*Hyptis spicigera*), 30

ibala farming systems (Zambia), 130–32, 133, 137, 138

Iban people (Borneo), 105–7, 110–15, 122n2; Baleh, 106, 111

Ifugao (Philippines), 16–20, 21, 97; Conklin on, 47, 49, 51; rice terraces in, 170, 171

illipe nut tree (*Shorea macrophylla*), 113

Imperata cylindrica grass, 7, 37, 174, 176n6

Inca Empire, 24, 169

income, 36–37, 194, 235; and innovation, 45; from migrant labor, 205; and savings, 182; wage, 35, 37, 220

India: and alternative agriculture, 257; biotechnology in, 264, 266; colonialism and, 219; crops in, 92, 225, 226, 227, 242; diversification of rural economy in, 235; domestication of crops in, 63; Green Revolution in, 222–30; intercropping in, 272; irrigation in, 173; and Islamic conquest, 184; land use systems in, 50; north, 222–28; population pressure in, 215; protoagriculture in, 68; shifting cultivation in, 105, 111; soil nutrients in, 94–95, 126; south, 228–29; summer crops in, 224, 226, 227, 236n7; and transgenics, 264; winter crops in, 224, 227. *See also* Karimpur; Shahidpur

Indian gum arabic, 68

indigenous technical knowledge. *See* local knowledge

indigo, 195

Indonesia: agroforestry in, 147; crops in, 5, 230; external intervention in, 219, 220; farmers in, 231, 241; immigrants from, 36; landraces in, 22n7. *See also particular countries*

infields, 51, 97

Information Centre for Low-External-Input and Sustainable Agriculture (ILEIA), 272, 280n1

Innis, D. Q., 272, 273

innovation: in agroforestry, 164; bursts of, 181–83, 220; diffusion of, 183; as element of diversity, 54–55; in Europe, 186–87; by farmers, xiv, 196, 198–217; and government policies, 282–83; in Japan, 194; in Mali, 201–6; in

Mediterranean region, 184, 185; in
Mindoro, 120; and organizational
diversity, 45; and private farms, 16; to
reduce risk, 215; and shifting cultivation,
103; technological, 216; top-down, 219.
See also technology
insecticides, 253, 258, 276; and transgenics,
250, 251–52, 269. *See also* pesticides
insects. *See* pests
Institute for International Tropical Agriculture
(IITA), 267
integrated pest management (IPM), 232, 253,
258, 269, 282
intensification of production, 54, 60, 214–16;
definition of, 200–201; super special
(*Supra Intensifikasi Khusus*), 233
Intensive Agricultural Development Programs
(IADPS; India), 225
intercropping, 48, 77, 221, 286; in Africa,
196n1, 201–2, 272; and agrobiodiversity,
272–74, 282; in Amazonia, 122n5, 144;
in Borneo, 112; and crop yields, 272–73;
in Cuba, 259; in Java, 235; in Karimpur,
224; and management diversity, 277; in
Mediterranean region, 183, 184; in
Mindoro, 118; with vegetables, 228, 235
International Agricultural Research Centers
(IARCs), 219, 246, 247, 251
International Board for Plant Genetic
Resources, 246
International Board for Soil Research and
Management (IBSRAM), 159
International Centre for Research in
Agroforestry (ICRAF), 147
International Crops Research Institute for the
Semi-Arid Tropics, 272
International Food Policy Research Institute
(IFPRI), 228
International Institute for Environment and
Development, 272
International Plant Genetic Resources
Institute (IPGRI), 248
International Rice Research Institute (IRRI),
219, 220, 231, 257
International Service for National Agricultural
Research (ISNAR), 271
Internet, 247, 250, 266, 280n1

investment: capital, 22n4, 184, 188, 201,
208; changes in, 229; as element of
diversity, 54–55; in irrigation, 184, 188,
190, 196; and labor, 214; in land
improvement, 182–83; and large
households, 205; in Mali, 201; in
managing degradation, 174–75; in new
technology, 200, 228; in organizational
skills, 174, 200; patterns of, 202; physical
capital, 188; by small farmers, 200; social,
188
Iquitos (Peru), 145
Ireland, 180
Irian Jaya, 72
ironwood (*Eusideroxylon zwageri*), 150
irradiation, 256
irrigation, xiii, 47, 229; in alternative
agriculture, 254; canal, 190, 226; in
Cuba, 259; and drought, 210; in Europe,
186; evolution of, 70–71; excessive, 222;
and field types, 51; government
intervention in, 225; and Green
Revolution, 221; investment in, 184, 188,
190, 196; in Japan, 189, 191, 192; in
Java, 230; in Peninsular Malaysia, 34, 37;
and labor, 191–92; as landesque capital,
55, 175; in Philippines, 17, 18, 19, 120;
and pollution, 260n2; and rice, 39n4,
114, 189, 227, 228; and salinization,
176n10; technology of, 78n8, 185; and
terraces, 168, 170, 211; winter, 223, 224.
See also wet-rice cultivation
Isoberlinia, 123
Israel, 175
Italy, 185
Ives, J. D., 176n9
Izikowitz, K. G., xv

jackfruit, 113
Jamaica, xi
Jana, S., 249
Janda Baik (Peninsular Malaysia), 33, 34, 35,
36, 37, 38
Japan, 188–94; alternative agriculture in, 254,
255; biological husbandry in, 257;
farmers in, 192, 193, 194; and foreign
trade, 189; guest soil (*kyakudo*), 190;

Japan (*Continued*)
  irrigation in, 189, 190, 191, 192; and
    Java, 194–96; population growth in, 189;
    rice in, 189, 191, 192, 193, 220; *samurai*
    (warrior class) in, 189; seventeenth-to
    twentieth-century, 183, 188–94;
    Tokugawa shogunate in, 188, 191, 192;
    trade in, 188–90; tree crops in, 64–65
Jatiluhur dam (Java), 230–31, 232, 233
Java, 222, 236n5; agrodiversity in, 38; and
    Borneo, 113, 151; change in, 196, 230;
    cluster of dry crops (*palawija*) in, 234,
    237n13; commercial demand on, 196;
    communal land in, 197n6; external
    intervention in, 218; farmers and the state
    in, 230–33; germplasm bank in, 151;
    Green Revolution in, 196, 230–31, 234;
    Gunung Sewu, 107, 169; immigrants
    from, 34; and Japan, 194–96; Jatiluhur
    dam, 230–31, 232, 233; mass guidance
    (*bimbimgan massal*; BIMAS) in, 231;
    monocropping in, 151; nonfarm work in,
    235; population of, 169, 171, 194; rice in,
    xiii, 22n6, 194, 195, 197n6, 230; stone
    lines in, 170–71; Sukahaji, 107, 234, 235,
    237n13; *talun-kebun* system of, 146;
    village forests in, 151
Jensen, E., 106
Job's tears (*Coix lachrymae-jobi*), 108, 112
Johnson, K., 167
Jones, J. A., 89
Jordan, Carl, 126
Jordan valley, 63, 69, 70
Julbernardia, 123
Juma, C., 236n4, 245, 247, 251

kaffir corn (*amasaka*), 129
Kala village (Mali), 201–6, 209, 214, 216
Kalimantan, 105, 106, 107, 110, 113, 114,
    121; agroforests in, 38, 149, 150, 151,
    182; old forests in, 146, 149
Kantu' people (Borneo), 106, 107, 110, 111,
    112, 114, 115; *vs.* Iban, 122n2
Karimpur (Uttar Pradesh, India), 222,
    223–28
Katanga (Congo), 133
Kates, R. W., 200

kava (*Piper methysticum*), 276
Kayan people (Borneo), 110, 112, 113, 114,
    115
Kayapó people (Amazonia), 143, 144, 152–53
Kenya, xi
Kenyah people (Borneo), 105–8, 110–15,
    121, 147
Kerridge, E., 186, 187
Kiome, R. M., 160, 214
Kissidougou prefecture (Guinea Republic,
    West Africa), 153, 154, 155, 161
Klock, J., 18, 20
knotweed (*Polygonum erectum*), 67
Kofyar (northern Nigeria), 206–11, 214, 216,
    217
Kooistra, L. I., 75
Korea, 255
Kuala Lumpur (Malaysia), 32–38

labor: and agrodiversity, 285; and alternative
    agriculture, 255, 275–76; bonded, 192;
    and capital, 200, 216; changes in, 187,
    188, 191–93, 195, 213; conscripted, 190,
    203; cooperative, 44, 118, 209, 214;
    corvée, 195, 203; and decision making,
    51; demand for, 6, 26, 54, 192, 200, 203;
    dependent, 192; of farmers, 182; gender
    division of, 44, 127, 130; hired, 44, 209,
    218; and investment, 214; and irrigation,
    18; management of, 58n1, 201, 208,
    209–10; of men, 133, 136; migrant, 11,
    162, 163, 202, 205; mobilization of,
    22n3, 44; and organizational diversity, 41,
    44; scarcity of, 130, 133, 136, 191, 193;
    wage, 121, 188; of women, 44, 127, 130,
    131, 134, 136. *See also* employment
Lahjie, A., 106
Lal, R., 157
land: access to, 109, 115–16; degradation of,
    53, 56, 57, 152, 157–61, 163–65, 169,
    171, 175, 211; and extensification, 200;
    and intensification, 200; investment in,
    171, 174, 192, 208; prices of, 35–36, 37;
    reclaiming, 155; rental of residences on,
    36; selection of, 109–11, 117; speculation
    in, 35–36, 39; types of, 7–8, 26; valuation
    of, 15, 19, 38, 191, 226

land equivalent ratio (LER), 272–73

land holding, xiv; absentee, 36; in Borneo, 115–16, 121; changes in, 14, 15, 213; control of, 210; diversity in, 116; flexible systems of, 15; individualization of, 183, 188; inequitable, 15; in Java, 195, 197n6; in Peninsular Malaysia, 32–33, 34; in Papua New Guinea, 4; in Philippines, 19, 117, 120; and population pressure, 217; reforms of, 24, 26, 188, 195, 197n5; and tenancy, 36, 120, 188, 192; in Vanuatu, 75; and Zande, 31

land management: in Andean region, 24–25; and biodiversity, 55, 113; commercial, 20, 194; evolution of, 69–70, 74; and farming spaces, 39, 187; and floods, 21, 78n6, 155–56, 168, 171, 172–73, 182, 223; innovation in, 16, 194; through land husbandry, 161; and population, 134; and soils, 21, 69–79; and technology, 70; in Vanuatu, 75; and water, 69–70

land use: in Andean region, 24; and catenas, 89; codification of, 50; complexity of, 183, 184; decision making in, 196; in Java, 195; in Peninsular Malaysia, 33; mapping of, 4, 48–49, 50, 52, 84, 85; market signals and, 215; patterns of, 13–15, 49, 138–39; and soil diversity, 96–99; in Zambia, 124–32

landraces: adaptability of, 248; in Andean region, 22n7, 24, 25, 26; defined, 22n7; and domestication of crops, 61; and field types, 52; and genetic diversity, 21, 245, 246, 249; stress resistance of, 221, 252; taxonomy of, 46

Lappé, M., 251, 270

Larcher, W., 90

Lawrence, D. C., 151

Leach, Melissa, 153, 154, 161, 175

Leaf, Murray, 223, 226, 227, 228

Lee, J., 121

legumes, 25, 123, 124; and nitrogen, 90–92, 95, 273; use of, 129, 131, 185. *See also* beans

Leipzig conference on Conservation and Sustainable Use of Plant Genetic Resources for Food and Agriculture (1996), 281

Lenné, J. M., 55, 221, 245–46, 248, 274

lentils, 63, 68, 92

Lewis, M. W., 20

ley farming, 186, 187

Lian, F. J., 110, 111, 112, 113, 115, 121; on Borneo, 105, 106

Life Sciences Knowledge Center, 250

Lipton, M., 45, 220

little barley (*Hordeum pusillum*), 67

livestock: in Africa, 202, 203, 204, 205, 207, 208–9, 212, 213; and agrodiversity, 60, 186; in Andean region, 24, 25; in Borneo, 108; cattle, 24, 120, 203, 204, 212, 213; in Cuba, 258; and decision making, 51; domestication of, 62, 63, 64, 68, 69; draft, 191; in Europe, 185, 187; fodder for, 92, 185, 188, 191, 227; and genetic erosion, 246; improved, 186; and income disparities, 45; integration of production, 235; in Java, 169; and mechanization, xiv, 227; in Mediterranean region, 183, 184; in Nepal, 167; in Philippines, 18, 119, 120; pigs, 4, 6, 7, 8, 9, 11–12, 18, 19; and plowing, 78n7; and soils, 26, 157; and *tapades*, 98, 99, 165; and transgenics, 283

Livingstone potato (*Coleus esculentus*), 132

local knowledge, 43, 57, 76, 87, 245, 277, 279, 282

loess, 82

Longhurst, R., 220

Lopez-Gonzaga, V., 118, 119

Louette, D., 248

low-external-input and sustainable agriculture (LEISA), 272

Lowrance, R., 285

lucerne (alfalfa), 188

Lun Dayeh people (Borneo), 88, 114

Lundy, R. T., 196n2

McGrath, D. G., 130

Machakos (Kenya): Akamba people in, 160, 201, 211–14, 217; external interference in, 211–12; pattern of change in, 212–14; population pressure in, 214–15

Magdoff, F., 95

magic, 28, 31, 39n2, 108–9

magnesium, 126, 151

Mahmud, Zaharah, 32

maize (*Zea mays*), xvi; in Africa, 29, 30, 31, 131, 162, 163, 207, 211, 212; in Amazonia, 122n5, 144, 173; in Andean region, 24, 25; apomixis in, 267; in Borneo, 111, 112; domestication of, 61; and drainage systems, 72; in Europe, 180, 185; and field types, 52; fields of (*faamu*), 136; and genetic conservation, 248; hybrid, 136, 219; in India, 224, 227; interplanting of, 92, 129, 273; landraces of, 22n7, 26; in Mindoro, 118, 119, 120; and monocropping, 274; and Mycorrhizal fungi, 92, 94; in Papua New Guinea, 4, 9; in Peru, 144, 169; phytoliths of, 79n10; in protoagriculture, 66, 67; spread of, 137; in *tapades*, 98; trade in, 26, 135; transgenic, 250

Majcherczyk, A., 95

Malang (Java), 107

Malaysia, 32–38, 51. *See also particular regions*

Mambwe people (Zambia), 131, 135, 137

management: and agrodiversity, 60–61; of agroforests, 55, 140, 141–51, 156, 277; and alternative agriculture, 255; authoritarian, 148; biological, 241, 260; of biophysical diversity, 275–76; cooperative, 277; of environment, 66; of fallow systems, 141, 144–49, 156n1, 233; by farmers, xviii, 149, 167–68, 183–85; of fields, 143, 149; integrated pest (IPM), 232, 253, 258, 269, 282; of irrigation, 18; of labor, 58n1, 201, 208, 209–10; of nutrients, 276; of plants, 140–41, 143, 152; of production, 180; of resources, 55, 56, 58n3, 60–61; skills of, 48, 201; of slope dynamics, 74–75, 165–70, 207, 275; successional, 141–51; time, 193; of water, 69–70, 113, 168, 170–73, 190, 192, 194, 205, 220, 230; of weeds, 162, 275; of wetlands, 70–72; of wild animals, 55, 66, 129, 140. *See also* land management; soil management

management diversity: adaptation in, 53; and agrodiversity, 42–43, 50, 59; and crop rotation, 43, 277; defined, 41; and disease, 43, 276; and drainage, 71; and farming spaces, 52; and household size, 205; and income disparities, 45; and local knowledge, 43, 57, 277; and pest control, 276, 277; and political economy, 285, 286; and pollution, 276; in protoagriculture, 69–75, 77; and resilience, 276–77; and soils, 43, 271, 276; and sustainability, 276–78

Manandhar, S., 167

mango (*Mangifera indica*), 29, 94, 98, 113, 207

Mangyan. *See* Hanunóo people

manioc (*Manihot esculenta*), xvi, 69, 137, 154; in Africa, 29, 30, 31, 32, 129, 130, 131, 133, 135, 163, 180, 208; in Amazonia, 122n5, 144, 173; in Borneo, 111, 112; in Indonesia, 180; in Java, 169; in Mindoro, 119; and Mycorrhizal fungi, 94; in *tapades*, 98; trade in, 135; and weed management, 162

Marga Tani (Java), 107, 232

marketing. *See* cash crops; trade

marsh elder (*Iva annua*), 67

mashua (*Tropaeolum tuberosum*), 25

mass wasting, 165, 166, 176n8

Matengo system (Tanzania), 131

matrilineality, 33, 75

Maure people (Mali), 201, 203

Mauritania, 72

Mayans, 142

Mayer, E., 24, 25

maygrass (*Phalaris caroliniana*), 67

Mead, R., 273

Mearns, R., 161, 175

mechanization, 187, 194, 220, 226; and alternative agriculture, 256; in Cuba, 258; and livestock, xiv, 227; and soil erosion, 243

medicinal plants, xixn2, 18, 28, 68, 76, 98, 119, 151, 249, 260n4

Mediterranean region, 71, 78n9, 159, 183–85, 188

Meerut district (northwest Uttar Pradesh), 229–30

Meggitt, M. J., 217

Meiji Noho (Meiji agricultural ways), 191

Mellon, M., 270

men's houses (Papua New Guinea), 8, 9, 11

Mesopotamia, 176n10

Messerli, B., 176n9

Mexico, xi, 72, 142, 219, 220, 274, 278; farming practices in, 47, 171, 182; genetic conservation in, 247, 248; irrigation systems in, 71, 171; land degradation in, 56

microenvironments, 28, 31, 43, 44, 54, 81

Midwest (United States), 67, 73

migration: in Africa, 203, 206–8; and agrodiversity, 116, 285; in Borneo, 108, 115, 121, 174; and domestication of crops, 63; in Indonesia, 33, 34, 36; in Papua New Guinea, 6, 7–9, 8; in Philippines, 20, 119–20

millet: in Borneo, 108, 112; bullrush (*Pennisetum glaucum*), 29; domestication of, 62, 63; eleusine (*Eleusine indica*), 31, 32, 129, 131, 133, 135; finger (*Eleusine coracana*), 28, 29, 30, 31, 129, 130, 131; fonio (*Digitaria exilis*), 98, 207, 208, 209, 211; and Green Revolution, 221; in India, 224, 228, 230; long-cycle, 204; pearl (*Pennisetum glaucum*), 63, 64, 72, 92, 202, 207

Milne, G., 124

Minangkabau people (Peninsular Malaysia), 33

Mindoro (Philippines), 116–20

Mintima (Chimbu, Papua New Guinea), 3–16

miombo, 123, 124, 133–38

mitanda (temporary houses, in Zambia), 136

Mitchell, P. B., 82

Miyamoto, M., 117, 120

mobile topsoil, 82, 84, 89–90, 93

modernist agriculture, xvii, 242, 261n8, 276, 282, 285; and biological husbandry, 257; costs of, 243; in Cuba, 258, 260; *vs.* alternative agriculture, 244, 253, 255–56

Modjeska, N., 22n4

Mombasa (Kenya), 211

monoculture, xvii, 158, 221, 274–75; and

agrodiversity, 285; and alternative agriculture, 256; in cash crops, 260n1; of cotton, 31, 260n2; in Cuba, 258, 259; and farmers, 242; in Java, 151; of rice, 235

Monsanto Corporation, 250, 264, 279

Mool, P. K., 176n8

Mooney, P. R., 247

Moore, H. L., 125, 127, 128, 129, 130, 133, 135, 137

Moore, R. H., 192

Moran, E., 116, 153

Mortimore, M., 53, 160, 212, 214, 215

Moss, R. P., 49, 96

mounding practices, 155, 235; *mputa* (in Zambia), 130, 131, 209

Mubyarto, 234

multinational corporations, 242, 251; and biotechnology, 265, 267, 268, 280, 284; and genetic erosion, 245; and patents, 265, 267, 268; resistance to, 278–79; seed-chemical, 262–65; and transgenics, 252

Muslims, 183–85, 187

MVS. *See* seeds: modern varieties

Myanmar (Burma), 148

Mycorrhizal fungi, 90, 91–92, 94, 124, 191, 275

Nabta playa (Egypt), 62

Nairobi (Kenya), 211, 213

Naregu tribe (Chimbu, Papua New Guinea), 3–16; changes in land use of, 12–15; farming practices of, 9–11; migrations of, 6, 7–9

natural farming, 254. *See also* alternative agriculture; organic farming

Near East, 75, 77n1, 245, 274; domestication in, 63, 64, 65; protoagriculture in, 67–68. *See also particular countries*

neo-Malthusians, 200

neo-Marxism, xv

Nepal, 166–68, 173

Nepalese alder (*Alnus nepalensis*), 148

Netherlands, 75, 186, 188

Netting, Robert, xiii–xiv, xvii, 44, 49, 58n1, 201, 206, 208–10

New Economic Policy (Malaysia), 35
New England (U.S.), 66, 67
Ngo, T.H.G.M., 106
Nibbering, Jan, 169
Nigeria, 201, 204, 208, 211, 241, 272;
    agrodiversity in, 44, 210; Hausa kingdoms
    in, 206; Jukun state in, 206; soils in, 82,
    96. *See also* Kofyar
Nile valley, 70
Nishida, M., 65
nitrogen, 87, 90–92, 147, 151, 165, 213; and
    bacteria, 95, 99n3, 124; and compost,
    228; in fertilizers, 243; and fire, 125, 126,
    127, 128, 145; fixation of, 124, 259,
    261n8, 273, 276; and Mycorrhizal fungi,
    94; and nutrient cycles, 93, 148
noria (Persian wheel), 185, 226, 236n8
Norman, D. W., 272, 273
North America, 250; alternative agriculture
    in, 255; biological husbandry in, 257;
    control of seed use in, 252; eastern,
    65–68, 72–73, 74; and genetic
    conservation, 245, 247; herbicide resistant
    crops in, 269; modernist agriculture in,
    xiii; multinational corporations in, 251;
    transgenics in, 270; use of flood plains in,
    75. *See also* United States
Northern Grassland (Mambwe) system, 137
nutmeg, 150, 151
nutrients: cycles of, 93; evidence of, 74; and
    fire, 125–26, 127; and hillside terracing,
    169, 214; and Japanese forests, 190–91;
    management of, 276; mineral, 261n7; in
    soils, 90–95, 124, 125, 159, 160, 165,
    168, 271; storage of, 94–95; and
    transgenics, 265; from trees, 127, 128,
    145
Nye, P. H., 60, 61, 86, 126, 158, 272

oats (*Avena* spp), 62
oca (*Oxalis tuberosa*), 25
Odum, E. P., 271, 285
oil crops, 30, 121, 228, 269
oil palms (*Elaeis guineensis*), 121, 162, 164,
    207
Okigbo, B. N., 48, 272
Okura Nagatsune, 193

Oldeman, L. R., 158
Olson, E. A., 167
Orang Asli people (Peninsular Malaysia), 33,
    34, 37
organic farming, 253, 254, 282. *See also*
    alternative agriculture
organizational diversity, 41–42, 44–45, 50,
    53, 201, 205, 236
Orinoco River, 69, 72
Östberg, W., 81
outfields, 51, 97
overgrazing, 212
Owusu-Bennoah, E., 164
oxen, 203, 204, 223, 226, 229, 259
oxen area (Paucartambo, Peru), 24–26

Pacey, A., 161
Pacific island region, xvi, 159
paddy. *See* wet-rice cultivation
Padoch, Christine, xii, 42, 88, 147, 156n4; on
    Amazonia, 55, 173; on Borneo, 105, 106,
    108, 110, 111, 112, 114, 115, 150
Pagiola, S., 282
Pakistan, 226
paleobotany, 61, 67, 68, 73–75, 141; and
    bioturbation, 81–82
palm wine, 163
palynology (pollen analysis), 73
pandanus, 9, 13, 14
papaya, 119, 207
Papua New Guinea, xi, xii, 48, 156n1;
    Chimbu, 3–16, 22n3, 32, 44, 49, 88,
    166, 170; compared to Philippines, 20;
    crops in, 9–10; drainage systems in,
    71–72; Enga, 9, 217; highlands of, 181;
    Indonesian, 166; monocropping in, 274;
    women in, 22n4
Parker, E., 152, 153
Parsons, Helen, xii
patent rights, 247, 251, 260n4; and
    agrobiodiversity, 284; and biotechnology,
    265–66, 267, 268; and control of seed
    use, 252
Paton, T. R., 82
Paucartambo (Peru), 23–27, 32, 50, 51
peas, 63, 92
Peluso, N., 105, 106, 108, 146, 147

Pentecost island (Vanuatu), 74–75

People, Land Mangement and Environmental Change project (PLEC), xi–xii, 40, 164, 214, 271; and farmers, 278, 279, 282

Pereira, W., 257

Persian wheel. *See* noria

Peru, xi, 86, 171; Amazonian region of, 144–45; Paucartambo, 23–27, 32, 50, 51; terraces in, 168, 169, 170

pesticides, 158, 220, 221, 242, 243; and alternative agriculture, 256; biological, 43, 259; in Cuba, 258, 259; in Indonesia, 231, 232, 233, 234; and transgenics, 251–52

pests: brown planthopper (*Nilaparvata lugens*), 231, 232, 234; and farmers, 231–32, 233; integrated management of (IPM), 232, 253, 258, 269, 282; and intercropping, 273; and management, 21, 43, 276, 277; and modernist agriculture, 243; and monocultural systems, xvii; and resilience, 277; resistance to, 245, 251–52; in rice, 22n6, 231, 232; white rice stem borer (*Scirpophaga innotata*), 232, 237n12

Peters, Charles, 132, 133, 134, 143, 146, 150

pH , 85, 99, 151, 165, 176n6; and fire, 126, 127, 128, 143

Phaseolus beans, 67, 92

Philippines, 148, 219, 231; Hanunóo people of, 116–20, 140, 156n1; Ifugao, 16–20, 21, 47, 49, 51, 97, 170, 171; terraced fields in, 19–20, 170, 171

phosphorus, 71, 87, 95, 124, 131, 151, 213; and alternative agriculture, 256; in ancient soils, 168; concentration of, 145; in fertilizers, 243; and fire, 126, 127, 128; in manufactured soils, 97; and Mycorrhizal fungi, 94; recycling of, 148

phytoliths, 74, 78n10

pigeon pea, 68, 119

Pimentel, D., 158

pineapples, 144

Pinedo-Vasquez, M., 155, 172

Pingali, P. L., 220

pistachio (*Pistacia atlantica*), 64

pit-pit (*Saccharum spontaneum*), 3–4

plants: breeding of, 220, 241, 249–53, 268; diversity of, 77; domestication of, 5, 62, 63, 76, 77n1, 132; macroremains of, 74; management of, 140–41, 143, 152; medicinal, xixn2, 28, 68, 76, 98, 119, 151, 249, 260n4; nutrients needed by, 87–88; and soils, 90–95, 154–55, 159; tissue cultures of, 249, 250

plants, wild, 76, 156n1, 162, 274; and domestication of crops, 62–63; and genetic conservation, 246, 248, 249; in protoagriculture, 64–67; and resource management, 55; sacred, 119; and shifting cultivation, 111, 113, 210

plowing: deep, 191, 243; with draft animals, 212, 213, 214, 232; foot-, 25, 85; history of, 78n7, 175n2; in India, 225; investment in, 203, 204, 214; mechanized, 158, 175n2; in Mindoro, 120; and modernist agriculture, 244. *See also* tillage

Plucknett, D. L., 77

Polanyi, K., 236n1

Poleman, T. T., 229

politics: of biotechnology, 262; and change, 180, 228; and genetic conservation, 247; and leadership, 11, 22n3; and shifting cultivation, 106; and small farmers, 21, 278–79

pollution, 243, 260n2, 276

polyculture. *See* intercropping

pond fields. *See* wet-rice cultivation

population, xvii; in Africa, 154, 201, 208, 210, 211; and agrodiversity, 59–60, 116, 222, 285; in Amazonia, 141–42; in Borneo, 105, 106, 150; and colonialism, 158, 195; decline of, 116, 190; growth of, 59–60, 63–64, 71, 158, 163, 189, 194, 195, 199, 212, 214, 215, 217, 237n13, 244, 263, 285; in Japan, 189, 190, 193; in Java, 169, 171, 194; and land, xiv, 49, 56, 157; in Mediterranean region, 184; in Papua New Guinea, 4, 6, 22n3; in Philippines, 119; shifts of, 180; in South America, 116; and transgenics, 263; and wet-rice production, 114–15

Porol Range (Papua New Guinea), 5, 6, 7

Portararo, A. V., 200

Porter, P. W., 212

Portugal, 20, 185

Posey, Darrell, 143, 152, 153

Postma, A., 117

potassium, 87, 94, 151, 243, 256; and fire, 126, 127, 128

potatoes, 52, 180, 211, 225, 228; in Andean region, 24, 25, 26; landraces of, 22n7, 25, 26; transgenic, 250. *See also* sweet potatoes

Potkanski, T., 169

Potrykus, I., 250

Potter, L., 106, 121

Prescott-Allen, C., 245

Prescott-Allen, R., 245

Pretty, J. N., 161

Pringle, R., 106

production: costs of, 186, 241; efficiency and, 199, 200; of food, 230, 242; of grain, 191; improvements in, 180, 186, 193, 200, 206; in Japan, 194; of leguminous crops, 221; management of, 180; to meet social demands, 215; of milk, 227; new combinations of, 181–82, 188; of sugar, 195; vegetable, 211, 213, 235

production zones. *See* farming spaces

protoagriculture, 63–75, 77; management diversity in, 69–75; sites of, 68–69; and wild plants, 64–67

public sector: and change, 218; and genetic engineering, 265–67; and research, 251, 256

pulses, 77n1, 131

pumpkins, 66, 129

Punan people (Borneo), 113

qanats, 78n8 '

Qualset, C. O., 248

Querol, D., 247

quinoa (*Chenopodium quinoa*), 24, 25

rainfall: in Africa, 204, 209; in Borneo, 106; and diversity, 17; and economy, 205; and flooding, 171, 176n7, 223; monsoonal, 166, 167, 228; in Nepal, 166, 167; and soils, 81, 82, 84

rainforests, 85, 86, 93, 94, 103

Ramakrishnan, P. S., 105, 111, 126, 237n11

Rao, M. R., 272

rattans, 18, 19

Reining, C. C., 27, 29, 30, 39nn2,3

Reintjes, C., 272

religion, 11–12, 44, 106, 117–18, 119, 122n4; Christianity, 117, 122n4. *See also* magic

research: in Amazonia, 126, 141–42; on biotechnology, 246, 263, 264, 267; farming systems (FSR), 58n2, 132–39, 221; international centers for agricultural, 147, 159, 219, 228, 246, 247, 251, 272; public sector, 251, 256; on rice, 219, 220, 231, 257. *See also particular agencies and organizations*

resilience: and agrodiversity, 59, 211, 271–72, 280, 286; and alternative agriculture, 255; and climate change, 285; of farmers, 279; and management diversity, 276–77; and resource management, 56; and soil management, 95; and sustainability, 277–78

Rhizobium spp., 90, 91, 92

rice: in Africa, 208; in alternative agriculture, 254; in Amazonia, 144, 172, 173; *basmati*, 227; in Borneo, 107–8, 109, 111, 117; in Cuba, 258; domestication of, 62, 63, 76; double cropping of, 229; and drought, 230; fields of, 146; genome maps of, 266; Green Revolution in, 218, 219, 220, 223, 228, 236; and GURTS, 264; hybrid, 252; in India, 224, 228–29; in Japan, 189, 191, 192, 193, 220; *japonica*, 236n3; in Java, 169, 195, 197n6; *javanica*, 17, 231, 236n3; landraces of, 22n7; *lokal*, 233, 234; in Peninsular Malaysia, 37, 51; monoculture of, 235; in Nepal, 167; in Philippines, 118, 119; price of, 234; in protoagriculture, 68; and rubber, 36–37; savanna land and, 154; and silt bars, 172; summer-grown, 227; swamp (*padi paya*), 109, 113–15; taxation of, 189, 191, 192; *tinawen* types of, 17; trade in, 135; transgenic, 250, 265;

transplantation of, 222; upland, 122n3; and wetlands, 71; yields of, 37, 120, 194. *See also* wet-rice cultivation

Richards, Audrey, 44, 58n3, 125, 129, 130, 132, 133, 135, 136, 137

Richards, Paul, xiv, 21, 149, 196n1, 273

Richter, D. D., 86, 88

Rigg, J., 220

Rissler, J., 270

roads, 9, 11, 15, 36

Roberts, A. D., 135

Robinson, W., 199

Rochelau, D. E., 212

Rockefeller Foundation, 219, 220, 268

Roe, E. M., 161

Roman period, 75, 171, 175n2, 183–84

Roosevelt, A. C., 69, 72, 141

root crops: in Amazonia, 143, 144; in Borneo, 108; in Cuba, 258; domestication of, 77n1, 132; in Mindoro, 119; ridge cultivation of, 132; in *tapades*, 98; *vs.* grains, 76

root systems, 93, 94

Rosset, P. M., 243, 256

Rousseau, J., 106, 112, 115

rubber: in Borneo, 109, 113, 121, 150, 151; and fallow systems, 149, 151; in Peninsular Malaysia, 34, 35, 38, 51; Para (*Hevea brasiliensis*), 34; and rice, 36–37; wild, 162

Rural Advancement Foundation International (RAFI), 247, 250, 263, 264

Russia. *See* Soviet Union

Ruthenberg, H., 52, 104

rye (*Secale cereale*), 62

Saccharum edule (edible pit-pit), 9

Sachs, P., 95

sago palm, 113

Sahel (West Africa), 82, 201–5

Salih, Kamal, 35

salinization, 176n10

Sandor, J. A., 168, 171

Sarawak, 93, 105, 106, 111, 113–14, 126

Sauer, C. O., 22n7, 77n2

Sauer, J. D., xvi, 63–64, 245

savanna, 127, 152, 153, 154–55, 163, 208; *chipya*, 124. *See also* forests

sawah (wet-rice production), 114–15

Schelhaas, R. M., 93, 126, 128

Schultz, J., 138

Schumpeter, J. A., 181

Schutjer, W., 199

Scleria spp., 111

Scoones, I., 201

seeds: chemically dependent, 262–65; distribution of, 219, 227; macroremains of, 74; modern varieties (MVS), 219, 220, 221, 222, 223, 225, 229, 231, 234, 236n2; and multinational corporations, 252; saving of, 262–65; selection of, 52, 63, 77, 77n3, 246; stocks of, 164; storage of, 246–49; and terminator technologies, 263–65, 269

Senegal, 203

Serenje plateau (Zambia), 132, 133

sesame (*Sesamum indicum*), 30, 207, 224

settlement patterns, 11, 65, 137, 209; and external intervention, 29–30, 31, 32, 39

shadouf (levered bucket), 185

Shah, P., 161

Shahidpur (Punjab), 222–23, 226–28

Sharma, R., 229

shellfish, 18

Sherratt, Andrew, 61

shifting cultivation, xvii, 14, 103–22, 156; and adaptability, 134; in Africa, 32, 208, 210, 212; in Amazonia, 122n5, 172–73; in Borneo, 105–21, 146, 147, 174; definition of, 104–5; and drainage, 71; and fallow systems, 103, 104, 105, 187, 199; and field types, 51; and fire, 104, 105, 108, 123, 125; in history, 60–61, 75–76, 106, 186; and intensification, 199; and intercropping, 272; in Java, 169; in Peninsular Malaysia, 33, 34, 38; literature of, 138–39; and long-term change, 182; and monocropping, 274; observations of, 46, 47; in Papua New Guinea, 4; and population density, 116; and protoagriculture, 64; site selection for, 109–11, 115; and soils, 105, 123, 157,

shifting cultivation (*Continued*)
186; tools for, 78n7; *vs.* permanent
systems, 130; and wild plants, 111, 113,
210

Shiva, V., 247

Shorea javanica, 149

Sierra Leone, 44

Silberfein, M., 212

silk production, 189, 190

silt bars, 172, 173

Singer, A., 27

slash-and-burn cultivation, 104. *See also*
shifting cultivation

Slater, G., 237n11

slope dynamics, 17, 49; Himalayan, 176n8;
and landslides, 165–66, 167, 168, 176n7;
management of, 74–75, 165–70, 207,
275; and soils, 84, 89, 96, 165; and
terraced fields, 165–70, 182; in West
Indies, 176n7

smallholders, xiii–xiv, xvii. *See also* farmers:
small

Smith, B. D., 67

Smith, J., 215, 222

Smith, N.J.H., 77

Smith, T. C., 72, 190, 192, 196n2, 197n3

social conditions, 21, 142, 228; and
agrodiversity, 59; and farming practices,
38, 52, 53; and innovation, 45; and
relations, 6, 41, 109; and status, 15

Soedjito, H., 106, 147

Soeharto, 230

Soekarno, 230

Soil Association (British), 257, 261n7

soil erosion, 26, 56, 74; control of, 131, 156,
207, 208, 214; and external intervention,
212; and intercropping, 273; and
livestock, 157; and management, 158,
159, 276; and mass wasting, 165; and
mechanization, 243; and mobile topsoil,
90, 167, 168; and rainfall, 129, 176n7;
sheet, 160; and shifting cultivation, 118,
157; and slope dynamics, 165, 166, 167,
207; and soil formation, 86; and soil loss,
158, 159, 161; and soil structure, 96, 158;
and tillage, 158, 243, 275; and trash-lines,
160

soil fertility: and bioturbation, 81; and cereal
crops, 76; and ecological engineering,
152; and fire, 128; and floods, 173; loss
of, 26, 29, 130, 159, 164, 165, 214; and
management, 156, 186, 208; in Roman
period, 184; and shifting cultivation, 105,
186; variations in, 138

soil management, 43, 95, 227; and fertility,
156, 186, 208; and plant management,
140, 154, 155; and soil erosion, 158, 159

soils, 80–99; and agroforestry, 110, 273–74;
and alternative agriculture, 253, 255, 256;
and biophysical diversity, 41; biota in, 84,
89, 91, 93, 95, 96, 124, 155, 277; and
biotechnology, 284; bioturbation of, 74,
81, 89, 96; clay-rich, 226, 227;
compaction of, 243–44; conservation of,
26, 31, 161, 166, 168–70, 175, 213, 251,
275–77; in Cuba, 259, 260; degradation
of, 53, 56, 57, 97, 152, 157–61, 163–65,
169, 171, 175, 211; diversity of, xvii, 80,
86–88, 94, 96; effects of fire on, 105,
125–28; flooding of, 223; formation of,
82, 84–85, 88–90, 155; healthy, 243;
humus, 95, 125, 244, 253; improvement
of, 7, 138, 146, 148, 186, 208, 211; and
land management, 21, 69–70; as
landesque capital, 55, 175; local
knowledge of, 81, 87; manufactured, 39,
41, 55, 96–99, 141, 174, 175; mapping
of, 85; mobile topsoil, 82, 84, 89–90, 93;
modification of, 123–39, 175; nutrients
in, 82, 84, 85, 86, 87–88, 90–95; organic
matter in, 95; paleobotanical evidence
from, 10, 73–75; parsimonious, 5,
123–39; *plaggen,* 97; preparation of, 235;
and resilience, 277–78; and shifting
cultivation, 123; on sloping lands, 159;
stratigraphy of, 74; structure of, 95–96,
164, 168, 171, 226; taxonomy of, 82–84,
88, 90, 96, 124; in terraces, 19, 168–70;
terra preta do Indios, 97; texture of, 96;
tillage of, 251, 253, 280n4; tropical,
85–88, 121; types of, 7, 159; and wind,
81, 82, 126. *See also* catenas

Sombroek, W. G., 158

sorghum (*Sorghum bicolor*), 68, 92, 129, 221,

274; in Africa, 29, 30, 31, 208, 209, 211; domestication of, 62, 63; in India, 224, 228, 230

South America, xi, 70, 72, 73, 140; Andean region of, 23–27; crops from, 5, 20, 29, 92; floodplains in, 69, 75; genetic diversity in, 245; monocropping in, 274; phytoliths from, 79n10; population in, 116, 141; successional forest in, 141–44. *See also particular countries and regions*

Southeast Asia, xii, 20, 23, 219; agroforestry in, 147, 149; field types in, 51; shifting cultivation in, 103; soil diversity in, 88; successional management in, 146–51. *See also particular countries*

Southeast Asian Universities Agroecosystem Network (SUAN), 47

Soviet Union, 243, 246, 258, 260n2, 267

soya bean, 92; transgenic, 250, 252, 262

Spain, 20, 24, 184, 185

Spencer, J. E., 48, 104, 146, 168, 170

Spooner, B., 199

squash, 66, 67, 273

Sri Lanka, 93, 126, 128, 222

Steinberg, P. E., 212

Stemler, A.B.L., 62

Stocking, Michael, 40, 131, 157, 159, 160, 175

Stone, Glenn, 58n1, 201, 206, 208, 209, 210

Strømgaard, Peter, 127, 128, 131, 133, 134

Struik, Paul, 43, 80

Sudan, 27–32, 29–30

Suet, Ho Wai, 34

sugar cane, 4, 94, 135, 236n7; in Borneo, 112; in Cuba, 258; in India, 224, 225, 226, 227; in Java, 195

Sugiyama, Y., 131, 136

Sukahaji (Java), 107, 234, 235, 237n13

sulfur, 87, 95, 125

Sumatra, 33, 34, 38, 149, 151

sunflower: *Helianthus annuus*, 67; *Tithonia diversifolia*, 148

Sungai Pencala (Peninsular Malaysia), 33, 34, 35, 36, 37, 38

Sungai Serai (Peninsular Malaysia), 33, 34, 35, 36, 37, 38

Susanto, A., 114

sustainability, xvii, 56, 120; and agrodiversity, 284–86; and alternative agriculture, 255, 262; and biotechnology, 283; and modernist agriculture, 244; and resilience, 271, 277–78; and shifting cultivation, 103

Sutton, J.E.G., 72, 168, 169, 170

Swaminathan, M. S., 257, 285

Swaminathan Research Foundation, 257

sweet potatoes (*Ipomoea batatas*), 135, 163, 274; in Africa, 29, 30, 98, 207; in Amazonia, 122n5; in Borneo, 112; in Papua New Guinea, 4, 5–6, 7, 12; in Philippines, 18, 19, 20, 119

swidden cultivation, xv, 104. *See also* shifting cultivation

Swift, M. J., 124

Switzerland, 44

Syria, 175

Taiwan, 220

taluds (accumulation terraces), 75

Tamil Nadu (south India), 228–29

Tanzania, xi, 81, 131, 170

tapades (compound fields, Guinea), 97–99

taro, 75, 162, 276; in Africa, 98, 208; in Borneo, 112; *Colocasia esculenta*, xvi, 4, 5, 10; *Colocasia* spp., 71; in Philippines, 18, 19; *Xanthosoma*, 207

tarwi (*Lupinus mutabilis*), 25

taungya system of forest management, 148

taxes, 195, 202, 205, 218, 224

technology, 52, 199, 211; changes in, 193, 200, 226, 228; and domestication of crops, 63; and FSR, 58n2; intermediate, 284; and irrigation, 71; and management, 43, 70; mixed, 283; and modernist agriculture, 244. *See also* biotechnology; tools

Technology Protection System (TPS), 263

tenancy, 36, 120, 188, 192, 224

terminator technologies, 263–65, 269

terraced fields, 7, 39, 51, 212; accumulation (*taluds*), 75, 168; cross-valley, 171; and early field systems, 73; *fanya juu* (thrown up; Kenya), 170, 213, 214; investment in, 174, 214; and irrigation, 70, 168, 170,

terraced fields (*Continued*)
171, 211; in Java, 169, 195, 196, 237n13; as landesque capital, 55; and manufactured soils, 97; in Mediterranean region, 78n9; in Nepal, 173; in Philippines, 17–21, 170, 171; and soil conservation, 19, 168–70; in South America, 169, 170; and steep slopes, 165–70, 182; stone-walled, 168, 169, 207; and water, 170–71, 213, 237n13. *See also* wet-rice cultivation; Ifugao

Thailand, xi, 93, 126, 148, 159

Thirsk, J., 187, 254, 261n8

Thomas, P. J., 237n11

Thrupp, L. A., 161, 283

Tibet, 176n8

Tiffen, Mary, 160, 200, 212, 214, 215

tillage, 6, 135, 137, 155, 280n4; in Africa, 125, 203, 207; and erosion, 158, 243, 275; and fire, 127; minimal, 251, 253, 276, 282. *See also* cultivation; plowing

tobacco, 31, 66, 112, 169, 250, 258

tomatoes, 250, 266

tools: in Andean region, 25; in Borneo, 108, 109; disadvantages of, 280n4; evolution of, 78n7, 275; hoe, 78n7, 154, 164, 175n2, 203, 280n4; improved, 186, 187, 220; in India, 225; in Papua New Guinea, 9; in Philippines, 21; plow, 25, 78n7, 85, 232

Toulmin, C., 201, 202, 204, 205, 206

tourism, 12, 18, 20

trade: in Africa, 135, 202; and agrodiversity, xvii, 58n1, 145; and biotechnology, 265; in Borneo, 109, 117; in England, 188; and farmers, 145, 215; free, 242; and imports, 27, 54; in India, 224, 228, 229; and industry, 188; and innovation, 198; in Japan, 188–90, 197n4; in Papua New Guinea, 11; in Philippines, 120, 122n7

transgenic crops, 249–52; advances in, 265; and agrobiodiversity, 282, 283; and agrodiversity, 271; dangers of, 283; and GURTS, 264–65; opposition to, 262–63, 284; safety of, 268–70

Trapnell, Colin, 124, 125, 129, 131, 132, 134, 137

Treacy, J. M., 144

tree crops: in Africa, 98, 162, 163, 164, 207, 210, 212, 213; almond, 64; in Borneo, 109, 113, 121; as capital, 175, 214; cutting of, 127–28, 132, 133, 136, 143, 173; and defense, 154; domestication of, 65; and drainage systems, 72; fruit, 18, 24, 34, 74, 135, 144, 145, 150, 151, 154, 183; and genetic erosion, 246; in India, 224; introduction of, 77; in Peninsular Malaysia, 33; mycorrhizal, 94, 124; *Newbouldia laevis*, 164; in Papua New Guinea, 9; in Philippines, 119; and protoagriculture, 64–65; ring-barking, 143; shade, 10; and shifting cultivation, 104, 110, 123, 148, 208

Treide, B., 156

tropics, 93, 105, 152, 245, 271; soils of, 80, 85–88, 95

Tuareg people (Africa), 64

Tukolor people (Mali), 201

Turner, B. L. II, 200

Ucayali river, 172

Uchucamarca valley (Peru), 25

Uganda, xi, 261n10

ulluco (*Ullucus tuberosus*), 24, 25, 26, 52

umari (*Poraqueiba sericea*), 145

Union for the Protection of New Plant Varieties (UPOV), 252, 265

United Nations (UN), 266, 281, 282, 283

United Nations Development Programme (UNDP), 264

United Nations Environment Programme (UNEP), xi

United Nations University (UNU), xi

United States (U.S.): alternative agriculture in, 254, 255, 256; biotechnology in, 220; and Cuba, 257–58; multinational corporations in, 242; Northeastern, 65–68, 72–73, 74; occupation of Philippines by, 20; soil erosion in, 158, 243; Southeastern, 67, 73; transgenics in, 250, 262, 263; and USLE, 161

Universal Soil Loss Equation (USLE), 161

Unruh, J. D., 145

urbanization, 23, 32, 35

Uribe, R., 104

U.S. Agricultural Research Service (USDA-ARS), 267, 268

U.S. Department of Agriculture (USDA), 250, 263, 264

USLE. *See* Universal Soil Loss Equation

Vanuatu, 74–75, 166

Varro, 184

Vaughan, M., 125, 127, 128, 129, 130, 134, 135, 137

Vavilov, N. I., 61, 77n2, 246

vegetables, 76, 183; in Borneo, 112; as intercrops, 228, 235; in Peninsular Malaysia, 34, 38; and Zande, 30, 31. *See also particular species*

vegetation, 49, 152, 163, 174; slashing, 154, 164; successional, 140, 164; transformation of, 53, 183, 185

Verro, 184

vetch, 63, 92

Visser, L. E., 108

Voeks, R. A., 76

Vogt, J., 187, 188

wadis (ephemeral streams), 171

Wahgi valley (Papua New Guinea), 3, 5, 6, 7, 9

Walker, D., 5

walnut, 65

Wanmali, S., 222

Ward, R. E., 190

warfare, 6, 16, 22n3, 108, 116; civil, 23, 32

water: conservation of, 168, 213, 275, 276, 277; ground, 226, 228, 229, 237n9; in India, 223, 225, 227, 229; in Japan, 189; management of, 43, 69–70, 113, 168, 170–73, 190, 192, 194, 205, 220, 230; shortages of, 237n13; and soils, 89, 95–96, 158, 223; and terraced fields, 170–71, 213, 237n13. *See also* irrigation; wet-rice cultivation

water chestnuts, 65

Waters-Bayer, A., 272

Watson, A. M., 183, 185

Watters, R. F., 105

Watts, Michael, 218

weathering, 97–98, 176n7; and mobile topsoil, 90; and slope dynamics, 165, 167; and soil formation, 84, 85, 86, 88

weeds, 118, 122n5, 163, 172, 268–69, 273, 275

Weeks, J., 45

wells, 175, 203, 204, 205, 214; in India, 223, 225, 226, 229

West Indies, xixn1, 176n7

wetlands, 49, 70–74, 76

wet-rice cultivation: in Borneo, 113–14, 121, 122n3, 150; conversion to, 182; and cyanobacteria, 91; in India, 227; in Indonesia, 231; in Java, 169, 171, 195, 197n6; in Peninsular Malaysia, 33, 34, 35; and manufactured soils, 97; and monocropping, 274; and swamp rice, 114–15. *See also* terraced fields

Wetterstrom, W., 63

wheat: in Andean region, 24, 25, 26; domestication of, 61, 63; in Europe, 185, 187, 188; and genetic conservation, 247; Green Revolution in, 218, 219, 223; and GURTS, 264; high-yielding, 219–20, 225; in India, 224, 226, 227; in Japan, 189, 191; Mexican varieties of, 226, 236n3; and Mycorrhizal fungi, 92, 94; in Nepal, 167; in protoagriculture, 68; and rape, 269; transgenic, 250; wild, 274

White, B., 194

Wilken, G. C., 47, 56, 155, 168, 182, 274

Willey, R. W., 272, 273

Winarto, Yunita, 232, 233

Wiser, Charlotte, 223, 225, 236n6

women: in Africa, 30, 127, 130; in Borneo, 110; and history of agriculture, 68–69, 77, 210; labor of, 44, 127, 130, 131, 134, 136; in Papua New Guinea, 8, 10, 11, 12, 22n4; private gardens of, 205; and warfare, 6

Wood, D., 55, 221, 245–46, 248, 274

woodlots. *See* agroforests

World Bank, 212, 282, 283

World Trade Organization (WTO), 265

World War II, 119, 195

yams: in Africa, 29, 75, 162, 164, 207, 208, 209, 210; in Papua New Guinea, 4; in Philippines, 118
Young, A., 90

Zaire. *See* Congo
Zambia: independence of, 136; Lala people in, 132; mining industry in, 133, 136; northeastern, 123, 125, 127, 132, 134, 137, 138; northwestern, 127, 132
zamindari system of land holding in India, 224
Zande (Southern Sudan), 27–32, 50, 51
Zimbabwe, 133, 136
Zimmerer, Karl, 24, 25, 26, 47, 52
Zohary, D., 67, 92